Mobile Actuation Systems for Magnetic Small-scale Robots

Mobile Actuation Systems for Magnetic Small-scale Robots

Qianqian Wang

School of Mechanical Engineering, Southeast University, Nanjing, China
Jiangsu Key Laboratory for Design and Manufacturing of Precision Medicine Equipment, Southeast University, Nanjing, China

Li Zhang

Department of Mechanical and Automation Engineering, The Chinese University of Hong Kong, Hong Kong, China
Shenzhen Loop Area Institute, Shenzhen, China

Tiantian Xu

Shenzhen Institutes of Advanced Technology, Chinese Academy of Sciences, Shenzhen, China

Mingxue Cai

Shenzhen Institutes of Advanced Technology, Chinese Academy of Sciences, Shenzhen, China

IEEE Press
WILEY

Published by John Wiley & Sons, Inc., Hoboken, New Jersey.

Library of Congress Cataloging-in-Publication Data:

Names: Wang, Qianqian, 1992- author | Zhang, Li (Professor of mechanical and automation engineering) author | Xu, Tiantian (Professor of robotics) author | Cai, Mingxue author

Title: Mobile actuation systems for magnetic small-scale robots / Qianqian Wang, Li Zhang, Tiantian Xu, Mingxue Cai.

Description: Hoboken, New Jersey : Wiley-IEEE Press, 2026. | Includes bibliographical references and index.

Identifiers: LCCN 2026003504 | ISBN 9781394356515 hardback | ISBN 9781394356539 epdf | ISBN 9781394356522 epub | ISBN 9781394356546 oBook

Subjects: LCSH: Microrobots—Automatic control | Actuators | Magnetic devices

Classification: LCC TJ211.36 .W36 2026

LC record available at https://lccn.loc.gov/2026003504

Cover Design: Wiley
Cover Image: ipopba/Getty Images

To the Visionaries, Innovators, and Explorers Pioneering the Frontiers of Small-scale Robotics

Contents

About the Authors

Qianqian Wang is a Professor and Associate Dean at the School of Mechanical Engineering, Southeast University (SEU), Nanjing, China. He received his Ph.D. in Mechanical and Automation Engineering from The Chinese University of Hong Kong (CUHK). Before joining SEU, he was a Postdoctoral Fellow at CUHK. He has published more than 40 papers in peer-reviewed journals, including *Science Robotics, Science Advances, Nature Communications, IEEE Transactions series*, and presented his work at ICRA, IROS. etc. Several of his journal papers have been indexed as ESI Highly Cited Papers and Hot Papers (Top 0.1%) and have been highlighted in Nature News and international press reports. He has published two monographs with Springer Nature and CRC Press. His research focuses on the small-scale robotics field, including collective behaviors of micro/nanorobots, medical imaging-guided delivery, and precision medical systems. He has been selected into the National High-level Young Scholar Program.

Li Zhang is a Professor in the Department of Mechanical and Automation Engineering and a Professor by Courtesy in the Department of Surgery at The Chinese University of Hong Kong (CUHK). He also serves as a supervisor for joint training of PhD students at the Shenzhen Loop Area Institute (SLAI) and acts as a program leader in the Multi-scale Medical Robotics Center (MRC), InnoHK, at the Hong Kong Science Park. Dr. Zhang's primary research interests include small-scale robotics and their translational applications. He has authored or co-authored over 400 publications (H-index: 93), including Science Robotics (2017, 2021, 2024, 2025a, 2025b), Nature Machine Intelligence (2020, 2022, 2024), Nature Materials (2024), Nature Reviews Bioengineering (2024, 2026a, 2026b), Nature Synthesis (2025), Nature Biomedical Engineering (2025), as a corresponding author. Dr. Zhang is elected as Fellow of IEEE (FIEEE), American Institute for Medical and Biological Engineering (FAIMBE), American Society of Mechanical Engineers (FASME), Royal Society of Chemistry (FRSC), Asia-Pacific Artificial Intelligence Association (FAAIA), The Hong Kong Institution of Engineers

(FHKIE), a member of the Hong Kong Young Academy of Sciences (YASHK), and an Outstanding Fellow of the Faculty of Engineering at CUHK. He also serves as Senior Editor for IEEE T-ASE and IEEE T-RO, and as Associate Editor for Science Advances (AAAS).

Tiantian Xu is a Professor at the Shenzhen Institute of Advanced Technology, Chinese Academy of Sciences. She received her Ph.D. in Robotics from Sorbonne University. Her research focuses on medical micro- and nano-robotics, intelligent control of magnetically driven small-scale robots, and their biomedical applications. She has led five national-level projects, including those from the National Natural Science Foundation of China and the National Key R&D Program. Dr. Xu has published over 30 papers in IEEE journals, including four in IEEE Transactions on Robotics (T-RO), and more than 10 in high-impact journals like *Nature* and *Nature Biomedical Engineering*. She has received multiple awards, including the Best Application Paper at IROS and the Second Prize of Natural Science from the Chinese Institute of Electronics.

Mingxue Cai is an Associate Professor at the Shenzhen Institute of Advanced Technology, Chinese Academy of Sciences (SIAT), and a high-level talent in Shenzhen, China. He joined SIAT in July 2023. His primary research focuses on magnetically driven medical robots for targeted therapy, intelligent control, and sensing technologies. He has published over 20 papers in top IEEE journals, including IEEE Transactions on Robotics (TRO) and IEEE Transactions on Cybernetics (T-Cyber). His work has been recognized with the Best Conference Paper Award at RCAR 2025 and the Best Student Paper Nomination at ICIRA 2025. Dr. Cai has led several projects, including those funded by the National Key R&D Program, the National Natural Science Foundation of China (NSFC), and various provincial and municipal research programs in Guangdong and Shenzhen.

Preface

The recent rise of small-scale robots that can convert external energy into mechanical movement and/or forces has been a fascinating topic for both academic and industrial communities, as well as the public. Over the years of development, magnetically driven small-scale robots have been proposed and applied in various fields, especially in biomedical applications, since magnetic fields are compatible with the human body and various medical imaging systems, allowing researchers to build imaging-guided small-scale robotic systems. Current books or book chapters focus more on the small-scale robot itself, while we believe the systematic design methodology of magnetic actuation systems requires more attention, which drives the topic of this book: mobile actuation systems for magnetic small-scale robots.

The control and manipulation strategies of small-scale robots have evolved from manual operation, through semi-automation and fully automated systems, toward an adaptive, intelligent approach, thanks to the rapid development of control theory and deep learning techniques. Thus, magnetic actuation systems play an essential role in controlling these small-scale robots for conducting complex tasks, such as remote sensing, active targeted delivery, small-scale manipulation, and sensing. Recent research has focused on the development of mobile magnetic actuation systems because of their advantages, including large workspaces, quickly variable fields, multiple system configurations, great scalability, and bandwidth, compared with stationary systems. Currently, mobile magnetic actuation systems have been used in various scenarios, including tasks both ex vivo and in vivo. Thus, in the next 5–10 years, we believe mobile magnetic actuation systems will become more intelligent and will be used for conducting more complex tasks in a self-adaptive manner.

This book focuses on mobile magnetic actuation systems for manipulating magnetic small-scale robots, aiming to summarize recent developments in the design, control, validation, and optimization of magnetic actuation systems. By reading this book, readers will grasp the key research challenges and opportunities of magnetic actuation systems, including system design methods, modeling logic, optimization strategies, and experimental validation demonstrations. Readers will have a deeper understanding of how to apply these systems for practical applications, such as biomedical applications, especially for audiences in the fields of robotics, control, imaging, and mechanical/electrical engineering. For practitioners, the improved understanding of automation and intelligence will broaden their horizons and encourage them to seek collaboration opportunities with experts in the small-scale robotics field and intelligent control field.

The goal of this book is to convey recent advances in the design, construction, and control of mobile magnetic actuation systems and to promote filling the gap between their theoretical investigation and practical applications. The book is suitable for scientists, engineers, and students who are engaged in the investigation of small-scale robotics. We have tried to make the book a systematic introduction to the current state-of-the-art and future research directions of magnetic small-scale robots, which may help researchers and practitioners have a better understanding and further stimulate the development of such an exciting field.

The material of this book is composed of 10 chapters. Chapter 1 is devoted to briefing the background, introducing the basic concepts, actuation principles, and the historical development of magnetic actuation technology. Chapter 2 presents the design and real-time optimization of actuation systems, followed by a learning-based control framework in Chapter 3. Chapters 4–5 introduce the design and control of reconfigurable electromagnetic actuation systems. Chapters 6–7 cover multimode control, mechanism design, and parameter optimization of mobile actuation systems. Chapters 8–9 are devoted to practical implementation in medical and engineering contexts, with an emphasis on vascular intervention. Finally, Chapter 10 summarizes the design methodologies and integration strategies between imaging and actuation systems and discusses future opportunities and challenges in translational biomedicine and interdisciplinary research.

The book is written with the kind help of many organizations and individuals. The authors would like to acknowledge support from the National Natural Science Foundation of China with project Nos. 52205590 and 52575652; the Natural Science Foundation of Jiangsu Province with project No. BK20220834; the start-up Research Fund of Southeast University (SEU) with project No. RF1028623098; the Chinese University of Hong Kong (CUHK) internal grants; the Shenzhen Institutes of Advanced Technology (SIAT)-CUHK Joint Laboratory of Robotics and Intelligent Systems; the Multiscale Medical Robotics Center (MRC) at the Hong Kong Science Park; the CUHK Chow Yuk Ho Technology Centre for Innovative Medicine; and the CUHK T Stone Robotics Institute.

We are very grateful to Prof. Bradley Nelson from ETH Zurich, Switzerland; Prof. Yu Sun from the University of Toronto; Prof. Zhonghua Ni and Prof. Yunfei Chen, Co-directors of Jiangsu Key Laboratory for Design and Manufacturing of

Precision Medicine Equipment, SEU; and the other collaborators, as well as the colleagues from the School of Mechanical Engineering at SEU, Faculty of Engineering at CUHK, and the SIAT-Chinese Academy of Sciences (CAS), for their long-term guidance and generous support to our group's research on mobile magnetic actuation systems. We appreciate all the laboratory ex- and current members, and many other researchers for contributing their excellent research work to this book, including Prof. Xingzhou Du, Prof. Jiangfan Yu, Prof. Lidong Yang, Prof. Dongdong Jin, Prof. Xinyu Wu, Dr. Zhengxin Yang, Dr. Moqiu Zhang, Dr. Shiwei Du, Zhaoyang Qi; and we would also like to thank Zhiqiang Chen, Junjian Zhou, Xinyu Gu, Ziyang Qi, and Hao Xie for their assistance in editing and proofreading. The editorial and production staff of Wiley-IEEE Press are gratefully acknowledged for the kind arrangement in writing this book that enables its publication. Thanks to the great and fascinating efforts achieved by numerous researchers all over the world, the continuous discoveries keep pushing the forefront of mobile small-scale robotics, and we sincerely hope that this book enhances readers' comprehension of small-scale robotics and inspires further exploration in this exciting area of research. May the contents within these pages provide valuable references and inspire new perspectives for researchers venturing into this domain.

Finally, we would like to take this opportunity to send our best greetings for celebrating the 110th anniversary of the School of Mechanical Engineering, Southeast University in 2026.

Nanjing, China
December 2025

Qianqian Wang
(On behalf of all authors)

List of Acronyms

2D	Two-dimensional
3D	Three-dimensional
3-DOF	Three Degrees of Freedom
5-DOF	Five Degrees of Freedom
AA	Adjusted Angle
ADRC	Active Disturbance Rejection Control
ANN	Artificial Neural Network
APF	Artificial Potential Fields
ARMM	Advanced Robotics for Magnetic Manipulation
ASMC	Anti-slip Model Control
AV	Adjusted Velocity
BF	Base Frame
C-arm	C-arm X-ray Imaging System
CCS	Coil Configuration Space
CM	Cylindrical Magnet
CPU	Central Processing Unit
CT	Computed Tomography
D-EMF	Directional External Magnetic field
DOB	Disturbance Observer
DOF	Degree of Freedom
DOFs	Degrees of Freedom
DRL	Deep Reinforcement Learning
Dyn-NN	Dynamic Neural Network
EM	Electromagnet/Electromagnetic
EMA	Electromagnetic Actuation
EMF	Electromagnetic Field
EMS	Electromagnetic System
ESO	Extended State Observer
FEM	Finite Element Method

FLC	Fuzzy Logic Controller
FLM	Fuzzy Logic Mapping
FoV	Field of View
FPS	Frames Per Second
FPP	Flow Pulsation Pump
FS	Front Sight
GCA	Guidewire/Catheter Advancer
GI	Gastrointestinal
GPU	Graphics Processing Unit
GRU	Gated Recurrent Unit
GUI	Graphical User Interface
HB	Helical Bullet
HC	Helmholtz Coil
HCs	Helmholtz Coils
HMMs	Helical Micromachines
HOGF	Histogram of Oriented Gradients Feature
LQR	Linear Quadratic Regulator
LS	Left Sight
MAD	Mean Absolute Deviations
MC	Maxwell Coil
MCGRS	Magnetically Controlled Guidewire Robotic System
MCRs	Magnetic Continuum Robots
MCs	Maxwell Coils
MDG	Magnetic Driller-tipped Guidewire
MFW	Magnetic-feasible Workspace
MLESAC	Maximum Likelihood Estimator Sample Consensus
MMR	Magnetic Micro/Millirobot
MMRs	Magnetic Micro/Millirobots
MNMs	Micro-/Nanomachines
MOO	Multi-objective Optimization
MPC	Model Predictive Control
MRI	Magnetic Resonance Imaging
MSC	Mesenchymal Stem Cell
MSCs	Mesenchymal Stem Cells
NdFeB	Neodymium-Iron-Boron
NSGA-II	Nondominated Sorting Genetic Algorithm II
OCP	Overall Cross-sectional Picture

OCT	Optical Coherence Tomography
PI	Proportional-integral
PID	Proportional-integral-derivative
PM	Permanent Magnet
PMCS	Parallel-mobile-coil System
PRF	Pulse Repetition Frequency
PRP	Platelet-rich Plasma
PSO	Particle Swarm Optimization
PVC	Polyvinyl Chloride
PVD	Peripheral Vascular Disease
PWM	Pulse-width Modulation
R-EMF	Rotating Electromagnetic Field
RAM	Random Access Memory
RBC	Red Blood Cell
REMA	Reconfigurable Electromagnetic Actuation System
RH	Rotary Holder
RGB-D	Red-green-blue-depth
ROI	Region of Interest
RRT	Rapidly exploring Random Tree
RS	Right Sight
RWD LSI	Real-time Widefield Laser Speckle Imaging System
S-RBCs	Simulated Red Blood Cells
SAC	Soft Actor-critic
SD	Standard Deviation
SMC	Sliding Mode Controller
SMMR	Soft Magnetic Miniature Robot
STL	StereoLithography
STM32	STMicroelectronics 32-bit Microcontroller
TC	Teleoperation Console
TCP/IP	Transmission Control Protocol/Internet Protocol
U-Net	U-shaped Convolutional Neural Network
UR	Universal Robot (UR10)
US	Ultrasound
VIS	Vascular Interventional Surgery
VMTK	Vascular Modeling Toolkit
WMM	Wired Magnetic Microrobot
YOLO-V4	You Only Look Once Version 4

1

Introduction to Mobile Magnetic Actuation System

1.1 Introduction

In recent years, miniature robots have increasingly captured the focus of researchers and practitioners, primarily owing to their substantial potential in medical and bioengineering fields [1–3]. The definition of a "miniature robot" pertains to a device that is controllable and has dimensions spanning from micrometers to millimeters. Due to their small size, these robots can access intricate and confined areas within the human body in a minimally invasive fashion, including the gastrointestinal (GI) tract, vasculature, brain, and eye. Furthermore, they possess the capability to carry out a variety of tasks, such as targeted drug delivery [4], precision surgical operations [5, 6], and comprehensive medical examinations [7]. Besides, microsized robots hold promise for in vitro applications, like cell manipulation and tissue engineering, because they can manipulate subcellular entities with high precision and repeatability.

Given their small size, the newly developed tiny agents hold great promise for transforming medical diagnosis and treatment via minimally invasive techniques. Nevertheless, incorporating conventional onboard components, including actuators, processors, and power sources, poses significant difficulties in submillimeter robots due to their constrained volume. Consequently, researchers have directed their efforts toward remote actuation methods to achieve wireless control of untethered robots. The proposed approaches encompass optical [8], acoustically controlled [9, 10], chemical reaction, and magnetic field-based methods [11].

Optical tweezers, as highly sensitive and versatile tools, are capable of trapping cells and microparticles. However, their penetration depth within the body is limited, resulting in insufficient force for practical applications. Acoustically controlled methods offer a non-harmful driving force for specially engineered microrobots inside the human body, yet precise control over their direction and trajectory remains elusive, with research in this area still in its infancy. Chemical reaction methods can propel microrobots forward, but managing the chemical changes induced in the body proves to be a formidable challenge. In contrast, the magnetic field method enables wireless actuation of magnetic robots without signal attenuation during penetration through the human body. Magnetically controlled microrobots can be precisely guided to navigate within living organisms, addressing a multitude of medical issues. However, magnetic actuation is not without its drawbacks, including a confined workspace, restricted magnetic flux density for actuation, and the bulkiness and costliness of the designed system. Therefore, the development of magnetic actuation systems (MASs) has been geared toward overcoming these limitations.

The inherently interdisciplinary nature of MASs in clinical settings underscores the need for coordinated efforts among a diverse group of experts, encompassing roboticists, clinical practitioners, regenerative medicine researchers, and biomedical engineers. Clinical specialists blend conventional approaches with engineering insights to devise innovative solutions to prevailing challenges [12].

Simultaneously, regenerative medicine scientists have been refining stem cell modifications to elicit therapeutic outcomes at designated target locations. Within the realm of biomedical engineering, magnetic micro/millirobots (MMRs) have been harnessed as delivery vehicles to achieve pinpoint drug administration through closed-loop control mechanisms [13, 14]. Notably, the development of MASs for clinical use by roboticists and engineers serves as the cornerstone that supports and enables all these collaborative initiatives [15–17]. The fusion of specialized knowledge and concerted efforts across these disciplines drives progress and fosters innovations that hold immense promise for the future landscape of medical robotics.

The traditional paradigm of medical treatment commonly involves introducing high-dosage medications into patients via the bloodstream, relying on passive transport. Nevertheless, this approach has inherent drawbacks. A substantial amount of the medication can be taken up by nontargeted organs or cells, which may lead to side effects and harm to the body. Consequently, there is a pressing need for a more accurate and controlled drug delivery mechanism. MASs offer the capability of precise, wireless control over magnetic robots, facilitating targeted delivery and controlled movement within the body. This unique feature unlocks a wide array of clinical applications. In addition to precision drug delivery, MMRs, when actuated by MASs, can be employed to deliver functional cells, such as stem cells, to specific sites for repairing damaged cartilage. Moreover, MASs are relevant in various scenarios requiring precise delivery, such as removing blood clots from the cardiovascular system and performing vitreoretinal or GI endoscopy. Addressing these diverse applications requires the development of suitable MASs and MMRs that are tailored for clinical use. Significant attention has been devoted to advancing the MAS and MMR designs to satisfy the specific needs of clinical applications, emphasizing their potential to revolutionize the field of medicine.

The development of actuation systems for clinical applications is accompanied by inherent challenges. A critical issue stems from the intrinsic properties of MASs. From an energy perspective, the natural instability of the dipole in all directions poses a significant obstacle to the effective control of permanent MASs. Additionally, the magnetic field strength decreases rapidly as it moves away from the source, following a cubic distance decay pattern. This rapid attenuation renders the field unsuitable for many applications where sustained strength is required. Furthermore, even when the magnetic field is amplified at the center of the workspace, the steep gradients near the electromagnet limit the effective operational area to a region smaller than the full 3D environment. This constraint adds complexity to the design of MASs for real-world practical use.

In response to the challenges inherent in MAS design, researchers have adopted the closed-loop flux path method to boost magnetic field intensity and gradient. When integrated with appropriate mapping strategies, this system can function as a magnetic catheter guidance platform. A thorough examination of magnetic field distribution across the workspace and between electromagnets is critical for calculating the precise current levels needed to accomplish specific tasks [18]. Strict current regulation is necessary to prevent excessive heating, which could compromise wire integrity. Moreover, the effective operational workspace for many practical applications remains constrained. To address these limitations, recent studies have focused on innovations such as optimized wire configurations and advanced coil designs [19]. These advancements aim to elevate manipulation capabilities and extend the boundaries of magnetic actuation in intricate clinical scenarios.

The efficient control of MMRs is paramount to their operational performance, particularly when these robots collaborate to execute intricate tasks. A robot's morphological design profoundly influences its control dynamics and functional capabilities. Presently, significant research efforts have been directed toward understanding core locomotion mechanisms—such as slipping, rolling, walking, and climbing—as these foundational modes can be integrated to achieve diverse operational objectives [20]. A critical consideration in MMR design is material selection, which must prioritize magnetizable properties while ensuring biocompatibility (non-cytotoxicity). Notably, recent advancements have centered on developing MMRs using materials that are both biodegradable and magnetizable, aiming to create robots capable of completing tasks and subsequently degrading safely within the body [21, 22]. The objective is to enable precise navigation and propulsion by integrating multiple locomotion strategies. Beyond robot design, the development of MASs capable of generating sufficiently strong magnetic field gradients or torque within expansive workspaces is equally essential. This aspect has been a focal point in the literature, with studies emphasizing the need for MAS designs that ensure accurate, high-gradient magnetic fields during MMR navigation [23, 24]. Collectively, these advancements form the foundation for medical interventions leveraging magnetic manipulation techniques.

Beyond the technical optimization of MASs and MMRs, the operational success of magnetic manipulation systems hinges critically on real-time feedback derived from advanced imaging modalities. Among these, photoacoustic imaging has emerged as a particularly promising in vivo imaging technique [25], garnering significant interest for its utility in providing precise feedback signals for contemporary research applications. Depending on the application context, alternative imaging approaches such as fluoroscopy or magnetic resonance imaging (MRI) may also be employed. However, these methods are accompanied by notable limitations, including potential health risks associated with prolonged exposure (e.g., ionizing radiation in fluoroscopy) and the inherently time-consuming nature of MRI protocols. Collectively, these factors underscore the need for careful selection and integration of imaging technologies to balance feedback quality with practical constraints in clinical settings.

The control of magnetic robots can be broadly categorized into three paradigms: individual manipulation, coordinated multi-agent systems, and large-scale swarm control [26, 27]. The incorporation of machine learning algorithms and optimization strategies further elevates operational efficiency and precision [28]. These advanced methodologies have yielded encouraging outcomes, particularly in enabling accurate drug delivery within intricate three-dimensional (3D) environments—a critical capability for targeted medical interventions.

1.2 Magnetic Actuation Systems

1.2.1 Magnetic Actuation Principle

Magnetic actuation operates by exerting forces or torques on magnetic entities—either embedded with permanent magnets or composed of magnetizable materials—through externally controlled magnetic fields (see Figure 1.1). Notably, the absence of electrical currents within the operational environment allows the quasistatic magnetic field to be rigorously modeled using Maxwell's equations, as established in [29]. This framework provides a foundational basis for analyzing and optimizing magnetic manipulation systems.

$$\nabla \cdot \mathbf{B} = 0 \quad \nabla \times \mathbf{B} = 0 \tag{1.1}$$

where $\mathbf{B}$ is the external magnetic field and ∇ is the gradient operator. The equation implies that the gradient matrix of $\mathbf{B}$ is symmetric and trace free. A magnetic torque τ acts on a magnetic object m when misalignment occurs between the magnetization of the object and the orientation of the magnetic field as follows:

$$\tau = \mathbf{m} \times \mathbf{B} = \begin{bmatrix} 0 & B_z & -B_y \\ -B_x & 0 & B_x \\ B_y & -B_x & 0 \end{bmatrix} \begin{bmatrix} m_x \\ m_y \\ m_z \end{bmatrix} \tag{1.2}$$

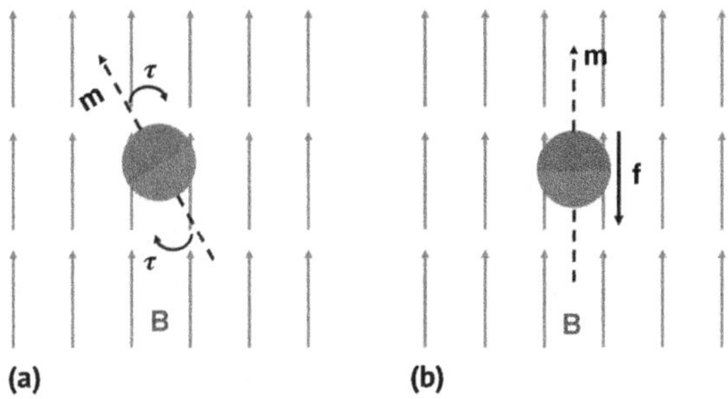

Figure 1.1 Diagram of magnetic interaction. (a) pure torque under a uniform magnetic field; (b) pure force under a nonuniform magnetic field.

Moreover, a magnetic force **f** exists, when a magnetic object is in a nonuniform magnetic field as follows:

$$\mathbf{f} = (\mathbf{m} \cdot \nabla)\mathbf{B} = \begin{bmatrix} \frac{\partial B_x}{\partial y} & \frac{\partial B_y}{\partial y} & \frac{\partial B_y}{\partial z} \\ \frac{\partial B_x}{\partial z} & \frac{\partial B_y}{\partial z} & -\left(\frac{\partial B_x}{\partial x} + \frac{\partial B_y}{\partial y}\right) \end{bmatrix} \begin{bmatrix} m_x \\ m_y \\ m_z \end{bmatrix} \tag{1.3}$$

Either effect can be deployed individually or in concert to achieve six-degree-of-freedom motion. In the analytical framework of Eqs. (1.2) and (1.3), each robot is treated as a magnetic dipole, an assumption justified when its size is dwarfed by the workspace. With full field programmability, the dipole model caps controllable degrees of freedom (DOFs) at five—three for translational force and two for torque—since rotation about the intrinsic magnetization axis is unavailable. A force couple generated by tailored multimagnet architectures reinstates the absent degree and thereby secures complete six-axis guidance [30]. Altogether, the magnetic field offers eight independent coordinates: three field components together with five gradient components.

To characterize the energy field in an electromagnetic manipulation system, the first analytical step relies on the four Maxwell equations, which collectively establish the theoretical basis for modeling both electric and magnetic field behaviors. These equations form the cornerstone for understanding field interactions and energy distribution in such systems.

$$\nabla \cdot \mathbf{E} = \frac{\rho}{\varepsilon_0} \tag{1.4}$$

$$\nabla \times \mathbf{E} = -\frac{\partial \mathbf{B}}{\partial t} \tag{1.5}$$

$$\nabla \cdot \mathbf{B} = 0 \tag{1.6}$$

$$\nabla \times \mathbf{B} = \mu_0 \mathbf{J} + \mu_0 \varepsilon_0 \frac{\partial D}{\partial t} \tag{1.7}$$

where **E** represents the electric field strength, related to the electric displacement vector **D** via $\mathbf{E} = \frac{1}{\varepsilon_0}\mathbf{D}$, with ε_0 denoting vacuum permittivity; **B** denotes the magnetic flux density, linked to the magnetic field strength **H** (in A/m) through $\mathbf{B} = \mu_0 \mathbf{H}$, where $\mu_0 = 4\pi \times 10^{-7}$H/m is the vacuum permeability. The electric charge density is symbolized by ρ, while **j** corresponds to current density (zero in the manipulation region but non-zero within magnets). Alternative formulations of Maxwell's equations are derived from the foundational Eqs. (1.4)–(1.7).

The magnetic dipole model is widely adopted to describe the spatial variation of magnetic flux density around a single pole of a permanent magnet. For such configurations, the scalar magnetic potential is typically expressed using a Taylor series expansion, which provides a systematic framework for approximating the field's behavior. This approach enables analytical modeling of the magnetic potential's dependence on spatial coordinates near the magnet's pole region.

$$\mathbf{\Phi}(r) = \frac{1}{4\pi} \sum_{n=0}^{\infty} \frac{1}{\|r\|^{n+1}} \oint_S \rho^n \mathrm{P}_n(\mathbf{r} \cdot \rho)(\mathbf{n} \cdot \mathbf{M})\, \mathrm{da} \tag{1.8}$$

where ρ denotes the distance from the magnetic volume's center to the integration point, $\mathrm{P}_n()$ represents Legendre polynomials, **M** is the magnetization vector, **n** is the normal unit vector, and **r** is the position vector from the magnetic volume's center to the point of interest (expressed as $\mathbf{r} = {}^{\mathbf{i}}P_{\mathbf{B}} +^{\mathbf{i}} \mathbf{P}_{\mathbf{m}}$, as illustrated in Figure 1.2a). This formulation simplifies to a multipole expansion when the analysis points lie outside the minimum bounding sphere—the smallest sphere entirely containing the magnet during magnetic flux evaluation.

The magnetic flux density can be represented with an expansion form as:

$$\begin{aligned} \mathbf{B}\{\mathbf{r}, \mathbf{m}\} &= \mu \mathbf{H} = -\mu \nabla \Phi \\ &= \left(\frac{\mu}{4\pi \|\mathbf{r}\|^2} \mathbb{P}_2\{\hat{\mathbf{r}}\} + \frac{\mu}{4\pi \|\mathbf{r}\|^3} \mathbb{P}_3\{\hat{\mathbf{r}}\} + \frac{\mu}{4\pi \|\mathbf{r}\|^4} \mathbb{P}_4\{\hat{\mathbf{r}}\} \right. \\ &\quad \left. + \frac{\mu}{4\pi \|\mathbf{r}\|^5} \mathbb{P}_5\{\hat{\mathbf{r}}\} + \ldots \right) \mathbf{m} \end{aligned} \tag{1.9}$$

where μ denotes the permeability of the surrounding medium, which is $\mu_0 = 4\pi \times 10^{-7}\ \mathrm{T \cdot m \cdot A^{-1}}$ in free space. This value can vary depending on the medium in which the MMR is placed. The dipole moment **m** is defined as the integral of the magnetization **M** over the volume **V**.

Based on the properties outlined in Eq. (1.6), the even-order terms ($\propto \frac{1}{r^2}$, $\propto \frac{1}{r^4}$, etc.) are nullified in the expanded version of Eq. (1.9). Consequently, the initial

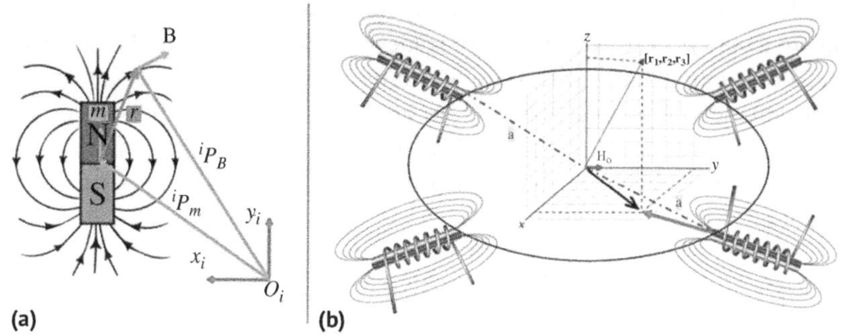

Figure 1.2 Magnetic field illustration. (a) The representation of magnetic flux density around a permanent magnet (represented as a magnetic dipole), where the original coordinate is $x_iO_iy_i$, $\mathbf{m}$ represents the magnetic moment, $\mathbf{B}$ is the magnetic flux density at the position away from the center of the magnetic dipole with a distance $\mathbf{r}$. (b) The multiple electromagnets' field distribution, where the $[\mathbf{r}_1, \mathbf{r}_2, \mathbf{r}_3]$ represents the position under global coordinate, $\mathbf{H_0}$ is the unit magnetic field strength, and $\mathbf{a}$ is the radius of the workspace.

nonzero term in the polynomial expansion is characterized by a decay rate of $||r||^{-3}$ (the dipole term with $n = 2$), followed by subsequent terms that diminish at a rate of $||r||^{-5}$, and so forth.

The magnetic dipole model is frequently employed to estimate magnetic flux density values, particularly when the magnet's geometry and internal magnetization distribution are unknown. This simplified approximation serves as the most commonly referenced approach under such conditions, enabling practical estimation of field strength based on measurable external parameters.

$$\mathbf{B}_{\text{dipole}}(\mathbf{r}) = \frac{\mu_0}{4\pi}\frac{1}{||\mathbf{r}||^3}\left(3\hat{\mathbf{r}}\hat{\mathbf{r}}^T - \mathbb{I}\right)\mathbf{m}; \tag{1.10}$$

where $\hat{\mathbf{r}}$ represents the unit vector of $\mathbf{r}$ and $\mathbb{I} \subseteq \mathbb{R}^{3\times3}$ denotes the identity matrix. As per Eq. (1.10), the dipole field's magnitude diminishes cubically with distance. In terms of geometry optimization for permanent magnets in static MAS, research indicates that cylinders with rectangular cross-sections (with a diameter-to-length ratio of $\sqrt{\frac{4}{3}}$) and cubes are the most effective shapes.

From an energy standpoint, the total energy U_b of a magnetic dipole with a moment $\mathbf{m}$ can be represented as follows.

$$U_b = -\mathbf{m}\cdot\mathbf{B} = -\begin{bmatrix} m_x & m_y & m_z \end{bmatrix}\cdot\begin{bmatrix} B_x \\ B_y \\ B_z \end{bmatrix} \tag{1.11}$$

Earnshaw's theorem elucidates the inherent instability of a magnetic dipole within an external magnetic field **B**. To achieve stable levitation of the magnetic dipole, it is essential that Eq. (1.8) possess a minimum point. This necessitates the Laplace equation of the total energy U_b to satisfy the condition $\nabla^2 U_b > 0$, ensuring the system's stability through a positive definiteness in the energy landscape.

Because both the divergence and curl of **B** are zero, the Laplace equation of **B** is zero, that is

$$\begin{aligned}\nabla^2 U_b &= -\frac{\partial^2 \mathbf{m} \cdot \mathbf{B}}{\partial x^2} - \frac{\partial^2 \mathbf{m} \cdot \mathbf{B}}{\partial y^2} - \frac{\partial^2 \mathbf{m} \cdot \mathbf{B}}{\partial z^2} \\ &= -m_x \nabla^2 B_x - m_y \nabla^2 B_y - m_z \nabla^2 B_z = 0\end{aligned} \tag{1.12}$$

Therefore, the dipole cannot reach the minimum or maximum energy points. The dipole cannot be stable in all directions at any point in free space under magnetic field **B**.

To address instability in electromagnetic systems, researchers have explored diverse strategies. One approach involves modifying system accessories to compensate for unstable directions during analysis, while another utilizes magnetic traps to confine objects at specific 2D surface locations [31]. However, these solutions lose efficacy in 3D environments when holding plates or accessories are removed, leading to renewed instability. A related technique in magnetic catheter-based guidance balances mechanical reaction forces with magnetic gradient-generated forces to stabilize objects, though such methods are impractical in complex in vivo clinical environments—a recognized challenge in existing studies [32]. Alternative approaches circumvent Earnshaw's theorem through dynamic magnetic field control [33]. By applying time-varying currents to electromagnets, researchers create transient trapping points, enabling adaptive applications like targeted radiation therapy and selective embolization in cancer treatment. Most clinical systems rely on open-loop preprogrammed controls, which lack adaptability to dynamic fields. A commonly adopted solution involves integrating perception feedback to enhance magnetic robot control accuracy, stabilizing 3D motion under MAS. The following section elaborates on open-loop and closed-loop control methodologies.

1.2.2 Electromagnetic Field and Force/Torque Analysis

The underlying basis for electromagnetic field generation is outlined in Eq. (1.7), illustrating how a magnetic field forms in response to an applied electric field. Notably, the second term on the right-hand side accounts for the hysteresis effect, which is linked to electric displacement.

The force applied to a magnetic object with magnetic dipole moment **m** in field **B** can be represented by the following equation.

$$\mathbf{F} = (\mathbf{m} \cdot \nabla)\mathbf{B} \tag{1.13}$$

At room temperature, **m** is linearly related to **B** for a superparamagnetic under-saturated magnetic robot. The moment can be represented by Eq. (1.14):

$$\mathbf{m} = V_m\mathbf{M} = V_m \frac{\chi_m}{(1 + \chi_m)\mu_0}\mathbf{B} \tag{1.14}$$

where χ_m is the magnetic susceptibility.

According to Eqs. (1.6) and (1.7), the conditions $tr(\nabla\mathbf{B}) = 0$ and the symmetry of $\nabla\mathbf{B}$ are satisfied within an electromagnetic actuation system's workspace for a magnetic object. Consequently, the magnetic force can be expressed accordingly.

$$\mathbf{F} = \begin{bmatrix} \frac{\partial B_x}{\partial x} & \frac{\partial B_x}{\partial y} & \frac{\partial B_x}{\partial z} \\ \frac{\partial B_x}{\partial y} & \frac{\partial B_y}{\partial y} & \frac{\partial B_y}{\partial z} \\ \frac{\partial B_x}{\partial z} & \frac{\partial B_y}{\partial z} & -\left(\frac{\partial B_x}{\partial x} + \frac{\partial B_y}{\partial y}\right) \end{bmatrix} \mathbf{m}$$

$$= \begin{bmatrix} m_x & m_y & m_z & 0 & 0 \\ 0 & m_x & 0 & m_y & m_z \\ -m_z & 0 & m_x & -m_z & m_y \end{bmatrix} \begin{bmatrix} \frac{\partial B_x}{\partial x} \\ \frac{\partial B_x}{\partial y} \\ \frac{\partial B_x}{\partial z} \\ \frac{\partial B_y}{\partial y} \\ \frac{\partial B_y}{\partial z} \end{bmatrix} = \mathcal{F}(\mathbf{m})\mathcal{G}(\nabla\mathbf{B}^{\mathrm{T}}) \tag{1.15}$$

In a simplified scenario where the magnetic field coordinates coincide with the global coordinates, resulting in $\frac{\partial B_x}{\partial y} = \frac{\partial B_x}{\partial z} = \frac{\partial B_y}{\partial z} = 0$, the magnetic force can be simplified further.

$$\mathbf{F} = \begin{bmatrix} m_x & m_y & m_z & 0 & 0 \\ 0 & m_x & 0 & m_y & m_z \\ -m_z & 0 & m_x & -m_z & m_y \end{bmatrix} \begin{bmatrix} \frac{\partial B_x}{\partial x} \\ 0 \\ 0 \\ \frac{\partial B_y}{\partial y} \\ 0 \end{bmatrix} = \begin{bmatrix} m_x \frac{\partial B_x}{\partial x} \\ m_y \frac{\partial B_y}{\partial y} \\ -m_z \left(\frac{\partial B_x}{\partial x} + \frac{\partial B_y}{\partial y} \right) \end{bmatrix} \tag{1.16}$$

Magnetic torque arises from the rotation of the magnetic dipole **m** around an axis, with its direction being perpendicular to both **m** and **B**. For a magnetic dipole **m** positioned at an angle θ_x from the direction of **B** relative to the x-axis, the magnetic torque τ_x is given by the following expression.

$$\tau_x = \frac{\partial(\mathbf{B} \cdot \mathbf{m})}{\partial \theta_x} = \frac{\partial(||B||||m|| \cos(\theta_x))}{\partial \theta_x} = -||B||||m|| \sin(\theta_x) \tag{1.17}$$

The total torque in three dimensions can be represented as:

$$\tau = \mathbf{m} \times \mathbf{B} = \mathrm{Sk}(\mathbf{m})\mathbf{B} = \begin{bmatrix} 0 & -m_z & m_y \\ m_z & 0 & -m_x \\ -m_y & m_x & 0 \end{bmatrix} \begin{bmatrix} B_x \\ B_y \\ B_z \end{bmatrix} \tag{1.18}$$

where Sk(**m**) denotes the skew-symmetric form of the vector **m**×. Unlike the magnetic force, which possesses 3-DOFs, the magnetic torque has only two controllable DOFs. This limitation exists because it is not possible to produce torque about an axis that is aligned with the direction of **m**.

In a MAS system that includes several electromagnets, as depicted in Figure 1.2b, there is a linear relationship between the current applied to the coils and the electromagnetic flux density. The magnetic field at a particular point **p** in the workspace is the aggregate of the magnetic flux densities produced by each individual electromagnet. The principle of superposition is applicable only if the magnetization does not reach saturation. The current is represented by $I = [I_1, I_2, \dots, I_i, \dots, I_n]^T$ (in amperes), and the overall magnetic flux density **B** (in Tesla) is given by the following expression.

$$\boldsymbol{B}(\boldsymbol{p}) = \sum_{i=1}^{n} \boldsymbol{B}_i(\boldsymbol{p}) = \sum_{i=1}^{n} \tilde{\boldsymbol{B}}_i(\boldsymbol{p}) I_i \tag{1.19}$$

where $\tilde{\boldsymbol{B}}_i$ is a unit-current vector expressed in Tesla per Ampere.

The partial derivative of $\boldsymbol{B}$ with respect to the x-coordinate at point $\boldsymbol{p}$ in the effective workspace can be represented as follows:

$$\begin{aligned} \frac{\partial \boldsymbol{B}(\boldsymbol{p})}{\partial x} &= \left[\frac{\partial \tilde{\boldsymbol{B}}_1(\boldsymbol{p})}{\partial x}, \ \dots, \ \frac{\partial \tilde{\boldsymbol{B}}_n(\boldsymbol{p})}{\partial x} \right] \left[I_1, \cdots, I_n \right]^T \\ &= \sum_{i=1}^{n} \frac{\partial \tilde{\boldsymbol{B}}_i(\boldsymbol{p})}{\partial x} I_i \end{aligned} \tag{1.20}$$

The initial term in $\mathcal{G}(\nabla \mathbf{B}^{\mathrm{T}})$ of Eq. (1.15), specifically $\frac{\partial B_x}{\partial x}$, can be expressed as $\sum_{i=1}^{n} \frac{\partial (\tilde{B}_{x_i}(\mathbf{p}))}{\partial x} I_i$. Correspondingly, the first term in **B** of Eq. (1.18), B_x, is represented as $\sum_{i=1}^{n} (\tilde{B}_{x_i}(\mathbf{p})) I_i$, where $(\tilde{B}_x)_i(\boldsymbol{p})$ denotes the magnetic flux density in the x direction generated by the ith electromagnet at position $\boldsymbol{p}$.

By incorporating Eqs. (1.15)–(1.20), a compact representation that accounts for a controllable magnetic field and its gradient can be derived, which is expressed as follows:

$$\begin{aligned} \begin{bmatrix} \tau \\ \boldsymbol{F} \end{bmatrix} &= \begin{bmatrix} \mathrm{Sk}(\boldsymbol{m}) & \boldsymbol{O} \\ \boldsymbol{O} & \mathcal{F}(\boldsymbol{m}) \end{bmatrix} \begin{bmatrix} \boldsymbol{B} \\ \mathcal{G}(\nabla \boldsymbol{B}^{\mathrm{T}}) \end{bmatrix} \\ &= \begin{bmatrix} \mathrm{Sk}(\boldsymbol{m}) & \boldsymbol{O} \\ \boldsymbol{O} & \mathcal{J}(\boldsymbol{m}) \end{bmatrix} \begin{bmatrix} \boldsymbol{B}(\boldsymbol{p}) \\ \nabla \boldsymbol{B}(\boldsymbol{p})^{\mathrm{T}} \end{bmatrix} \boldsymbol{I} \end{aligned} \tag{1.21}$$

where $\boldsymbol{O}$ represents the zero matrix, which is instrumental in analyzing the electromagnetic forces and torques in all three spatial dimensions. By knowing the magnetic moment of the object and the unit magnetic flux density in the three directions, the force and torque exerted on the object can be directly linked to the applied current.

The current applied is obtained from the voltage of the power supply. Its variation over time can be depicted as follows:

$$\frac{dI(t)}{dt} = \frac{V_c}{L_a} - \frac{R_a}{L_a} I(t) \tag{1.22}$$

where $I(t)$ denotes the current applied, while $\frac{dI(t)}{dt}$ signifies the rate at which the current changes. V_c, L_a, and R_a represent the voltage across the coil, the inductance, and the resistance of the coil, respectively. When multiple coils are activated concurrently, mutual induction may occur between adjacent coils, although it is typically negligible and often assumed to be zero in practical applications. Further analysis is provided in Section 1.2.3, with a visual representation in Figure 1.2b.

For particular applications, on-site calibration techniques have been established to simplify the process of determining magnetic field distributions. Analytical methods based on Maxwell's equations have been utilized to depict the magnetic scalar potential [34]. Polynomial regression, which incorporates the Laplace equation for flexible field mapping, has been applied in these calculations. The accuracy of these calculations was verified by comparing them with finite-element method simulation results, which can predict magnetic field distributions in new areas.

1.2.3 System Modeling and Control

In a simplified magnetic robot motion analysis, the quasistatic motion can be analyzed as [35]:

$$m\boldsymbol{v}(\dot{t}) = \boldsymbol{F}_m - \boldsymbol{c} \circ \boldsymbol{v}(t) + \Delta \tag{1.23}$$

where $\boldsymbol{v}(t)$ is the velocity of a controlled object, c is the total drag coefficient, $\circ$ denotes the element-wise product, and Δ represents the disturbance. Only $\boldsymbol{F}_m$ is controllable; the other variables vary with the surrounding environment and the controlled object.

In the electromagnetic system analysis, the dynamics of the magnetic field strength $\boldsymbol{H}(r)$ can be modeled based on the applied voltage $\boldsymbol{u}(t)$. As per Eq. (1.9), the magnetic field strength can be converted to magnetic flux density using the parameter magnetic permeability μ; thus, $\boldsymbol{H}(r)$ is utilized in the subsequent analysis.

To analyze the magnetic field strength $\boldsymbol{H}(r, t)$ in a circular region of radius a centered at the origin and surrounded by n electromagnets, as depicted in Figure 1.2b, the total magnetic field at position r and time t is given by $\boldsymbol{H}(r, t) = \sum_{k=1}^{n} \tilde{\boldsymbol{u}}_k(t)\boldsymbol{H}_k(r)$. Here, $\boldsymbol{H}_k(r) \in \mathbb{R}^3$ represents the magnetic field strength at position $r = [r_1, r_2, r_3]^T$ in 3D space, generated by the unit voltage from the center of the workspace.

In the analysis, only four electromagnets were taken into account, and a 2D space was utilized. $\boldsymbol{H}_0$ represents the magnetic field strength distribution around the lower right corner electromagnet and points to the right, while $\boldsymbol{H}_k(r)$ can be represented in global 2D coordinates using a rotation and translation matrix.

$$\boldsymbol{H}_k(\boldsymbol{r}) = -\boldsymbol{O}_k^T \boldsymbol{H}_0(a\boldsymbol{e} - \boldsymbol{O}_k \boldsymbol{r}) \tag{1.24}$$

where $\boldsymbol{e} = [1, 0]^T$ for 2D space analysis, and the rotation matrix $\boldsymbol{O}_k$ is represented as:

$$\boldsymbol{O}_k = \begin{bmatrix} \cos\left(\dfrac{2\pi(k-1)}{n}\right) & \sin\left(\dfrac{2\pi(k-1)}{n}\right) \\ -\sin\left(\dfrac{2\pi(k-1)}{n}\right) & \cos\left(\dfrac{2\pi(k-1)}{n}\right) \end{bmatrix} \tag{1.25}$$

The method described is a universal approach to depict the magnetic field distribution in a circular workspace surrounded by four electromagnets.

In high-gradient electromagnetic actuation systems, there is a presence of high resistance and inductance. The applied voltage is represented by the control input vector $\mathbf{u}(t) = [u_1(t), u_2(t), \dots, u_n(t)]^T$, and the voltage-state vector necessary for applying the magnetic field strength is denoted as $y_k(t) = R_k I_k(t)$.

The electromagnetic flux produced by the kth electromagnet is given by $\phi_k = \sum_{j=1}^{n} L_{kj} I_j$. Consequently, $y_k = u_k - \dot{\phi}_k = u_k - L\dot{I}$. Thus, the state space equation can be formulated as follows:

$$\begin{cases} \boldsymbol{I}(t) = \boldsymbol{R}^{-1}\gamma(t) \\ \dot{\gamma}(t) = -\boldsymbol{R}\boldsymbol{L}^{-1}\gamma(t) + \boldsymbol{R}\boldsymbol{L}^{-1}\boldsymbol{u}(t) \end{cases} \tag{1.26}$$

$$R = \text{diag}(R_1, R_2, \dots, R_n); \quad I(t) = [I_1(t), I_2(t), \dots, I_n(t)]^T \tag{1.27}$$

where the inductance matrix $\mathbf{L} \in \mathbb{R}^{n \times n}$ is a symmetric constant matrix, the diagonal elements are L_{kk} denoted as the inductance of the k_{th} electromagnet, and $L_{jk} - L_{kj}$ is denoted as the mutual inductance between two electromagnets j and k. State space Eq. (1.26) can also be derived from Eq. (1.22).

The notation in Eq. (1.26) can be streamlined by defining $\xi\mathbf{A} = \mathbf{R}\mathbf{L}^{-1}$. If k electromagnets possess identical inductance and are symmetrically arranged, then all L_{kj} values are zero, allowing $\mathbf{A}$ to be the identity matrix. Consequently, the state space equation can be characterized by a single parameter ξ.

$$\boldsymbol{H}(r, t) = \boldsymbol{H}(r)\gamma(t) \tag{1.28}$$

$$\boldsymbol{F}_m(r, t) = K\nabla\|\boldsymbol{H}(r, t)\|^2 \tag{1.29}$$

$$\boldsymbol{F}_m(r, t) = 2K\left\{\frac{\partial\,[\boldsymbol{H}(r)\gamma(t)]}{\partial r}\right\}^T \boldsymbol{H}(r)\gamma(t) = K_f g(\mathbf{r}(t), \gamma(t)) \tag{1.30}$$

where K is volume and magnetic permeability-related parameters.

K_f contains the parameters ξ and K. The electromagnetic field is only related to the real-time position $r(t)$ and real-time voltage $\gamma(t)$, with a mapping of g().

According to Eqs. (1.23) and (1.30), the motion analysis is simplified to:

$$m\dot{\mathbf{v}}(t) = K_f g(\mathbf{r}(t), \gamma(t)) - \mathbf{c}_{\mathbf{o}}\mathbf{v}(t) + \Delta \tag{1.31}$$

According to Eqs. (1.26), (1.27), and (1.31), the complete dynamics of the electromagnetic actuator and the plant are:

$$\dot{\gamma}(t) = -\xi A\gamma(t) + \xi A\mathbf{u}(t) \tag{1.32}$$

$$\mathbf{v}(t) = \dot{\mathbf{r}}(t) \tag{1.33}$$

$$\dot{\mathbf{v}}(t) = \sigma_m K_f \mathbf{g}(\mathbf{r}(t), \gamma(t)) - \sigma_m \mathbf{c}_{\mathbf{o}} v(t) + \sigma_m \mathbf{\Delta} \tag{1.34}$$

where $\sigma_m = \frac{1}{m}$.

Preprogrammed control strategies are extensively employed across various applications, particularly in clinical settings where feedback is challenging to obtain [36, 37]. In such scenarios, the trajectory of the magnetic object is regulated by predefining the time-varying position sequence of the permanent magnet or the current input sequence to the electromagnetic system. For electromagnetic actuation systems, the effectiveness of the proposed method is significantly influenced by the coil design, and recent studies have highlighted the importance of optimizing coil design, which is emerging as a promising direction in MAS design.

Real-time acquisition of the magnetic moment is challenging. The necessary magnetic field is associated with the desired moment, as depicted in Eq. (1.14), which serves as a general solution for magnetic force control. This relationship is expressed by $\mathbf{m} = \mathcal{D}(\mathbf{B})$.

To symmetrically analyze and control the electromagnetic field and the gradient, according to Eqs. (1.19) and (1.21), the control input and output can be further represented as:

$$\begin{bmatrix} \mathbf{B} \\ \mathbf{F} \end{bmatrix} = \underbrace{\begin{bmatrix} \mathbb{I} & \mathbf{O} \\ \mathbf{O} & \mathcal{F}(\mathcal{D}(\mathbf{B})) \end{bmatrix}}_{\text{Second}} \underbrace{\begin{bmatrix} \mathbf{B}(\mathbf{p}) \\ \nabla \mathbf{B}(\mathbf{p})^{\mathrm{T}} \end{bmatrix}}_{\text{First}} \underbrace{\mathbf{I}}_{\text{Input}} \tag{1.35}$$

where $\mathbb{I} \subseteq \mathbb{R}^{3\times 3}$, $\mathcal{D}$(B) maps the magnetic field applied to the moment of controlled objects.

Equation (1.35) describes the physical process where the magnetic field and its gradient are produced using the current as an input. Subsequently, the controlled object acquires a magnetic moment through interaction with this magnetic field.

The preprogrammed control of MAS is categorized into three types based on the DOFs they control. The first category is 3-DOFs heading or rotational control. The second involves 3-DOFs force control, where the object' moments are typically aligned with the applied magnetic field for superparamagnetism or constrained in a specific direction for magnetized ferromagnets. The third type encompasses 5-DOFs control, combining heading (2-DOFs) and force control (3-DOFs). In scenarios achieving full 6-DOFs control, the focus is on generating torque around the magnetized axis [38]. The challenge lies in controlling the magnetic torque, which is not aligned with the magnetization direction, under the assumption of nonuniformity.

The open-loop preprogrammed control method programs the necessary magnetic field—whether uniform, rotating, or oscillating—to achieve the desired motion or trajectory [39]. Additionally, this control strategy can manage swarm shape and shows potential in medical applications.

The field alignment force and torque control, as mentioned, are designed to direct the MMR toward the desired orientation. Disturbances caused by the attractive forces between MMRs and other factors can impact motion control. Consequently, the subsequent sections will concentrate on feedback-based control methods to improve the precision of preprogrammed control.

Figure 1.3a depicts the comprehensive control framework for the MAS, which includes three key components: the planning model, learning module, and control

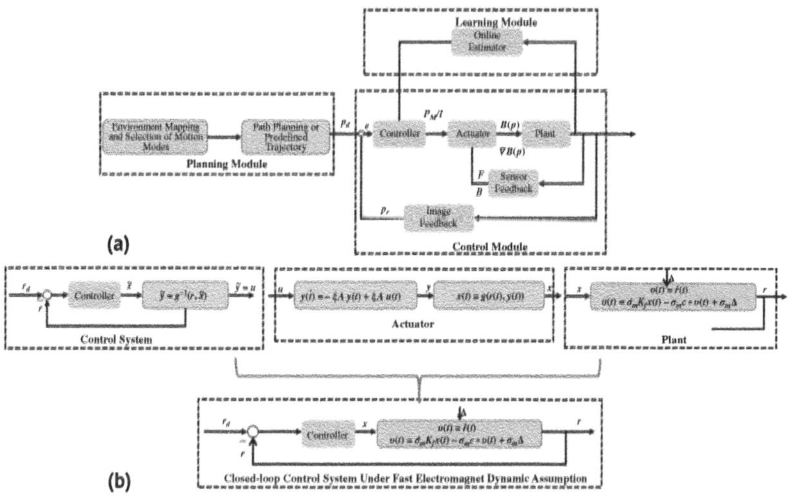

Figure 1.3 (a) General control scheme for MAS-based robot control with sensing feedback and online estimation. (b) The subsystems of the control system, actuation system, and plant; under the fast electromagnet dynamics assumption, the system can be simplified as the linear controller with the magnetic objects' motion dynamics.

module. The planning module serves as the starting point, tasked with responsibilities such as environmental mapping and the selection of suitable locomotion strategies. It then proceeds to path planning. In certain scenarios where the trajectory is predetermined, a preset trajectory sequence is utilized in lieu of path planning [40, 41].

The control module assumes the role of devising a controller based on the established model. Most controllers are aimed at position or velocity control. To achieve a controlled position or velocity, the requisite position ($\boldsymbol{P_M}$) for the permanent magnet or the required current ($\boldsymbol{I}$) for the electromagnetic system is output to regulate the MAS. The real-time imaging feedback is usually harnessed to continually minimize the disparity between the real-time position ($\boldsymbol{p_r}$) and the desired position ($\boldsymbol{p_d}$). In certain applications, feedback mechanisms like Hall effect sensors or force sensors are crucial for refining the magnetic flux density (**B**) or force (**F**) by providing precise feedback [42]. The learning module is essential for estimating unknown parameters during dynamic modeling or controller design, enabling the system to learn and adapt to the magnetic actuation principle.

The complete dynamic system, comprising the actuator and the plant as described by Eqs. (1.32)–(1.34), can be analyzed by dissecting the nonlinear aspects. By distinguishing the linear and nonlinear components in Eq. (1.34), the nonlinear component can be characterized as:

$$\mathbf{x}(t) = g(\boldsymbol{r}(t), \boldsymbol{\gamma}(t)) \tag{1.36}$$

where physical meaning of $\mathbf{x}(t)$ is the electromagnetic force.

Figure 1.3b presents the detailed equations for the block diagram of the controller, actuator, and plant. The applied voltage $u(t)$ serves as the input for the actuation system described by Eq. (1.32), while the state vector $\mathbf{x}(t)$ represents the electromagnetic force that connects the other two subsystems, as depicted by Eqs. (1.33) and (1.34).

Representing all vectors with state $\mathbf{x}$, the nonlinear system in Eq. (1.36) is represented as:

$$\mathbf{x} = g(\boldsymbol{r}, g^{-1}(\boldsymbol{r}, \boldsymbol{x})) \tag{1.37}$$

In practical applications, the dynamic system of electromagnetic actuation guarantees that the frequency of the subsystem described by Eq. (1.32) is much higher than that of the closed-loop control system dynamics. When A is an identity matrix, the system simplifies to the input $\boldsymbol{\gamma}(t)$ being asymptotically equivalent to the input $\boldsymbol{u}(t)$.

A nonlinear controller was proposed through feedback linearization with Eq. (1.38) as:

$$\hat{\gamma} = g^{-1}(\boldsymbol{r}, \tilde{\boldsymbol{x}}) \tag{1.38}$$

where $\tilde{\gamma}$ and $\tilde{\mathbf{x}}$ represent the estimated values of γ and $\mathbf{x}$, respectively. The controller design simplifies after cascading the nonlinear controller, as depicted by Eq. (1.38). Assuming rapid electromagnetic dynamics, the combined results are illustrated at the bottom of Figure 1.3b.

Many studies utilize a simplified system under the assumption of rapid electromagnetic dynamics to design controllers for position and velocity control. Classical controllers encompass linear types, such as proportional integral derivative control, linear-quadratic regulator control, and model predictive control [43, 44]. In certain applications, linear controllers and nonlinear filters are integrated into the system. Nonlinear controllers include adaptive and sliding mode control [45]. For autonomous control, the rapidly exploring random tree method is a common approach. In the learning model, iterative learning and deep learning are employed to estimate unknown parameters [46, 47], while a fuzzy logic-based control scheme is used for complex scenarios.

Preprogrammed control and dynamic magnetic field control are extensively used in clinical settings to accomplish tasks like selective embolization or stem cell delivery within high-viscosity joint fluid. Precise position control is crucial for accurate delivery applications, making feedback control particularly prevalent in intravenous drug delivery and endoscopy.

1.3 Development of Magnetic Actuation Systems

1.3.1 Development Phases and Trend of Magnetic Actuation Systems

In addition to systems with fixed electromagnets, reconfigurable systems have also been developed. These systems vary in their performance regarding magnetic actuation flexibility and precision: some excel in force control, while others show better torque control capabilities [48]. Some systems may encounter singularities, points where they lose the ability to generate specific forces or torques, or require excessive current near these points. To address these issues, adjustable MASs have been introduced. The BigMag system, for example, consists of six electromagnets placed in two fixtures. These fixtures can rotate around a common axis to avoid singularities and reduce total current, enabling the system to produce a strong magnetic field in any direction and at any point within the workspace (Figure 1.4a) [49, 50]. When all electromagnets are coplanar, the intersection angle between two adjacent ones is 60°. Chen et al. presented a system with four stationary and four movable electromagnets (Figure 1.4b) [51], similar to OctoMag but with independent adjustment of the upper set via robotic mechanisms. This design faces the challenge of balancing between a uniform magnetic field and a uniform field gradient.

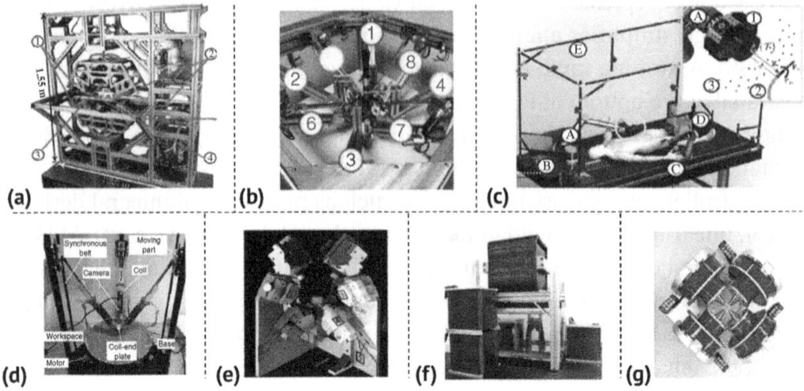

Figure 1.4 Magnetic actuation systems with movable electromagnets or novel designs: (a) BigMag system; (b) system with four stationary and four movable electromagnets; (c) ARMM system; (d) DeltaMag system; (e) BatMag system; (f) Omnimagnet system; (g) system comprising stepped cores. *Source*: Panel (a): Ref. [49] with permission of IEEE. Panel (b): Ref. [48] with permission of IEEE. Panel (c): Ref. [50] with permission of IEEE. Panel (d): Ref. [54] with permission of IEEE. Panel (e): Ref. [23] with permission of IEEE. Panel (f): Ref. [55] with permission of IEEE. Panel (g): Ref. [56] with permission of IEEE.

To address workspace constraints without enlarging electromagnet dimensions, driven mechanisms position mobile electromagnets closer to the region of interest (ROI). The Advanced Robotics for Magnetic Manipulation (ARMM) system pairs a cylindrical electromagnet with a 6-DOF UR10 robotic arm, generating precise magnetic fields at the ROI (Figure 1.4c) [52, 53]. Similarly, DeltaMag integrates three electromagnets into a parallel mechanism (Figure 1.4d), expanding workspace and reducing volume [54]. However, movable electromagnetic systems face challenges: electromagnets are bulkier and heavier than permanent magnets, and multi-coil setups with strong interaction forces are sometimes necessary. Integration with high-load, rigid, and compact drive mechanisms is crucial. Additionally, the total DOFs of drive mechanisms and coil currents can lead to multiple solutions and high nonlinearities, demanding robust, fast, and optimal field generation algorithms.

Several innovative designs have recently been introduced. Ongaro et al. developed the BatMag system, which uses nine electromagnets (Figure 1.4e) to independently control two devices in a 3D space by leveraging the inhomogeneity of strong magnetic fields [23]. Another design is a modular and reconfigurable MAS that uses multiple Omnimagents (Figure 1.4f) [55]. These can be temporarily rearranged during activation. Each Omnimagnet consists of a spherical core and three orthogonally arranged nested solenoids. The core's magnetization is determined by the currents in the three solenoids. Additionally, Li et al. developed a system with four electromagnets, each featuring a stepped core with a cone probe and a disk (Figure 1.4g) [56]. This optimized core shape enhances the magnetic field strength and gradient.

The use of MAS dates back to the 1920s [57], but researchers are still working to address the challenges that arise during its application by refining the system design. A permanent MAS creates a constant and uncontrollable magnetic field gradient. However, this feature restricts its application in certain scenarios where the magnets' positions are fixed due to the lack of controllability. To overcome this limitation, the orthogonal distributed electromagnetic system was developed. This system allows for controlled movement of magnetic objects in orthogonal directions and enables 3D position control through the use of six electromagnets. Despite these advantages, the system is limited by its working area, which makes clinical applications difficult.

The MAS design has evolved to incorporate nonorthogonal distributed electromagnets in response to the demand for greater controllability and enlarged working areas. This approach has gained mainstream status in MAS development. It addresses the limitations of the orthogonal design and enables more adaptable and accurate magnetic actuation for magnetic robots.

The permanent MAS, as detailed in Table 1.1, has some drawbacks, such as a constant and uncontrollable magnetic field gradient. However, it offers benefits like easily adjustable strength and lower costs, making it useful for certain applications. These applications include rotation control, localizing magnetic robots within defined workspaces [58], and particle sorting. Yet, its limitations, such as

Table 1.1 Advantages and disadvantages of different types of MASs.

Type	Advantages	Disadvantages
Permanent magnetic actuator	Easily adjustable magnitude; energy conservation and low cost; strong and stable magnetic field generation.	Limited DOFs; low frequency in a 3D environment; magnetic field cannot be immediately switched off.
Orthogonal distributed electromagnetic actuator	Controllable magnetic field; decoupled magnetic force in all dimensions.	Limited effective workspace; power consumption and heat dissipation.
Nonorthogonal distributed electromagnetic actuator	Larger effective workspace; higher magnetic field gradient.	Inflexible and bulky; power consumption and heat dissipation.
Robotic arm with a magnetic end effector	Extendable workspace; flexible, and easier to install.	Comparatively lower magnetic field gradient.

the unchangeable magnetic field gradient and limited control frequency, make it unsuitable for tasks that require precise positioning with quick on/off switching.

The orthogonal electromagnetic system allows for a manageable magnetic field gradient along orthogonal axes, enabling effective 3D position control with decoupling. However, its smaller operational range and increased heat production at high currents restrict its use in applications needing a bigger workspace or higher field gradient and torque.

The nonorthogonal electromagnetic system, unlike the orthogonal electromagnetic system, provides a bigger workspace, multidirectional controllability, and a higher field gradient or torque. This makes it suitable for surgical assistance, MMR delivery, and targeted drug delivery. However, these benefits are balanced by increased design and control complexity, higher energy consumption, and the risk of interference with other devices.

Recent studies have shown an increasing trend toward integrating the advantages of robotic arms and magnetic systems in MAS designs. This hybrid method can increase the system's workspace and flexibility, allowing for better control of MMRs in complex environments. One approach involves attaching a magnetic tip to a robotic arm to accurately position and manipulate MMRs [59]. Another method uses a robotic arm with multiple magnetic coils at its end to create a magnetic field gradient, helping to control microrobots or catheters with magnetic end effectors [60]. These hybrid systems greatly improve the precision and accuracy of magnetic robot control, making them ideal for clinical use. However, they are more complex than traditional MAS and may need advanced control algorithms and application conditions to optimize performance.

The classification of MAS types was based on papers documented from Google Scholar and the Web of Science databases between 2005 and 2024. After gathering the data, references in relevant publications were further screened. In the end, 42 high-quality papers and books were selected to illustrate the evolutionary trends of MAS types.

As depicted in Figure 1.5, systems with permanent magnet actuation were prevalent from 2005 to 2019. The combination of orthogonal and nonorthogonal electromagnetic systems, termed the electromagnetic actuation system, gained prominence from 2015 to 2024. Robotic arm-based systems saw significant development from 2020 to 2024. MAS research in clinical settings has evolved through several phases, with shifts in focus over time. Studies have progressed from system design to controller development, then to animal testing, and finally to human applications. Recent work on electromagnetic systems has centered on control theory and applications. For robotic arm-integrated systems, recent research has concentrated on system development, controller design, and application reporting within single studies.

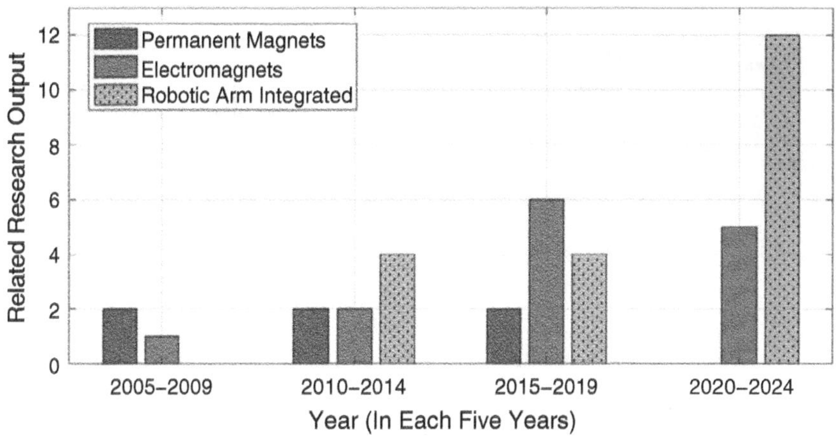

Figure 1.5 MAS development trend.

1.3.2 Development History of Electromagnetic Actuation Systems

The development timeline of electromagnetic actuation systems is illustrated on the right-hand side of Figure 1.6. Early designs were significantly influenced by the concepts of HC and MCs. Figure 1.6a presents a system that integrates HC and MCs to achieve 2D preprogrammed position control. However, clinical MAS applications require at least 3-DOFs control. The electromagnetic actuation system, equipped with multiple-DOF electromagnets, was designed to offer a large workspace and unconstrained DOFs (Figure 1.6b). The OctoMag was specifically developed to control intraocular microrobots for retinal surgery. An alternative method to increase the DOFs for magnetic robot manipulation involves using a dipole encircled by electrical wires, as depicted in Figure 1.6c.

Compared to traditional MAS designs, arranging multiple electromagnets around a workspace can generate a larger magnetic field gradient and workspace [61]. Figure 1.6d,e displays various nonorthogonal magnet arrangements. Adding an iron core greatly boosts the magnetic field strength and gradient from the electrical coils. In 2015, research indicated that a symmetrical system with a minimum of eight electromagnets could control all DOFs. This led to the development of an eight-electromagnet system with a back yoke. This enhanced catheter navigation system, with an enlarged workspace and magnetic field gradient, can perform ablation to treat cardiac arrhythmias. Experiments were done on a printed 3D model. Later, a similar design further expanded the workspace and enhanced the magnetic field gradient for use in multiple invasive therapies and diagnostics. After achieving system functionality and control, the focus shifted

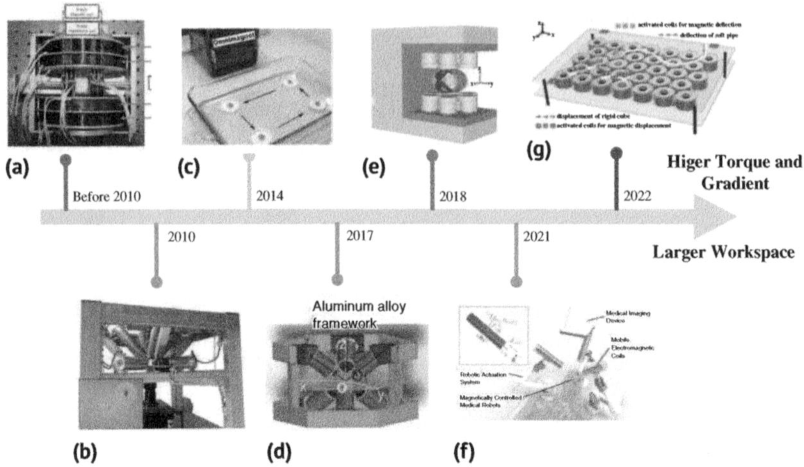

Figure 1.6 Development timeline for the electromagnetic actuators. (a) Two pairs of magnetic coil systems (HCs and MCs) to reach the locomotion in 2D. *Source*: Ref. [63] with permission of IOP Publishing. (b) OctoMag: 5-DOFs control under the electromagnetic field. *Source*: Ref. [64] with permission of IEEE. (c) Omnimagnet: a single stationary dipole source generates a fully controllable magnetic field and achieves 3D localization. *Source*: Ref. [65] with permission of IEEE. (d) Electromagnet-enhanced actuation system with a large workspace design for minimally invasive therapy. (e) Design of 18 copper coils distributed horizontally with a sufficiently large workspace for clinical applications. (f) Conceptual design of RoboMag in which three robotic arms are combined with three decoupled electromagnetic coils. (g) Planar electromagnetic array actuation system design applied in the deflection of soft pipes and displacement of rigid cubes.

to clinical-scale development. This system can be used for cardiac ablation in arrhythmia treatment or steering camera pills for GI diagnostic video acquisition. It has also been suggested for adaptive medical implants, like in adaptive radiation treatment for cancer therapy [62].

Table 1.2 presents a summary of the typical electromagnetic actuation systems intended for clinical use and their characteristics. The design standards took into account the controllable workspace, the maximum magnetic field intensity at the workspace's center when the maximum current is applied, the highest attainable magnetic field gradient, and the dimensions of the objects being controlled. The count of electromagnets served as a benchmark to assess whether complete control over the system was achievable.

As indicated in Table 1.2, despite meeting all the listed requirements, there are still limitations in the design of orthogonal and nonorthogonal distributed electromagnetic actuation systems, such as rigidity and bulkiness. To tackle these issues, different arrangements of electromagnets, as illustrated in Figure 1.6f,g, have been

Table 1.2 Typical electromagnetic actuat on systems and their properties.

	Workspace [mm]	Center B_{max} [mT]	∇B_{max}[Tm^1]	Current [A]	Number of electromagnets	Size of the controlled object
Octomag 2010	Hemisphere φ130	15	0.2	15	8	500 μm
Diller 2016	Cube 120 × 120 × 120	15	0.1	19	8	2 mm
Niu 2017	Sphere φ110	100	2.5	10	6	10 μm
Rahmer 2018	Cylindrical φ200	400	3	110–140	18	2 cm
Li 2019	Circle 2D φ50	139.2	20.5	10	4	100 μm
BatMag 2019	Sphere φ60 ≈ φ160	160	3.575	5	9	350 μm
ARMM 2019	Sphere φ1300	2350	45	20	1	Magnetically actuated catheters
RoboMag 2021	Hemisphere φ203	10	NA	5	3	500 nm microswarm
Array 2022	220.8 × 162.5 mm^2	20	1	Unit current	33	15 × 15 × 8 mm^3 cube and a soft magnetic pipe

proposed. Figure 1.4c presents the design of single or multiple robotic arms to achieve a more extensive workspace for applications in guiding surgical instruments within the human body. With this approach, the workspace can cover the entire human body, effectively addressing the problem of limited workspace.

Nonetheless, the magnetic flux intensity and gradient produced by a single electromagnet are constrained. When utilizing a single robotic arm along with an electromagnet, the magnetic gradient is fairly consistent within a radius of r/2 from the coil's center. Past this radius, the linearity of the gradient declines markedly. Hence, multiple electromagnets are integrated to intensify the magnetic field and gradient. As depicted in Figure 1.6f, it is recommended to place three electromagnets on top of three robotic arms positioned around the workspace. While the cooperative action of multiple electromagnets improves the working area and magnetic force, it also poses challenges for the analysis and control of the field distribution influenced by the movement of robotic arms. The application of a robotic arm-based system can be found in colonoscopy research [66], which was successfully executed in a pig. The time required for a single colonoscopy becomes shorter with autonomous magnetic manipulation. Level 3 autonomous inspection enhances the intelligence level in navigation within unstructured environments, bringing us one step closer to human application.

A different method, motivated by permanent magnet arrays, is to construct a 2D infinite space electromagnetic system, as shown in Figure 1.6g. Such a method addresses the constrained workspace issue in 3D space. Nevertheless, how to expand this kind of system into 3D space while preserving its features continues to be a difficult task.

1.4 Typical Applications of Magnetic Actuation Systems

The application of MAS in medical treatment dates back to 1985 with the use of an oscillating magnetic field for drug delivery. Significant advancements have been made in this field, particularly in applications targeting hard-to-reach areas within the human body that are promising for future use (see Figure 1.7).

This review focuses on developing novel medical treatments for various diseases. The benefits of MAS in surgery include improved precision, ease of use for novices, and reduced patient discomfort through minimally invasive and efficient procedures. Challenges in the operation of magnetic robots within the systems depicted in Figure 1.7 vary based on the disease characteristics and surgical environment.

Variety of environmental factors highlight the inherent differences in the system analysis of medical treatments across various regions of the human body.

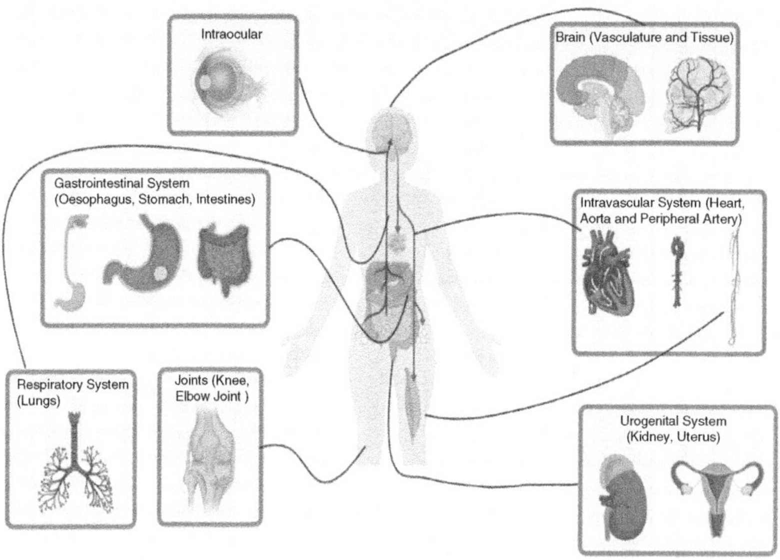

Figure 1.7 Potential environments for using MASs to control magnetic robots in hard-to-reach areas inside the human body.

For instance, intravascular applications require consideration of factors like blood flow speed and vessel size. The intra-ocular, GI, and respiratory systems, as well as human joints, offer natural pathways or accessible surfaces, making magnetic robots suitable for clinical access. Compared to naturally cavitated environments, the utility of MAS in medical treatment is evident in hard-to-reach areas where traditional robotic systems are inadequate. Notable examples include the brain, intravascular system, and urogenital system. Table 1.3 presents representative applications of seven different systems inside the human body, clinical application types, MASs employed, control methods, magnetic robots, imaging techniques, clinical stages, and tethering.

Beyond the typical examples, clinical applications in various human systems are clarified through experimental design, control algorithms, and validation. Control and actuation strategies differ according to the specific human systems or organs. To tackle the challenges of accessing these areas, studies have explored using untethered magnetized objects for drug or cell delivery, capsules with embedded magnets for endoscopy, and tethered continuum robots with magnetic end-effectors for surgery, supported by detailed clinical cases.

Table 1.3 Clinical applications of MASs.

Clinical applications	MASs	Control	Magnetic robot	Imaging	Clinical stage	Untethered
Brain tissue surgery	Permanent magnet array	Magnetic array trapping and motors	NdFeB with polymer housing	Fluoroscopic X-ray	Porcine brain	Y
Peripheral arterial disease surgery	Helmholtz coils	Magnetic sensor-based Preprogrammed control	Magnetic catheter	Digital microscope video	Vascular phantom	N
Aneurysm embolization	Robotic-assisted permanent MAS	Preprogrammed control	Self-adhesive microgels swarm	Fluoroscopy and US	Ex vivo human placenta	Y
Intravascular surgery	Helmholtz coils	Preprogrammed control	Claw-engaged microrobots	OCT catheter	Rabbit jugular vein	Y
Intraocular surgery	Electromagnetic actuation system	Preprogrammed control	Microrobot swarm	Camera/OCT	Bovine eye	Y
Knee cartilage regeneration	Electromagnetic actuation system	Precalibrated control	Medical microrobots	Optical microscope or arthroscope	Rabbit knee	Y
Gastrointestinal capsule endoscopy	Stereotaxis Niobe	Preprogrammed control	Video capsule	Fluoroscopy	Human	Y
Stomach ulcers treatment	Electromagnetic actuation system	Preprogrammed dynamic control	Soft film-like layers	Ultrasound imaging	Porcine stomach	Y
Lung inspection	8-coil MAS	Estimation-based catheter control	Magnetic endoscope	Camera	Lung phantom	N
Assisted urination	Permanent cubic magnet	Predefined field apply	Magnetic soft robotic bladder	Computed tomography	Porcine bladder	Y

1.4.1 Brain

The human brain is composed of the cerebrum, cerebellum, and brainstem, collectively termed brain tissue, with the remaining part being the brain's vascular surroundings. Research on MAS for animals (canine brains) dates back to 1990 [67] when a 5-mm-cylindrical permanent magnet was magnetically guided within the intraparenchymal brains of five adult canines to treat brain cancer, such as delivering drugs and taking biopsy samples. This study demonstrated the feasibility of remotely controlling untethered magnetized robots via a magnetic-driven system for clinical applications like placing hyperthermic heat lesions or targeted biopsies.

Recent studies have employed permanent magnet arrays to create a strong magnetic force trap, allowing robots to be drawn to traps without active control. A planar linear stage and one rotational axis were used to manipulate the microrobot's position within porcine brains. Ex vivo experiments on porcine brains, using fluorescent images for real-time tracking, highlighted the potential of MMRs in drug delivery, biopsies, hyperthermia, and cauterization, despite the study's limitation of a 6 cm workspace.

Controlling motion within the brain vasculature is another critical research area. Recent advanced research has developed cell-sized microrobots, equipped with real-time optoacoustic tracking for in vivo monitoring. The 3D real-time tracking of a microrobot within the murine cerebral vascular system has been accomplished for the first time. To simultaneously achieve imaging and magnetic response, the microrobot was coated with Ni and Au. Magnetic actuation was accomplished using a bar permanent magnet, with operations conducted manually. This successful experiment established a basis for safe and precise real-time procedures within the brain's vascular network.

1.4.2 Intravascular System

Deploying microrobots or catheters in human blood vessels faces a key challenge: navigating the varying vessel diameters to access the circulatory system. The human body has three main vessel types: arteries, veins, and capillaries, with diameters ranging from 25 mm in the aorta to 6 μm in capillaries. Large arteries, with diameters of 1–4 mm, are particularly important for intervention applications. Furthermore, their deeper location within muscle layers and thicker walls make arteries an ideal environment for interventions.

The successful use of endocardial catheters in medical interventions has been established for a decade [68], particularly in the milestone of human cardiac interventions using magnetic catheters for atrial fibrillation ablation. Myocardial ablation surgery, utilizing the commercial MAS named Stereotaxis Niobe to control the magnetic catheter tip under fluoroscopic imaging, is one of the few applications tested in humans. Other studies have documented related applications focusing on electrophysiological catheter guidance and control. Subsequent research has expanded to other heart diseases, including cardiac arrhythmias.

Current research initiatives are focused on developing solutions for a variety of diseases, such as peripheral arterial diseases and arterial aneurysms, by utilizing on-demand embolization techniques. In this context, microrobots have been developed to facilitate active retention during the process of drug delivery. Moreover, a continuum magnetic robot has been engineered with the aim of improving the access capability [69]. Simultaneously, the domain of soft robotics has been delving into the investigation of soft magnetic manipulation systems. The incorporation of soft magnetic actuation allows for greater flexibility in surgical procedures, which can be attributed to the implementation of remote magnetic control.

Peripheral arterial disease typically affects individuals aged 40 and older; it arises from narrowed blood vessels and may result in pain, impaired walking ability, and claudication. Percutaneous endovascular intervention stands as the principal treatment approach. The MAS plays a crucial role in steering the magnetic ring-tip catheter to the site of impairment and subsequently deploying a stent or balloon to unblock the vessel and restore its normal diameter. This method eliminates X-ray exposure through remote magnetic catheter control, with localization accomplished via catheter-tip sensing. A cylindrical configuration of Hall-effect sensors measures magnetic flux density in the workspace, permitting estimation of the magnetic catheter tip's location. Notwithstanding, X-ray imaging remains necessary for addressing bifurcations. The newly proposed magnetic-based localization algorithm enables real-time estimation of the magnetic tip's position. This approach has been practically applied using Helmholtz coils to manipulate the magnetic tip within the phantom.

Aneurysm is a condition where an artery's wall bulges outward. It can occur anywhere in the body, commonly in the aorta or brain. Due to their asymptomatic nature before rupture and the risk of severe internal bleeding, aneurysms have a high fatality rate. Embolization is an effective treatment strategy. Reference proposed a method of delivering microgels using magnetic-guided catheter assistance. This method combines C-arm fluoroscopy for real-time imaging with a magnetic robotic arm for catheter guidance. The self-adhesive microgel, delivered by the magnetic catheter and activated by an acidic stimulus to form connections, serves as an embolic agent. This study demonstrated the feasibility of controlling a swarm of microrobots in a dynamic human body environment to achieve effective embolization. The method was tested ex vivo on a human placenta. Ultrasound imaging confirmed effective aneurysm embolization, indicating that the microrobotic embolization platform is promising for application in human blood vessels.

Most intravascular applications rely on tethered magnetic catheters due to blood flow velocity. Steering untethered microrobots in the bloodstream poses a key challenge because of the circulatory system's rapid blood flow. To address this, a novel solution was proposed with an active retention structure. The microrobot features a claw to stay in place within blood vessels. This unique mechanism allows microrobots to remain positioned in the blood environment despite strong blood flow. This structure's successful use in a live rabbit's jugular vein has enabled the application of untethered microrobot swarms for treating vascular diseases and circulatory drug delivery via noninvasive methods.

To enhance the capability of a tethered catheter, researchers designed a novel untethered helical magnetic robot for better endovascular access. This robot features an articulating magnetic tip to improve steerability and a helical protrusion for rotation and forward motion energy generation. The system includes a central working channel for fluid injection. X-ray radiography monitors the robot's position in the vascular system in real-time. A commercial electromagnetic system called Navion controls the magnetic robot, achieving broad directional catheter control. An ex vivo experiment confirmed the helical structure's effectiveness in driving the robot forward. Further experiments using human placental blood vessels ex vivo and a porcine model in vivo validated the robot and system's effectiveness. Comparisons between the commercial and new catheters in the live pig study showed the new design's advantage in moving backward without vessel damage.

1.4.3 Intraocular System

In intraocular procedures, maintaining the stability of the surgeon's hand is crucial. As a result, significant research efforts have been dedicated to creating mechanical systems that can ensure a steady hand during surgical operations. Experiments conducted on chicken embryos and phantom models that mimic the human eye have contributed to this goal. Additionally, the development of robotic arms designed to reduce surgeons' hand tremors has been actively pursued.

Recent advancements in magnetically driven systems have sparked interest in exploring minimally invasive solutions for intraocular applications. In a 2010 study, localization was accomplished using a single camera, which demonstrated effectiveness in a model eye and laid the groundwork for future surgical procedures. Later, an in vivo rabbit eye experiment further confirmed the potential in this field [70]. Researchers developed a cylindrical microrobot, sized at 285 μm in diameter and 1.8 mm in length, to fit into a needle for easy injection into the eye. The Octomag system allows for intraocular surgery beneath the MAS, reserving space for imaging and surgical procedures upfront. Intraocular surgery offers imaging benefits due to its surface exposure. However, human trials are pending due to uncontrollably rapid motions during drug delivery. In material removal surgeries on the ocular globe, precise force control is essential, keeping its application in the animal trial phase. Recent research has explored targeted therapy via rotating-gradient magnetic fields. The success of this method in bovine eyes shows that magnetized microbots, driven by preprogrammed fields, can effectively deliver drugs in intraocular settings.

1.4.4 Knee and Elbow Joints

Knee and elbow joint cartilage damage is commonplace in athletes and the elderly, with a rising incidence among young adults in recent times. The standard approach to tackle this issue relies on arthroscopy for imaging. Regrettably, conventional therapeutic approaches do not provide means for fully restoring the damaged cartilage. Current treatments mainly focus on reducing inflammation and draining joint fluid to promote self-healing. The most cutting-edge method is platelet-rich plasma (PRP) injection, which delivers concentrated platelets to the injury site to alleviate pain and accelerate repair. However, the outcomes of this method are not uniformly remarkable, and its effectiveness varies significantly among individuals.

An alternative solution frequently suggested for treating serious cartilage injury is the utilization of stem cells. Mesenchymal stem cells (MSCs) can be easily obtained from placental tissue or umbilical cord blood. This novel method involves loading MSCs onto porous microrobots. These robots are injected into the affected joint via a needle and controlled by the MAS. Magnetic navigation enables precision medicine. As a 2020 study recorded [15], human adipose-derived MSCs can aid knee cartilage regeneration. Experiments employed a rabbit knee cartilage defect model. As the medical field increasingly accepts stem cell theory for therapy, this approach shows great potential for curing intra-articular damage clinically.

1.4.5 GI System

The GI system contains natural channels with relatively large spaces, making it easier for devices like millirobots, capsules, or catheters to navigate. Endoscopy is vital for GI malfunction diagnosis. Colonoscopy, gastroscopy, and esophagoscopy are used to examine the intestines, stomach, and esophagus respectively.

Early work on video capsules driven by Stereotaxis Niobe showed that navigating a 3D environment is central to GI endoscopy. Because the gut is much larger than intraocular spaces, the system must provide an extensive workspace while maintaining precise positioning. Researchers therefore combined robotic-arm actuators with refined control schemes: a rotary stage ending in a permanent magnet steered the capsule endoscope. The next year, further evaluation in a fluid-filled, mock-up stomach confirmed high robustness under reduced control frequency, localization errors, field misalignment, and singular configurations.

These two investigations established both the hardware layout and the algorithms required for complex tasks. As capsule endoscopy in the GI tract matured, later studies shifted toward smarter magnetic actuation, autonomy, and collaborative operation [71]. Reference examined colonoscopy with a focus on autonomy and intelligence. Higher autonomy not only lowers cost but also improves patient comfort, thereby promising broad clinical uptake of GI endoscopy.

Over the past few years, there has been a proposal to employ microrobots in stomach incision treatment, representing a nascent field of study that is still at the cellular slice stage. The design draws inspiration from the frictional anisotropy of micro-walkers, which mimics the process of walking. The most recent research has also highlighted this design approach, referred to as the asymmetrical friction effect, induced by magnetic torque. Enhancing the design of magnetic robots can considerably bolster the magnetic actuation methods applied in clinical settings. The rise of applications for precise drug delivery and surgical assistance in the stomach indicates that magnetic actuation technology, along with its control algorithms, has reached a certain level of maturity.

Recent research has explored using multiple microrobots to perform tasks. By dynamically controlling the magnetic field, researchers can manipulate a multilayer, film-like soft robot. The layers can separate under a predefined magnetic field and fix at different spots to treat stomach ulcers. These experiments were performed on a porcine stomach to address multiple ulcers.

1.4.6 Respiratory System

The respiratory system comprises the respiratory tract and lungs. The respiratory tract is divided into the upper portion (which includes the nose, mouth, pharynx, and throat) and the lower portion (which consists of the windpipe, bronchial tubes, and lungs). While the components of the upper respiratory tract are relatively easy to access, the lower respiratory tract, particularly the bronchial tubes and lungs, presents a complex situation. The subsequent discussion will primarily focus on the intricate areas of the respiratory system.

A position estimation method based on lung phantom experiments has been proposed [72]. When controlling an endoscope in the lung, positioning is the key difficulty. The approach uses images detected by the endoscope to determine the catheter's current position. The device's kinematics are captured by comparing input with visual feedback. This novel approach allows for real-time positioning within the human lung structure.

Bronchoscopy is vital for detecting diseases in the bronchial tubes. Optimizing the navigation trajectory is essential for smooth and autonomous detection. This has led to the development of a soft magnetic catheter, activated by two robot-driven permanent magnets. The goal is to enhance the navigation path for autonomous and atraumatic detection within the bronchial tubes, as highlighted by Pittiglio. By deploying a specially designed soft catheter robot, autonomous disease detection within the bronchial tree phantom has been achieved, with significant implications for medical diagnostics and interventions. This atraumatic autonomous endoscopy is possible through two robotic arms collaborating to control a patient-specific magnetic catheter. The robotic arms are programmed to follow a predefined trajectory, minimizing contact during navigation. A comparative study of four catheters using a leader-follower algorithm showed greatly improved tracking accuracy and reduced targeting errors.

1.4.7 Urogenital System

The urogenital system is composed of two parts: the urinary and genital tracts. The urinary tract includes key components such as the kidney, ureter, bladder, and urethra. One common ailment in the urogenital system is kidney stones. Retrograde intrarenal surgery is a procedure used to remove these stones. The endoscope passes through the urine storage area, enters the kidney, locates the stone, and then breaks it up using a laser.

Flexible magnetic continuum robots have been developed for use in the urogenital system. A catheter with a magnetic tip is used to navigate through a kidney phantom. The control method can predict the nonlinear behavior of the continuum robot in a heterogeneous magnetic field. In 2022, a novel application demonstrated that a specially designed magnetic soft robotic bladder could enhance contraction through magnetic actuation [73]. This innovation has potential for assisted urination. Its safety was tested in a porcine model over two weeks. The genital tract is divided into the female and male reproductive systems. In the male reproductive system, physical contraception is a key application. Research has led to the development of wireless robots for this purpose. Their potential has been shown through applications in a tubular phantom with ultrasound guidance.

1.5 Scope and Layout of the Book

The primary scope of this book is to bridge the existing gap in current literature, where the focus has predominantly been on the fabrication and individual design of small-scale robots, often overlooking the systematic design methodology of the external actuation systems required to manipulate them. Consequently, this text is dedicated to the comprehensive study of mobile MASs, summarizing the latest advancements in their design, control strategies, experimental validation, and performance optimization. Unlike stationary systems, the mobile MASs discussed herein are characterized by their ability to provide large workspaces, rapidly varied magnetic fields, and versatile system configurations with high scalability and bandwidth. The content is structured to guide readers—ranging from postgraduate students to researchers in robotics, control, and biomedical engineering—through the transition from manual and semi-automated operations toward fully adaptive, intelligent manipulation suited for complex tasks such as active targeted delivery and remote sensing.

The layout of the book is organized into 10 chapters that progress logically from fundamental principles to advanced system designs and specific applications. The text begins in Chapter 1 by establishing the necessary background, introducing the basic concepts, actuation principles, and the historical development of magnetic actuation technology. Following this foundational introduction, the book immediately addresses modern design challenges in Chapter 2, which focuses on the design and real-time optimization of actuation systems with enhanced flexibility, demonstrating capabilities such as autonomous navigation under ultrasound Doppler imaging. To tackle the complexities of controlling soft miniature robots, Chapter 3 introduces a deep reinforcement learning framework, highlighting the intersection of modern artificial intelligence with magnetic manipulation control.

The middle chapters of the book provide an in-depth technical analysis of specific system configurations and control methodologies. Chapters 4 and 5 explore reconfigurable electromagnetic actuation systems designed for large workspaces, detailing multi-objective optimization methods for coil configuration generation as well as performance-guided strategies for field and force control. This is followed by a comprehensive examination of parallel-mobile-coil systems in Chapters 6 and 7, which cover multimode control, mechanism design, and parameter optimization. These chapters also present advanced applications of these systems, such as motion planning for autonomous microrobotic navigation in dynamic flowing conditions, thereby validating the theoretical models through rigorous simulation and experimentation.

The final section of the book emphasizes practical implementation in medical and engineering contexts. Chapter 8 presents a magnetically controlled guidewire robotic system specifically designed for vascular intervention, including mathematical modeling and in vivo validation of multistage embolization strategies. Chapter 9 extends the discussion to wired magnetic microrobots utilizing parallel-mobile-coil systems. The book concludes in Chapter 10 with a summary of design methodologies and integration strategies between imaging and actuation systems. It offers a forward-looking perspective on the future of the field, predicting that mobile MASs will become increasingly intelligent and self-adaptive over the next five years, further expanding their utility in translational biomedicine and interdisciplinary research.

References

1 S. Palagi and P. Fischer. Bioinspired microrobots. *Nature Reviews Materials*, 3(6): 113–124, 2018.

2 M. Sitti. Miniature soft robots—road to the clinic. *Nature Reviews Materials*, 3(6): 74–75, 2018.

3 G.-Z. Yang, et al. The grand challenges of science robotics. *Science Robotics*, 3(14): eaar7650, 2018.

4 S. Jeon, et al. Magnetically actuated microrobots as a platform for stem cell transplantation. *Science Robotics*, 4: eaav4317, 2019.

5 S. Lee, S. Lee, S. Kim, C. Yoon, H. Park, J. Kim, and H. Choi. Fabrication and characterization of a magnetic drilling actuator for navigation in a three-dimensional phantom vascular network. *Scientific Reports*, 8(1): 3691, 2018.

6 Y. Kim, G. A. Parada, S. Liu, and X. Zhao. Ferromagnetic soft continuum robots. *Science Robotics*, 4(33): eaax7329, 2019.

7 D. Son, H. Gilbert, and M. Sitti. Magnetically actuated soft capsule endoscope for fine-needle biopsy. *Soft Robotics*, 7(1): 10–21, 2020.

8 M. Xie, A. Shakoor, Y. Shen, J. K. Mills, and D. Sun. Out-of-plane rotation control of biological cells with a robot-tweezers manipulation system for orientation-based cell surgery. *IEEE Transactions on Biomedical Engineering*, 66(1): 199–207, 2018.

9 Y. Deng, A. Paskert, Z. Zhang, R. Wittkowski, and D. Ahmed. An acoustically controlled helical microrobot. *Science Advances*, 9(38): eadh5260, 2023.

10 A. Aghakhani, O. Yasa, P. Wrede, and M. Sitti. Acoustically powered surface-slipping mobile microrobots. *Proceedings of the National Academy of Sciences*, 117(7): 3469–3477, 2020.

11 G. Kosa and P. Hunziker. Small-scale robots in fluidic media. *Advanced Intelligent Systems*, 1(7): 1900035, 2019.

12 L. Daneshmandi, et al. Emergence of the stem cell secretome in regenerative engineering. *Trends in Biotechnology*, 38(12): 1373–1384,2020.

13 J. Li, et al. Reconfigurable magnetic microrobot swarm: multimode transformation, reconfigurable swarm, and intensive killing of cancer cells. *Science Robotics*, 3(19): eaat8829, 2018.

14 T. Wei, et al. Development of magnet-driven and image-guided degradable microrobots for the precise delivery of engineered stem cells for cancer therapy. *Small*, 16(41): 1906908, 2020.

15 G. Go, et al. Human adipose–derived mesenchymal stem cell–based medical microrobot system for knee cartilage regeneration in vivo. *Science Robotics*, 5(38): eaay6626, 2020.

16 L. Zheng, Y. Jia, D. Dong, W. Lam, D. Li, H. Ji, and D. Sun. 3D navigation control of untethered magnetic microrobot in centimeter-scale workspace based on field-of-view tracking scheme. *IEEE Transactions on Robotics*, 38(3): 1583–1598, 2021.

17 D. Dong, L. Xing, L. Zheng, Y. Jia, and D. Sun. Automated 3-D electromagnetic manipulation of microrobot with a path planner and a cascaded controller. *IEEE Transactions on Control Systems Technology*, 30(6): 2672–2680, 2021.

18 Q. Boehler, S. Gervasoni, S. L. Charreyron, C. Chautems, and B. J. Nelson. On the workspace of electromagnetic navigation systems. *IEEE Transactions on Robotics*, 39(1): 791–807, 2022.

19 L. Yang, Z. Yang, M. Zhang, J. Jiang, H. Yang, and L. Zhang. Optimal parameter design and microrobotic navigation control of parallel-mobile-coil systems. *IEEE Transactions on Automation Science and Engineering*, 21(1): 855–867, 2022.

20 H. Xie, et al. Reconfigurable magnetic microrobot swarm: multimode transformation, locomotion, and manipulation. *Science Robotics*, 4: eaav8006, 2019.

21 D. Ahmed, A. Sukhov, D. Hauri, D. Rodrigue, G. Maranta, J. Harting, and B. J. Nelson. Bioinspired acousto-magnetic microswarm robots with upstream motility. *Nature Machine Intelligence*, 3: 311–320, 2021.

22 X. Yang, R. Tan, H. Lu, T. Fukuda, and Y. Shen. Milli-scale cellular robots that can reconfigure morphologies and behaviors simultaneously. *Nature Communications*, 13(1): 4156, 2022.

23 F. Ongaro, S. Pane, S. Scheggi, and S. Misra. Design of an electromagnetic setup for independent three-dimensional control of pairs of identical and nonidentical microrobots. *IEEE Transactions on Robotics*, 35(1): 174–183, 2019.

24 Z. Li, et al. Human mesenchymal stem cell derived miniature joint system for disease modeling and drug testing. *Advanced Science*, 9(21): 2105909, 2022.

25 P. Wrede, et al. Real-time 3D optoacoustic tracking of cell-sized magnetic microrobots circulating in the mouse brain vasculature. *Science Advances*, 8(19): eabm9132, 2022.

26 Q. Wang and L. Zhang. External power-driven microrobotic swarm: from fundamental understanding to imaging-guided delivery. *ACS Nano*, 15(1): 149–174, 2021.

27 J. Yu, L. Yang, and L. Zhang. Pattern generation and motion control of a vortex-like paramagnetic nanoparticle swarm. *The International Journal of Robotics Research*, 37(8): 912–930, 2018.

28 L. Yang, J. Jiang, X. Gao, Q. Wang, Q. Dou, and L. Zhang. Autonomous environment-adaptive microrobot swarm navigation enabled by deep learning-based real-time distribution planning. *Nature Machine Intelligence*, 4(5): 480–493, 2022.

29 A. J. Petruska and B. J. Nelson. Minimum bounds on the number of electromagnets required for remote magnetic manipulation. *IEEE Transactions on Robotics*, 31(3): 714–722, 2015.

30 C. R. Thornley, L. N. Pham, and J. J. Abbott. Reconsidering six-degree-of-freedom magnetic actuation across scales. *IEEE Robotics and Automation Letters*, 4(3): 2325–2332, 2019.

31 M. Parker. Magnetic traps help robots navigate soft tissue. *Nature Electronics*, 4: 770, 2021.

32 B. Shapiro, S. Kulkarni, A. Nacev, S. Muro, P. Y. Stepanov, and I. N. Weinberg. Wiley interdiscip. *Wiley Interdisciplinary Reviews: Nanomedicine and Nanobiotechnology*, 7: 446, 2015.

33 L. Xing, D. Li, H. Cao, L. Fan, L. Zheng, L. Zhang, and D. Sun. High-efficiency automatic catheter navigation system with a 3D magnetic-field-drive mechanism. *Advanced Intelligent Systems*, 4: 2100214, 2022.

34 Y. Xing, Y. Jia, Z. Zhan, J. Li, and C. Hu. Modeling and control of a soft magnetic micro-swimmer. In *2021 International Conference on Robotics and Automation (ICRA)*, pages 7281–7287, Piscataway, NJ, 2021. IEEE.

35 D. Son, M. C. Ugurlu, and M. Sitti. Permanent magnet array–driven navigation of wireless millirobots inside soft tissues. *Science Advances*, 7(43): eabi8932, 2021.

36 D. Jin, et al. Swarming self-adhesive microgels enabled aneurysm on-demand embolization in physiological blood flow. *Science Advances*, 9(19): eadf9278, 2023.

37 T. Li, et al. Bioinspired claw-engaged and biolubricated swimming microrobots creating active retention in blood vessels. *Science Advances*, 9(18): eadg4501, 2023.

38 A. J. Petruska. Open-loop orientation control using dynamic magnetic fields. *IEEE Robotics and Automation Letters*, 5(4): 5472–5476,2020.

39 T. Xu, C. Huang, Z. Lai, and X. Wu. Independent control strategy of multiple magnetic flexible millirobots for position control and path following. *IEEE Transactions on Robotics*, 38: 2875, 2022.

40 Q. Zou, X. Du, Y. Liu, H. Chen, Y. Wang, and J. Yu. Dynamic path planning and motion control of microrobotic swarms for mobile target tracking. *IEEE Transactions on Automation Science and Engineering*, 20: 2454–2468, 2022.

41 M. Salehizadeh and E. D. Diller. Path planning and tracking for an underactuated two-microrobot system. *IEEE Robotics and Automation Letters*, 6: 2674, 2021.

42 J. Wu, K. Yu, I. Lopez, A. A. Izquierdo, H. Saber, F. Alambeigi, and L. Zhou. Integrated magnetic location sensing and actuation of steerable robotic catheters for peripheral arterial disease treatment. *IEEE Robotics and Automation Letters*, 8: 5656, 2023.

43 J. Liu, T. Xu, and X. Wu. Optimal planning of internet data centers decarbonized by hydrogen-water-based energy systems. *IEEE Transactions on Automation Science and Engineering*, 20: 1577, 2023.

44 J. Jiang, L. Yang, and L. Zhang. Closed-loop control of a Helmholtz coil system for accurate actuation of magnetic microrobot swarms. *IEEE Robotics and Automation Letters*, 6: 827, 2021.

45 S. O. Demir, U. Culha, A. C. Karacakal, A. Pena-Francesch, S. Trimpe, and M. Sitti. Task space adaptation via the learning of gait controllers of magnetic soft millirobots. *The International Journal of Robotics Research*, 40: 1331, 2021.

46 J. Liu, X. Wu, C. Huang, L. Manamanchaiyaporn, W. Shang, X. Yan, and T. Xu. 3-D autonomous manipulation system of helical microswimmers with online compensation update. *IEEE Transactions on Automation Science and Engineering*, 18: 1380, 2020.

47 T. Xu, J. Liu, C. Huang, T. Sun, and X. Wu. Discrete-time optimal control of miniature helical swimmers in horizontal plane. *IEEE Transactions on Automation Science and Engineering*, 19: 2267, 2021.

48 R. Chen, D. Folio, and A. Ferreira. Performance metrics for a robotic actuation system using static and mobile electromagnets. In *2019 International Conference on Robotics and Automation (ICRA)*, page 2474, Montreal, 2019. IEEE.

49 J. Sikorski, I. Dawson, A. Denasi, E. E. Hekman, and S. Misra. Introducing BigMag – a novel system for 3D magnetic actuation of flexible surgical manipulators. In *2017 IEEE International Conference on Robotics and Automation (ICRA)*, page 3594, Singapore, 2017. IEEE.

50 J. Sikorski, A. Denasi, G. Bucchi, S. Scheggi, and S. Misra. Vision-based 3-D control of magnetically actuated catheter using BigMag—an array of mobile electromagnetic coils. *IEEE/ASME Transactions on Mechatronics*, 24: 505, 2019.

51 R. Chen, D. Folio, and A. Ferreira. Study of robotized electromagnetic actuation system for magnetic microrobots devoted to minimally invasive ophthalmic surgery. In *2019 IEEE International Symposium on Medical Robotics (ISMR)*, page 1, Atlanta, GA, 2019. IEEE.

52 J. Sikorski, C. M. Heunis, F. Franco, and S. Misra. The ARMM system: an optimized mobile electromagnetic coil for non-linear actuation of flexible surgical instruments. *IEEE Transactions on Magnetics*, 55: 1, 2019.

53 C. M. Heunis, Y. P. Wotte, J. Sikorski, G. P. Furtado, and S. Misra. The ARMM system-autonomous steering of magnetically-actuated catheters: towards endovascular applications. *IEEE Robotics and Automation Letters*, 5: 704, 2020.

54 L. Yang, X. Du, E. Yu, D. Jin, and L. Zhang. DeltaMag: an electromagnetic manipulation system with parallel mobile coils. In *2019 International Conference on Robotics and Automation (ICRA)*, page 9814, Montreal, 2019. IEEE.

55 A. J. Petruska, J. B. Brink, and J. J. Abbott. First demonstration of a modular and reconfigurable magnetic-manipulation system. In *2015 IEEE International Conference on Robotics and Automation (ICRA)*, page 149, Seattle, WA, 2015. IEEE.

56 D. Li, F. Niu, J. Li, X. Li, and S. Dong. Gradient-enhanced electromagnetic actuation system with a new core shape design for microrobot manipulation. *IEEE Transactions on Industrial Electronics*, 67: 4700, 2020.

57 A. Heilbronn. Eine neue methode zur bestimmung der viskosität lebender protoplasten. *Jahrbücher für Wissenschaftliche Botanik*, 61: 284, 1922.

58 C. Watson and T. K. Morimoto. Permanent magnet-based localization for growing robots in medical applications. *IEEE Robotics and Automation Letters*, 5: 2666, 2020.

59 J. S. Barajas-Gamboa, et al. First in-human experience with a novel robotic platform and magnetic surgery system. *International Journal of Medical Robotics and Computer Assisted Surgery*, 17: 1, 2021.

60 X. Du, et al. Real-time navigation of an untethered miniature robot using mobile ultrasound imaging and magnetic actuation systems. *IEEE Robotics and Automation Letters*, 7: 7668, 2022.

61 J. Rahmer, C. Stenging, and B. Gleich. Remote magnetic actuation using a clinical scale system. *PLoS One*, 13: e0193546, 2018.

62 J. Rahmer, C. Stehning, and B. Gleich. Spatially selective remote magnetic actuation of identical helical micromachines. *Science Robotics*, 2: eaal2845, 2017.

63 H. Choi, J. Choi, G. Jang, J. Park, and S. Park. Two-dimensional actuation of a microrobot with a stationary two-pair coil system. *Smart Materials and Structures*, 18(5): 055007, 2009.

64 M. P. Kummer, J. J. Abbott, B. E. Kratochvil, R. Borer, A. Sengul, and B. J. Nelson. OctoMag: an electromagnetic system for 5-DOF wireless micromanipulation. *IEEE Transactions on Robotics*, 26(6): 1006–1017, (2010).(b) *IEEE Transactions on Robotics*(c) *IEEE Transactions on Robotics*(d)

65 A. J. Petruska, A. W. Mahoney, and J. J. Abbott. Remote manipulation with a stationary computer-controlled magnetic dipole source. *IEEE Transactions on Robotics*, 30(5): 1222–1227, 2014.(b) *IEEE Transactions on Robotics*(c)

66 J. W. Martin, B. Scaglioni, J. C. Norton, V. Subramanian, A. Arezzo, K. L. Obstein, and P. Valdastri. Enabling the future of colonoscopy with intelligent and autonomous magnetic manipulation. *Nature Machine Intelligence*, 2(10): 595–606, 2020.

67 M. S. Grady, M. A. III Howard, J. A. Molloy, R. C. Ritter, E. G. Quate, and G. T. Gillies. Nonlinear magnetic stereotaxis: three-dimensional, *in vivo* remote magnetic manipulation of a small object in canine brain. *Medical Physics*, 17: 405, 1990.

68 F. Carpi and C. Pappone. Stereotaxis Niobe® magnetic navigation system for endocardial catheter ablation and gastrointestinal capsule endoscopy. *Expert Review of Medical Devices*, 6: 487, 2009.

69 M. Richter, M. Kaya, J. Sikorski, L. Abelmann, V. Kalpathy Venkiteswaran, and S. Misra. Magnetic soft helical manipulators with local dipole interactions for flexibility and forces. *Soft Robotics*, 10: 647–659,2023.

70 F. Ullrich, et al. Mobility experiments with microrobots for minimally invasive intraocular surgery. *Investigative Ophthalmology & Visual Science*, 54: 2853, 2013.

71 S. Kim, S. Bae, W. Lee, and G. Jang. Magnetic navigation system composed of dual permanent magnets for accurate position and posture control of a capsule endoscope. *IEEE Transactions on Industrial Electronics*, 71: 739, 2024.

72 J. Edelmann, A. J. Petruska, and B. J. Nelson. Estimation-based control of a magnetic endoscope without device localization. *Journal of Medical Robotics Research*, 3: 1850002, 2018.

73 Y. Yang, et al. Magnetic soft robotic bladder for assisted urination. *Science Advances*, 8(34): eabq1456, 2022.

2

Design and Real-time Optimization for a Magnetic Actuation System with Enhanced Flexibility

In this chapter, a magnetic actuation system based on three mobile electromagnetic coils is designed, and a control strategy for the system is proposed. Enhanced flexibility combined with optimization algorithms enables the system to satisfy various requirements in applications, such as avoiding collision between the coils and the obstructions within the workspace, placing the coils in optimal positions to enhance energy efficiency, generating wide varieties of magnetic fields for complex tasks, and tracking the location of the robot with an enlarged workspace.

2.1 Introduction

Minimally invasive procedures have garnered significant interest due to the capabilities of medical robots with remote actuation [1]. The magnetic field stands out as a leading actuation method, prized for its safety and ability to penetrate tissues [2]. Lately, robots actuated by magnetic fields have been engineered across scales from millimeters [3–5] to micrometers or nanometers [6], tailored for applications like targeted delivery [7], on-site diagnosis [8], and medical interventions [9]. Essential to these robots are magnetic actuation systems that produce the requisite magnetic fields for remote control [8] and additional practical uses [10, 11].

Magnetic actuation systems, utilizing multiple stationary electromagnetic coils, have been crafted to generate the necessary magnetic fields for medical applications. Helmholtz coils [12], individually or in conjunction with Maxwell coils, are prevalent for delivering magnetic torque and force. The incorporation of soft iron cores in these systems bolsters the magnetic field and enables sophisticated five-degree-of-freedom (5-DOF) [13, 14] manipulation. However, the expansion of such stationary coil systems for surgical applications presents significant challenges. Their substantial size [15] can lead to increased inductance, which in turn restricts the operational bandwidth of the magnetic fields. This necessitates the use of high-powered current sources and requires cooling mechanisms to manage the heat generated by the currents, consequently driving up the system's cost.

Mobile magnetic actuation systems offer an alternative, capable of generating the necessary magnetic fields over a broader operational area when compared to their stationary counterparts. These systems utilize either permanent magnets or electromagnetic coils to produce magnetic fields. In the case of permanent-magnet-based mobile systems, the magnetic field is modulated by altering the magnet's relative position to the robot [5], enabling dynamic magnetic field [16] adjustments and 5-DOF control [3, 17] through mechanical magnet repositioning or by incorporating additional magnets. A key advantage of these systems is the absence of heat generation, which is a concern with stationary systems that require more magnets for enhanced fields. However, the persistent strong magnetic field and the magnets' inertia can limit the operational bandwidth and the diversity of achievable magnetic fields. Electromagnetic-coil-based mobile systems, on the other hand, create magnetic fields through movable coils, facilitating an expanded working area. These systems are advantageous for their scalability and broad control bandwidth.

Compared with previously reported mobile-coil-based systems [18, 19], our magnetic actuation system with decoupled mobile coils will be able to further improve the energy efficiency due to the enhanced flexibility (see Figure 2.1). The system has the following features:

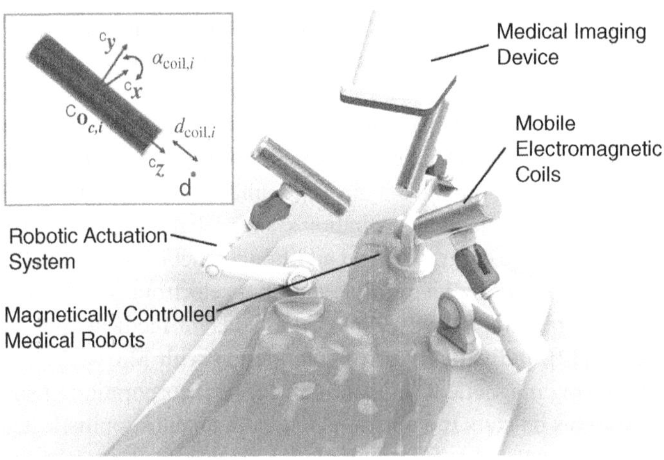

Figure 2.1 Concept of the proposed system with decoupled mobile electromagnetic coils. The enhanced flexibility enables the system to deal with complex obstructions inside the workspace (e.g., surface of the patient's body, other medical devices, etc.) and increase energy efficiency by moving the coils to optimal positions.

1. The electromagnetic coils are capable of moving within a spatial area, thereby enabling the workspace to be effectively expanded by continuously tracking the position of the controlled robot.
2. The system's high flexibility, combined with its collision detection algorithm, enables it to effectively prevent potential collisions between the coils and any obstructions within the workspace.
3. The system can position the coils optimally during swimmer actuation to reduce current, due to the real-time optimization algorithm and the coils' high flexibility.

Moreover, untethered small-scale robots offer great promise for medical applications in complex biological environments. However, challenges remain in the control and medical imaging of a robot for targeted delivery inside a living body, especially in flowing conditions. In Section 2.4, we propose a strategy to autonomously navigate a miniature helical robot in dynamic blood flow under ultrasound Doppler imaging guidance to overcome the existing challenges. So the main work covered in this chapter is as follows:

1. Construct a magnetic field model for the system by utilizing coordinate transformation and linear superposition.
2. Calculate the optimal positions of the coils based on this model through implementation of a particle swarm optimization (PSO) based algorithm.
3. Execute the demanded currents in each coil and the required mechanical platform motions, which are calculated using the optimal coil positions, via actuators.
4. Fabricate a prototype system called RoboMag and conduct simulations and experiments to validate the algorithm's feasibility.
5. Based on the aforementioned mobile electromagnetic coil system, we employ an autonomous navigation strategy using ultrasound Doppler imaging to conduct validation experiments for the targeted thrombolysis of a miniature helical robot.

2.2 System Design, Modeling, and Optimization

2.2.1 Modeling of System

For a robot positioned at a spatial location denoted as $\mathbf{d}$, the magnetic torque τ_d and force $\mathbf{F}_d$ exerted on it can be determined using the following equations:

$$\tau_d = \mathbf{m}_d \times \mathbf{B} \tag{2.1}$$

$$\mathbf{F}_d = (\mathbf{m}_d \cdot \nabla)\mathbf{B} \tag{2.2}$$

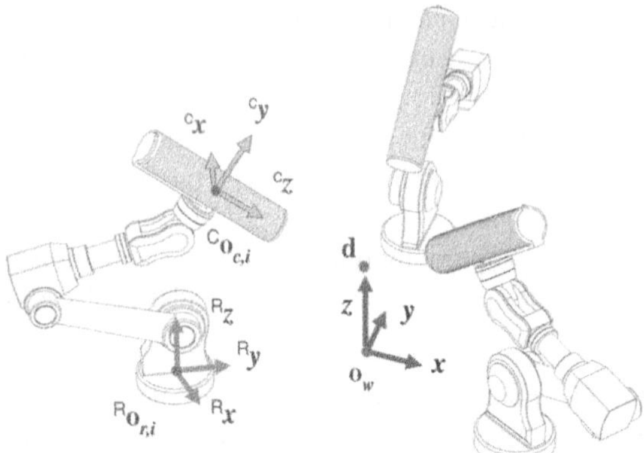

Figure 2.2 Three types of frames are defined for system modeling: world frame (with no superscript), coil's frame (**C**), and robotic arm frame (**R**).

where **B** is the magnetic flux density at the robot's location and $\mathbf{m}_d$ is the magnetic moment of the magnet or magnetized material in the robot. To generate demanded magnetic torque and force at a location **d**, a model is proposed as follows.

The modeling procedure relies on coordinate transformations and the principle of linear superposition. Three distinct coordinate systems are established (refer to Figure 2.2): the world frame (without a superscript), the robotic arm frame (denoted by **R**), and the coil's frame (denoted by **C**). In the world frame, the center of the workspace is designated as the origin (marked as $\mathbf{o}_w$ in Figure 2.2). For the robotic arm frame, the base of each arm is set as the origin (labeled as ${}^{\mathbf{R}}\mathbf{o}_{r,i}$ in Figure 2.2). For the frame of the ith coil, its origin is positioned at the center of the coil (labeled as ${}^{\mathbf{C}}\mathbf{o}_{c,i}$ in Figure 2.2), with the z-axis aligned along the coil's axial direction. The axes of these coordinate systems adhere to the right-hand rule.

Throughout this document, scalar values are represented by italic lowercase letters, vectors or coordinate points are represented by bold lowercase letters, and matrices are represented by bold uppercase letters. However, the magnetic flux density **B**, force **F**, and currents **I** are represented by bold uppercase letters due to standard conventions.

For a magnetic medical robot positioned at coordinates $\mathbf{d} = (x_d, y_d, z_d)$ within the global reference frame, the relative position with respect to the ith coil (where $i = 1, \ldots, n$) can be determined by transforming **d** into the ith coil's coordinate system using the rotation matrix ${}^{\mathrm{C}}_{\mathrm{W}}\mathrm{T}$ and the translation vector ${}^{\mathrm{C}}_{\mathrm{W}}\mathrm{t}$. The matrix ${}^{\mathrm{C}}_{\mathrm{W}}\mathrm{T}$ accounts for the rotational component necessary to switch from the global frame to the ith coil's frame, while ${}^{\mathrm{C}}_{\mathrm{W}}\mathrm{t}$ represents the translational component. These values are derived from the real-time position $\mathbf{o}_{c,i}$ and orientation $\hat{\mathbf{r}}_{c,i}$ of the coil. The position of **d** in the coil's frame is represented as ${}^{\mathrm{C}}\mathbf{d}_i$.

To determine the magnetic field at ${}^{C}\mathbf{d}_i$, a data-driven model of the magnetic field for a single coil is constructed and calibrated. This model is based on data from the finite element method and is calibrated using the measured magnetic flux density [20]. Using this model, the magnetic flux density at ${}^{C}\mathbf{d}_i$ with a current of $I_i = 1$ A can be computed and is denoted as ${}^{C}\widehat{\mathrm{B}}_i$. The computed magnetic flux density ${}^{C}\widehat{\mathrm{B}}_i$ is then transformed into the world frame as $\widehat{\mathrm{B}}_i$. Consequently, the required currents in the coils $\mathbf{I} = [I_1, I_2, \dots, I_n]^T$ for generating the magnetic flux density $\mathbf{B}$ at point $\mathbf{d}$ can be calculated.

$$\mathbf{I} = \mathbf{M}^{-1} \cdot \mathbf{B} \tag{2.3}$$

where $\hat{\mathbf{M}}$ consists of three linearly independent base vectors: $\hat{\mathbf{M}} = [\hat{\mathbf{B}}_i, \hat{\mathbf{B}}_j, \hat{\mathbf{B}}_k]$, where $i, j, k = 1, \dots, n$ and $i \neq j \neq k$.

In the context of magnetic force actuation, the computation of the base vectors follows a methodology analogous to that used for determining magnetic flux density.

$$\begin{aligned}\hat{\mathbf{F}}_i &= {}^{C}_{W}\mathbf{T}^{-1} \cdot {}^{C}\hat{\mathbf{F}}_i = {}^{C}_{W}\mathbf{T}^{-1} \cdot ({}^{C}\mathbf{m}_d \cdot \nabla){}^{C}\hat{\mathbf{B}}_i \\ &= {}^{C}_{W}\mathbf{T}^{-1} \cdot {}^{C}\hat{\mathbf{G}}_i \cdot {}^{C}_{W}\mathbf{T} \cdot \mathbf{m}_d\end{aligned} \tag{2.4}$$

where $\mathbf{m}_d$ is the magnetic moment of the robot under the world frame, and ${}^{C}\hat{\mathbf{G}}_i$ is a matrix that represents the gradient of magnetic flux density [21] at position ${}^{C}\mathbf{d}$ when $I_i = 1$ A. The required currents for generating force $\mathbf{F}$ at $\mathbf{d}$ can be solved by

$$\mathbf{I} = \hat{\mathbf{H}}^{-1} \cdot \mathbf{F} \tag{2.5}$$

where $\hat{\mathbf{H}}$ is made of three linearly independent base vectors: $\hat{\mathbf{H}} = [\hat{\mathbf{F}}_i, \hat{\mathbf{F}}_j, \hat{\mathbf{F}}_k]$, where $i, j, k = 1, \dots, n$ and $i \neq j \neq k$. For the modeling in this work, $\hat{\mathbf{M}}$ and $\hat{\mathbf{H}}$ are related to the real-time position and orientation of each coil and the location of the robot.

2.2.2 System Optimization

The magnetic field model serves as the foundation for optimizing the positioning of electromagnetic coils. Magnetic-torque actuation is prevalent in magnetic actuation techniques due to its superior efficiency over force actuation, particularly in the context of microrobot applications. A variety of magnetic-torque-actuated robots have been engineered, including capsule endoscopes [4, 16], magnetic catheters [5, 18], microrobots [8], and microrobotic swarms [6, 7]. Consequently, the primary focus of this article is on magnetic-torque actuation. Within this section, a real-time optimization algorithm grounded in PSO is introduced to determine the optimal mobile coil positions in real-time during operations. The goal of this algorithm is to fulfill several criteria:

1. To prevent any collisions between the electromagnetic coils and the surrounding obstacles;
2. To produce the desired magnetic field at a specified spatial location;
3. To minimize the electrical currents applied, thereby reducing heat generation;
4. To decrease the mechanical motions required during the operation.

To enhance system [22] efficiency and expedite the optimization process, a strategy has been adopted: the mobile coils consistently maintain the robot along their axial direction and monitor the robot's movements. Consequently, the parameters defining each coil's position can be reduced to two: the distance from the coil's tip to the robot $d_{\text{coil},i}$, and the tilt angle between the coil and the horizontal plane $\alpha_{\text{coil},i}$ (see the inset of Figure 2.2). The proposed optimization algorithm aims to identify the optimal parameter set $\mathbf{p} = \{d_{\text{coil},i}, \alpha_{\text{coil},i}\}$ $(i = 1, \dots, n)$ that meets the aforementioned objectives, which are encapsulated by the cost function of the optimization.

To avert collisions, an algorithm (Algorithm 1) founded on the principle of triangle-ray intersection [21] is deployed for detecting potential impacts between the system and non-uniform environmental obstacles when the parameter set p is put into action. The outcome of this collision detection will constitute a component within the objective function of the optimization procedure.

In specific operational scenarios, barriers can be captured using RGB-D imaging techniques and subsequently saved as three-dimensional (3D) models in StereoLithography (STL) format with the aid of 3D reconstruction algorithms [23, 24]. Within STL files, the 3D model of the barrier is delineated by J triangular facets in space, with the coordinates of each triangle's three vertices denoted as $\mathbf{v0}_j$, $\mathbf{v1}_j$, and $\mathbf{v2}_j$ for $j = 1, \dots, J$. Assuming the ith mobile coil is positioned at $\mathbf{o}_{c,i}$ and $\mathbf{q}_i$ is a point on the outer surface of the coil, the vector $\overrightarrow{\mathbf{q}_i\mathbf{o}_{c,i}}$ points from $\mathbf{q}_i$ to $\mathbf{o}_{c,i}$. Here, $\mathbf{q}_i$ and $\mathbf{o}_{c,i}$ are functions of the parameter set $\mathbf{p}$. The point $\mathbf{q}_i$ can be selected as a location where collisions are probable, such as the edge of the coil (refer to $\mathbf{q}_{i,1}, \dots, \mathbf{q}_{i,k}$ in Figure 2.3).

In this work, $\mathbf{q}_i$ is chosen to be the center of the surface at the tip of the coil for rapid computation suitable for real-time optimization. Concurrently, the boundary of the obstruction is expanded outward to create a new boundary for the point $\mathbf{q}_i$, with the discrepancy between the original and new boundaries exceeding the coil's radius to preclude potential collisions at the coil's edges. Vectors $\overrightarrow{\mathbf{v0}_j\mathbf{v1}_j}$ and $\overrightarrow{\mathbf{v0}_j\mathbf{v2}_j}$ are defined in a similar manner, establishing the plane $\{\mathbf{v0}_j\mathbf{v1}_j\mathbf{v2}_j\}$ on which the jth triangle resides (refer to Figure 2.3). The vector $\mathbf{e}(\overrightarrow{\mathbf{q}_i\mathbf{o}_{c,i}})$ is a unit vector originating from $\mathbf{q}_i$ and aligned with $\overrightarrow{\mathbf{q}_i\mathbf{o}_{c,i}}$. For every triangle constituting the 3D model of the boundary, the parameters t_j, u_j, and v_j must be resolved.

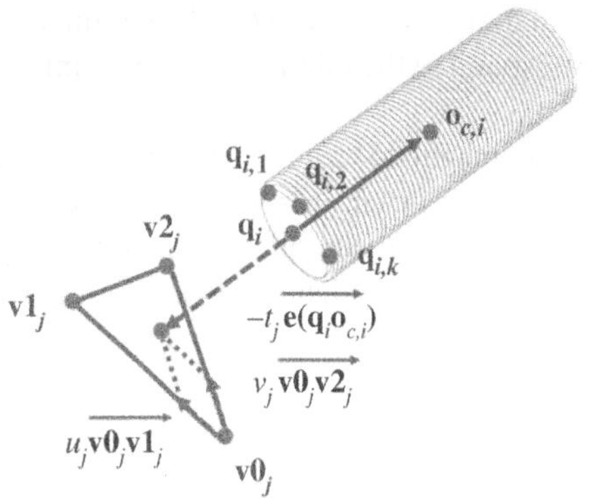

Figure 2.3 Schematic depiction of the collision detection algorithm based on triangle-ray intersection.

$$\begin{bmatrix} -\mathbf{e}(\overrightarrow{\mathbf{qo}_{c,i}}) & \overrightarrow{\mathbf{v0}_j\mathbf{v1}_j} & \overrightarrow{\mathbf{v0}_j\mathbf{v2}_j} \end{bmatrix} \begin{bmatrix} t_j \\ u_j \\ v_j \end{bmatrix} = \mathbf{q}_i - \mathbf{v0}_j \tag{2.6}$$

The distance t_j represents the interval between $\mathbf{q}_i$ and the intersection point on the plane defined by vertices $\begin{bmatrix} \mathbf{v0}_j & \mathbf{v1}_j & \mathbf{v2}_j \end{bmatrix}$. The coordinates u_j and v_j denote the barycentric coordinates of this intersection point. If the conditions $t_j \leq 0$, $u_j \geq 0$, $v_j \geq 0$, and $u_j + v_j \leq 1$ are met, it indicates that the intersection point is in the same direction as the vector $\overrightarrow{\mathbf{q}_i\mathbf{o}_{c,i}}$ and lies within the boundaries of triangle j. Consequently, we can infer that a ray extending from $\mathbf{q}_i$ in the direction of $-t_j\mathbf{e}(\overrightarrow{\mathbf{q}_i\mathbf{o}_{c,i}})$ intersects with triangle j. Considering the 3D barrier composed of J triangles, if any triangle j' intersects with the ray $-t_{j'}\mathbf{e}(\overrightarrow{\mathbf{q}_i\mathbf{o}_{c,i}})$ where $t_{j'} \geq 0$ and $t_{j'} \in \mathbb{R}$, then a collision with the electromagnetic coil is imminent. This collision detection process can be formulated as follows:

$$Q(\mathbf{p}) = \begin{cases} 1 & \text{if } \exists j \in [1, J] \text{ so that } t_j \leq 0 \wedge u_j \geq 0 \wedge v_j \geq 0 \wedge u_j + v_j \leq 1 \\ 0 & \text{otherwise} \end{cases} \tag{2.7}$$

where $Q(\mathbf{p})$ is the result of collision detection if parameter set $\mathbf{p}$ is implemented. 1 means collision will happen, and 0 means no collision.

Algorithm 2 (Refer to Appendix) delineates the procedure for the real-time optimization algorithm. The algorithm undergoes a series of iterations spanning G generations, with each generation comprising S distinct sets of parameters. Within this context, the parameter set denoted as $\mathbf{p}_s^g$ represents the sth set within the gth generation, where s ranges from 1 to S and g ranges from 1 to G. The parameters are bounded by $\mathbf{p}_{\min}$ and $\mathbf{p}_{\max}$, which correspond to the minimum and maximum limits, respectively.

To reduce the current requirement during magnetic-torque actuation, where a periodic alternating magnetic field is typically utilized, it is necessary to determine the amount of current demanded over each period of the magnetic field. Here, we define $I_i(k/mf, d_{\text{coil},i}, \alpha_{\text{coil},i}, \mathbf{d})$ as the current demanded in the ith coil at time $t = k/mf$, which can be computed via Eq. (2.3). The parameters $d_{\text{coil},i}$ and $\alpha_{\text{coil},i}$ are used to ascertain the position of the coil, while f represents the frequency of the required periodic dynamic field and d denotes the current position of the robot. Consequently, if the parameter set $\mathbf{p}_s^g = \{d_{\text{coil},i}, \alpha_{\text{coil},i}\}$ (for $i = 1, \ldots, n$) is applied, the total current demanded by the n coils of the system over one period of the dynamic field can be calculated as follows:

$$I(\mathbf{p}_s^g) = \sum_{i=1}^{n} \sum_{k=1}^{m} \left(I_i \left(\frac{k}{mf}, d_{\text{coil},i}, \alpha_{\text{coil},i}, \mathbf{d} \right) \right)^2 \tag{2.8}$$

Afterward, we can calculate the cost function for optimization, which is defined as:

$$C(\mathbf{p}_s^g) = w_1 I(\mathbf{p}_s^g) + w_2 \|\mathbf{p}_s^g - \mathbf{p}_{\text{pre}}\| + w_3 Q(\mathbf{p}_s^g) \tag{2.9}$$

In the given equation, the term $\|\mathbf{p}_s^g - \mathbf{p}_{\text{pre}}\|$ represents the variation between the current set of parameters and the previously determined optimal set, effectively quantifying the extent of required mechanical adjustments. $Q(\mathbf{p}_s^g)$ denotes the outcome of the collision detection process for the parameter set $\mathbf{p}_s^g$, as outlined in Eq. (2.7). The equation comprises three main components that address the required currents, the mechanical displacement, and the collision avoidance, with w_1, w_2, and w_3 being the corresponding weights for each component. Typically, w_3 is set to approach infinity, while w_1 and w_2 are assigned based on specific requirements.

Typically, the weight $w_3 \to \infty$, while w_1 and w_2 are adjustable based on specific requirements. An increased value of w_1 results in a reduced current demand, thereby improving system efficiency; conversely, a higher w_2 contributes to the minimization of mechanical adjustments by decreasing the frequency of parameter alterations, which in turn broadens the control bandwidth. In this context, the values assigned are $w_1 = 1$ and $w_2 = 2$.

The real-time optimization process is executed concurrently with the system's operation. It takes into account the instantaneous position of the magnetic medical robot $\mathbf{d}$, as well as the characteristics and parameters defining the magnetic field, such as the type of field (rotating or oscillating), pitch angle, and direction angle. The outcome of this process is the determination of the optimal parameter set. Upon any alteration in input, the optimization algorithm is triggered, and the resulting optimized parameters are stored as $\mathbf{p}_{\text{pre}}$. This stored set is then utilized in subsequent optimization cycles to expedite the convergence process.

The optimization procedure is divided into two main stages: initialization and search, consistent with the description in Algorithm 2. Regarding the initialization phase, an innovative approach has been devised to enhance computational efficiency. At the commencement of actuation and during the initial run of the optimization, the starting values for the parameter sets are ascertained through this method.

$$\mathbf{p}_s^0 = \text{rand}(\mathbf{p}_{\min}, \mathbf{p}_{\max})(s = 1, \dots, S) \tag{2.10}$$

The function rand(**a**, **b**) is utilized to create a random vector within the range defined by the lower limit **a** and the upper limit **b**. Conversely, during the actuation phase, the initialization of the current optimization's parameter sets is based on the outcomes of the previous optimization cycle, denoted as $\mathbf{p}_{\text{pre}}$. This approach is illustrated in the subsequent equation:

$$\begin{cases} \mathbf{p}_s^0 = \mathbf{p}_{\text{pre}} & (s = 1) \\ \mathbf{p}_s^0 = \mathbf{p}_{\text{pre}} + w\text{rand}(\mathbf{p}_{\min} - \mathbf{p}_{\text{pre}}, \mathbf{p}_{\max} - \mathbf{p}_{\text{pre}}) & (s = 2, \dots, S) \end{cases} \tag{2.11}$$

in which w is a manually assigned constant between 0 and 1.

Once the initial set of S parameters has been established, the subsequent steps involve computing the magnetic field and the collision detection function using these parameters, which is then followed by the evaluation of the cost functions. The parameter set that yields the most favorable cost is then designated as the initial global optimum, $\mathbf{p}_{\text{best}}$. Building upon these initial parameters, the subsequent phase entails an iterative search for the optimal solution across G generations. For each generation g (where g ranges from 1 to G), the S parameter sets are ascertained according to the subsequent equations:

$$\begin{cases} \mathbf{v}_s^g = c_0\mathbf{v}_s^{g-1} + c_s r_s(\hat{\mathbf{p}}_s - \mathbf{p}_s^{g-1}) + c_g r_g(\mathbf{p}_{\text{best}} - \mathbf{p}_s^{g-1}) \\ \mathbf{p}_s^g = \mathbf{p}_s^{g-1} + \mathbf{v}_s^g & (s = 1, \dots, S) \end{cases} \tag{2.12}$$

In this context, $\mathbf{v}_s^g$ denotes the velocity of change for the sth parameter set during the gth iteration. This velocity comprises three components: inertia, the impact of the individual's best performance up to the gth generation represented by $\hat{\mathbf{p}}_s$, and the impact of the global best performance up to the gth generation denoted by $\mathbf{p}_{\text{best}}$. The inertia weight is denoted by c_0, while c_s and c_g are coefficients that quantify the influence of the personal best and global best, respectively. These coefficients are assigned the constriction values $c_0 = 0.730$, $c_s = 1.496$, and $c_g = 1.496$. r_s and r_g represent random variables within the range of 0–1. Following G iterations, the determined global best $\mathbf{p}_{\text{best}}$ is forwarded to both the magnetic field and mechanical actuation systems for implementation, with $\mathbf{p}_{\text{pre}}$ being updated to reflect the current $\mathbf{p}_{\text{best}}$ for subsequent optimization rounds. For the purposes of this study, the parameters are established with $G = 200$ and $S = 50$ in the absence of $\mathbf{p}_{\text{pre}}$, or $S = 30$ otherwise. Overall, the fast calculation speed can be guaranteed by the following:

1. The optimization will be executed when the change of the inputs reaches a specific extent that can be defined by the user.
2. Within the optimization process, the iteration of the main loop will stop if the number of loops reaches G, or the value change of the best cost $\triangle C(\mathbf{p}_{\text{best}})$ within the most recent five iterations is smaller than ϵ and, meanwhile, the cost value is smaller than $w3$.
3. Fewer particles in each optimization.
4. Every time the optimization executes, searching of the new optimal parameter sets will start from the result of the last optimization process.

2.3 System Prototype and Validation

2.3.1 System Prototype

To validate the performance of the proposed design, a prototype named RoboMag has been constructed. This RoboMag system is composed of several key components: a host computer, a magnetic field generation system, a mechanical actuation system, a magnetically actuated robot, and localization equipment. In this architecture, computations are allocated across various subsystems to minimize interference and enhance processing speed (refer to Figure 2.4).

The prototype is depicted in Figure 2.5a. The host computer, equipped with an Intel(R) Core (TM) i5-7400 (3.0 GHz) CPU and 8 GB of RAM, features a user interface developed using LabView from National Instruments Co., Austin, TX, USA, for operator command input. The positioning of the robot and the optimization of coil positions are also performed on the host computer.

The coordinates optimal for the positioning of the coils are transmitted to the mechanical actuation system for subsequent operations. A myRIO microcontroller from National Instruments Co., Austin, TX, USA, computes the angles required for each motor, which are then actuated by three four-axis robotic arms provided by SDCQ Motors Company Ltd., Beijing, China. Each of these robotic arms comprises four articulations: one capable of rotation within the horizontal plane, with the remaining three having rotation planes that are perpendicular to the x–y plane (refer to Figure 2.5b).

The determination of the coordinates ${}^{\mathbf{R}}\mathbf{o}_{c,i}$ is based on the parameters obtained from the optimization process, with the angles of the joints being derived using

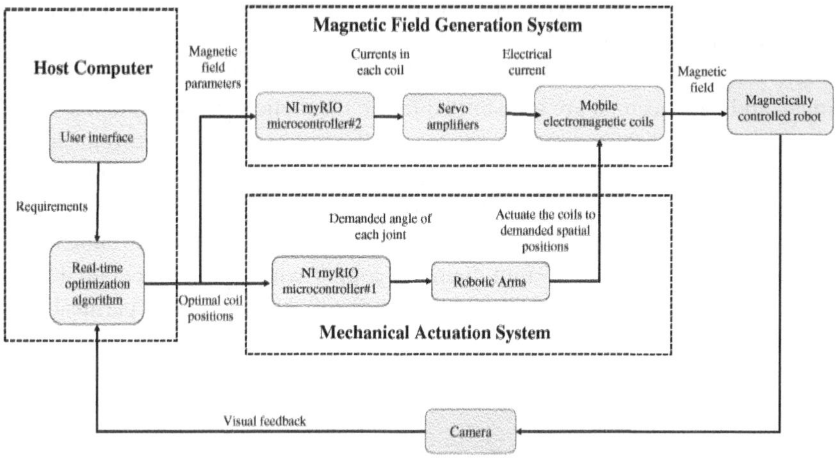

Figure 2.4 Hardware structure of the prototype RoboMag system. The system consists of three main modules, including the host computer, magnetic field generation system, and mechanical actuation system.

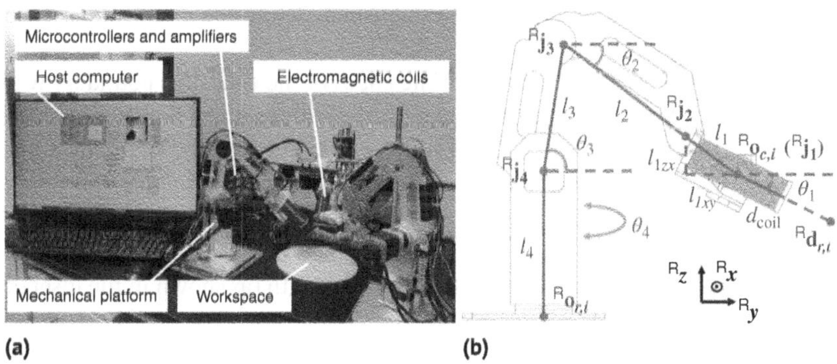

Figure 2.5 (a) Prototype RoboMag system. A host computer with a user interface is implemented for operating and conducting main calculations. The mechanical platform consists of three four-axis robotic arms that can move the three coils in a hemispherical space with a diameter of 203 mm. (b) A four-axis robotic arm that is implemented in the prototype RoboMag system for actuating the mobile electromagnetic coils.

the Denavit-Hartenberg parameterization method. The robotic arms facilitate the movement of the three mobile coils within a hemispherical region that has a diameter of 203 mm. Localization of the microrobot is achieved through the use of two cameras with orthogonal placements; one equipped with an IMX291 sensor from Sony Co., Tokyo, Japan, and featuring a 90° visual angle lens, captures the x-y coordinates, while the other secures the z coordinate. A LabView IMAQ pattern matching function is utilized to identify the microrobot and ascertain its position within the image frame.

The magnetic field parameters, which include amplitude, frequency, pitch angle, and base vectors, are transmitted to the magnetic field generation system. A separate myRIO microcontroller generates the necessary digital current signals in accordance with these parameters. These signals are then relayed to three servo amplifiers (ESCON 70/10, Maxon Inc.) to energize the coils. A 1000 W power supply is employed to deliver the required energy. The coils are custom-wound, comprising 830 $\pm$ 10 turns of 0.9 mm diameter copper wire, with cores crafted from 1J21 alloy—a soft magnetic material known for its high magnetic permeability. Each coil is capable of producing a magnetic field of 10 mT at a distance of 50 mm with a current of 5 A.

2.3.2 Experimental Validation

This experiment was designed to evaluate the efficacy of the RoboMag system and a newly proposed real-time optimization algorithm by observing the magnetic propulsion of a helical swimmer along a 3D path. A workspace boundary mimicking the shape of a human head was established, delineated by a white acrylic frame (refer to Figure 2.6a).

To prevent potential collisions with the edges of the coils, this boundary was extended by 10%. Within this setup, a flexible tube with an inner diameter of 8 mm was utilized, filled with a 90% glycerol solution to simulate a low Reynolds number environment. Replicating the 3D trajectory from the simulation, the tube's shape was adjusted accordingly. The helical swimmer, measuring 6 mm in diameter by 8 mm in length, featured a cylindrical NdFeB magnet affixed to one end, with dimensions of 1 mm in diameter by 2 mm in length (as shown in the inset of Figure 2.6a).

The outcomes of the experiment are depicted in Figures 2.6b through 2.7d. Figure 2.6b illustrates the actuation trajectory of the helical swimmer. The entire actuation process spanned 40 s, with no incidences of collision between the coils and the acrylic frame, as observed in Figure 2.6c. The system successfully tracked the swimmer's position, enabling the computation of optimal coil positions in real time. The currents through each coil were logged, and the root-mean-square values were determined for each feedback point of the swimmer's position, aligning well with the simulation data as shown in Figure 2.7a. Given potential velocity discrepancies between the experiment and the simulation, the horizontal axis represents the distance from the starting point (−15, −50, 28) to the swimmer's location. The measured currents were significantly lower than those in the control groups, by 62.7 and 45.0%, respectively. The magnetic field strength at key points along the trajectory was also verified against theoretical calculations, with an average deviation of 0.39 mT, as detailed in Figure 2.7b.

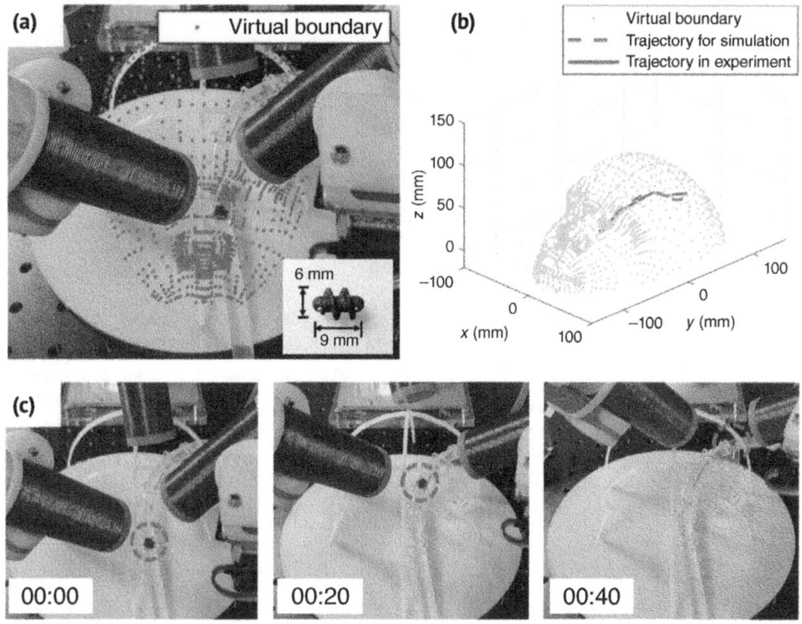

Figure 2.6 Actuation of a helical swimmer using a rotating field. (a) Experimental setup. (b) Experimental trajectory compares with the trajectory for simulation. (c) Actuation process of helical swimmer at 0, 20, and 40 s. The inset shows the image and dimension of the helical swimmer.

The paths traced by the tips of the coils are presented in Figure 2.7c, with the corresponding optimized parameters displayed in Figure 2.7d. These paths did not intersect with the expanded boundary, allowing the coils to be positioned close to the boundary for improved energy efficiency. A supplementary video provides a visual representation of the actuation process. Collectively, these results substantiate the effectiveness of the proposed algorithm in enhancing actuation efficiency and preventing collisions during rotating-field actuation.

In addition to evaluating the system's performance with a rigid-bodied swimmer, the actuation capabilities with a soft-bodied swimmer utilizing an oscillating magnetic field were also assessed. This soft swimmer measures 1 mm in diameter by 14 mm in length, as depicted in the inset of Figure 2.8a. The robot is equipped with a 1 mm diameter by 1 mm length permanent magnet at its head, with the magnetic moment aligned along the robot's axis. The swimmer's tail, composed of polydimethylsiloxane and colored black, is soft. The oscillating field used for actuation is configured with parameters identical to those used in the simulation. The experimental arrangement mirrors that used for the helical swimmer.

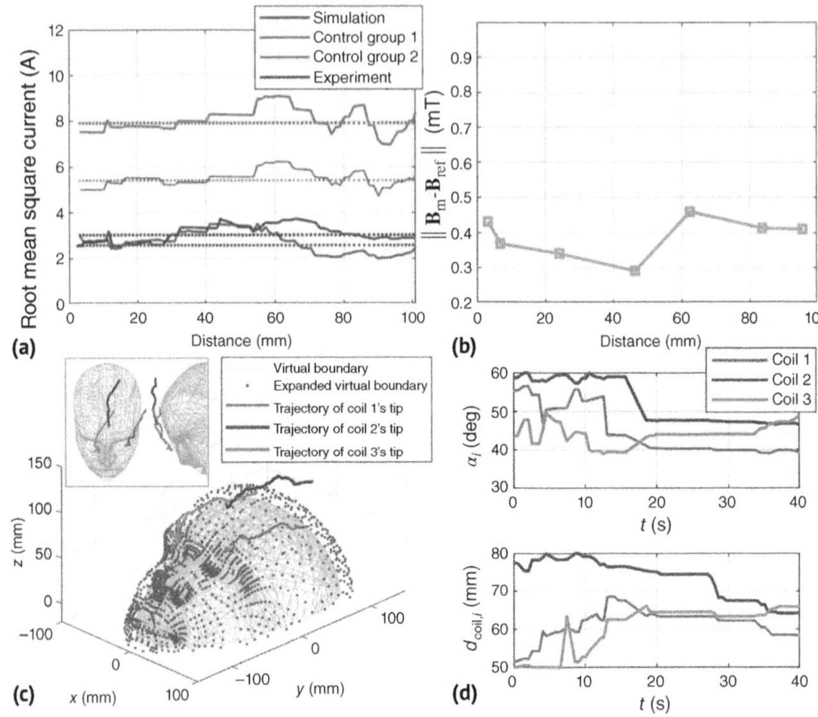

Figure 2.7 (a) Currents in the experiment are recorded, and the root-mean-square values between each position feedback are calculated for easier comparison. The dotted line shows the mean current. (b) Error of magnetic field on the trajectory during actuation. (c) Trajectories of the coils' tips during actuation. (d) Optimized parameters that control the coils' positions during the actuation.

The experimental outcomes are presented in Figures 2.8 and 2.9. The actuation of the swimmer along its path required 48 s. Throughout this procedure, the coils successfully tracked the swimmer's movement and prevented any collisions with the acrylic frame, as illustrated in Figure 2.8a. The experimentally applied currents on the coils were in close agreement with the simulated results, as depicted in Figure 2.8b, thereby validating the accuracy of the simulation. The currents recorded in these experiments were found to be 61.9 and 43.5% less than those in the two comparative control groups. Each coil's tip remained sufficiently near to, yet did not cross, the expanded boundary. The findings confirm the efficacy of the proposed real-time optimization algorithm.

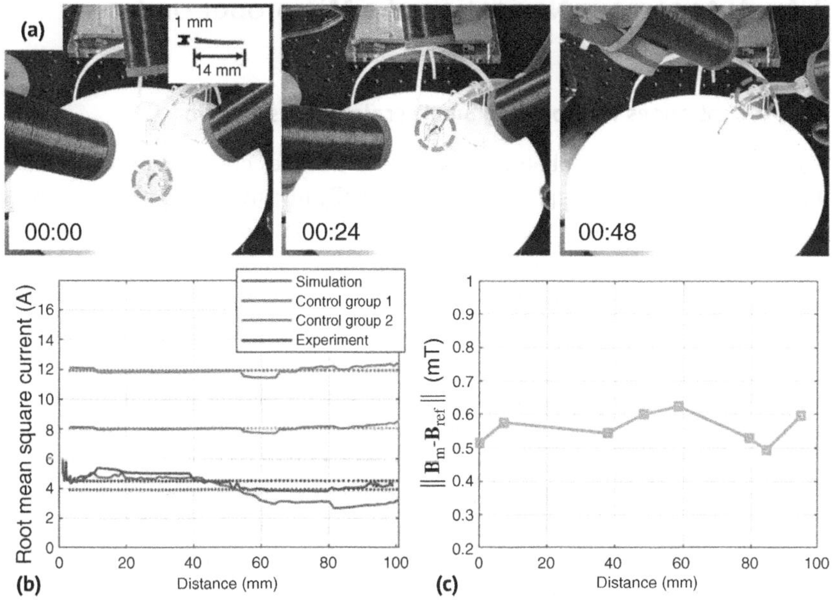

Figure 2.8 Actuation of a swimmer with a soft tail using an oscillating field. (a) The actuation process takes 48 s in total, and no collision happened between the mobile coils and the acrylic boundary. The inset shows the soft swimmer. (b) Recorded currents in the experiments compared with the simulation results. (c) Error of magnetic field on the trajectory.

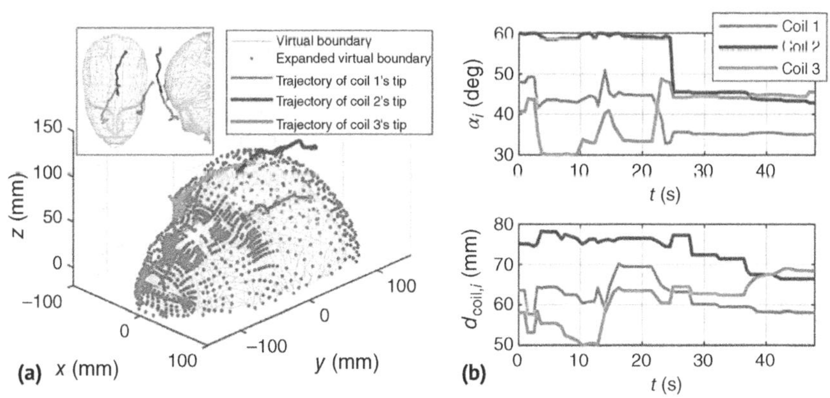

Figure 2.9 (a) Trajectories of the coils' tips are close to the expanded boundary, whereas they have no intersections. (b) Optimized parameters for the coils' positions during actuation.

2.4 Autonomous Navigation of a Microrobot Under Ultrasound Doppler Imaging

2.4.1 Challenges for Applying Small-scale Robots In Vivo

Small-scale, untethered robots that navigate in complex environments are emerging as valuable tools for applications in minimally invasive surgery, image-guided delivery, sensing, and targeted therapies within the human body [25–27]. Propelled by diverse energy sources, including self-propulsion [28], external fields [29–31], or hybrid mechanisms [32, 33], these robots with adjustable motion and steering capabilities can perform tasks in challenging media such as vitreous humor and gastrointestinal tract and in hard-to-reach areas like blood vessels and bile ducts [34, 35]. The integration of small-scale robots in medical procedures promises enhanced targeting, as seen in active delivery [36, 37], thrombolysis [38, 39], and the breaching of biological barriers [40].

Despite various proposed actuation strategies and improvements, the maneuverability of small-scale robots in stagnant or low-flow-rate fluids is restricted. The drag forces from significant fluid flows can undermine the robots' propulsion and control, leading to adverse net motion or uncontrollable locomotion in viscous biofluids [41, 42]. This can result in uncontrollable locomotion velocities and unpredictable motion trajectories, thereby reducing the efficiency and accuracy of targeted delivery to the desired region. Achieving controllable navigation in intricate biofluidic flows is a challenge that necessitates further exploration into actuation strategies to bolster the propulsion and navigation precision of small-scale robots under flow conditions.

An additional challenge in the in vivo deployment of small-scale robots is related to imaging challenges. The positional data obtained from imaging systems is crucial for robot localization, a technique that has been utilized for guiding robots and for developing closed-loop control systems to execute tasks autonomously [43, 44]. To date, a range of imaging techniques have been employed to track the position of small-scale robots, encompassing ultrasound imaging [45–47], magnetic resonance imaging [48], photoacoustic imaging [49–51], and positron emission tomography. Among these, ultrasound imaging is particularly promising for localizing mobile small-scale robots due to its high temporal resolution, which allows for rapid image acquisition and can be utilized as real-time feedback [52–54].

However, conventional B-mode ultrasound imaging often struggles with a low signal-to-noise ratio, making it difficult to distinguish small-scale robots from the surrounding biological tissues, thus compromising the accuracy of real-time tracking. In contrast to conventional B-mode imaging, Doppler ultrasound imaging detects the velocity of blood flow induced by the robot by reflecting high-frequency sound waves off red blood cells (RBCs). The resulting colored areas on the Doppler

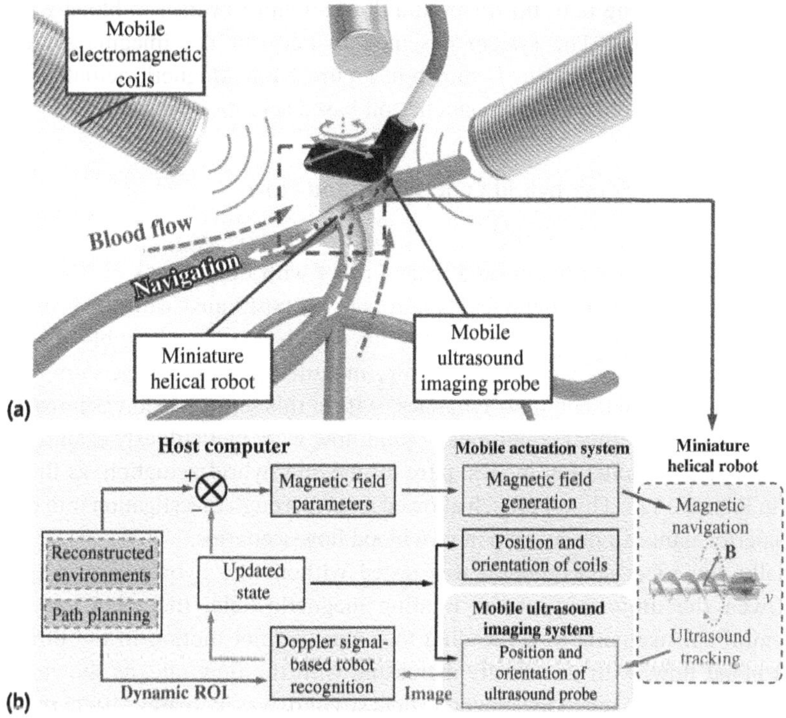

Figure 2.10 Schematic of the ultrasound Doppler imaging-guided autonomous navigation of a miniature helical robot in a vascular environment. (a) The torque and force-hybrid control is conducted using a mobile actuation system to overcome blood flow. (b) Overview of the Doppler imaging-based autonomous control strategy.

images pinpoint the robot's position, which aids in distinguishing the robot from its surroundings and facilitates real-time tracking. Hence, it is valuable to explore a strategy that combines control methods with feedback from Doppler ultrasound imaging for autonomous navigation in changing environments.

Addressing the challenges outlined earlier, we introduce an autonomous navigation strategy that leverages ultrasound Doppler imaging to facilitate the real-time deployment of a microrobot designed for precise thrombolysis, as depicted in Figure 2.10a. Our approach integrates a magnetic torque and force-driven actuation method to maneuver the helical robot through dynamic fluidic conditions. A specialized optimization algorithm is developed to produce magnetic field gradients in specified orientations when utilizing rotating magnetic fields. The Doppler signals generated by the rotating helical robot are analyzed within blood flow scenarios, enabling real-time localization of the robot. We have conducted experiments to assess real-time navigation and tracking of the robot within vascular settings, encompassing both upstream and downstream traversal in blood with varied flow velocities. The system designed to perform the imaging-guided autonomous navigation strategy is shown in Figure 2.10b. Magnetic actuation is conducted using a mobile electromagnetic coil-based actuation system.

2.4.2 Torque and Force-hybrid Control in Blood Flow

To evaluate the effectiveness of the hybrid control methodology, an experimental setup was established featuring a helical robot with dimensions of 7.30 mm in length and 2.15 mm in diameter. This robot was positioned within a 6-mm-diameter tube. The tube was subsequently filled with porcine whole blood and attached to a TJP-3A programmable pump, manufactured by LongerPump in China, to regulate the blood flow dynamics. Within this setup, the forces impacting the helical robot under conditions of fluid flow were meticulously examined under two distinct actuation modes: pure torque and hybrid actuation, as illustrated in Figure 2.11a. This approach allowed for a thorough investigation into the robot's performance in mimicking in vivo blood flow scenarios.

Initially, the robot's locomotion was tested without the hybrid control system. Given the uniformity of the rotating magnetic field, the robot experienced minimal magnetic force, leading to a negative net motion in the direction of blood flow, with its velocity increasing with the flow rate, as shown in Figures 2.11b, c. Even when the magnetic field strength was increased (10–15 mT), the passive net motion velocity remained between 0 and −2 mm/s, suggesting that the robot's operation was suboptimal at a field strength of 1–5 mT. Upon raising the flow rate to 20 mL/min, the negative net motion velocity further increased with the magnetic field strength. The increased blood flow distance caused parts of the robot to move away from the boundary, reducing the friction force. Consequently, the robot's magnetic dipole could not remain coplanar with the rotating field, leading to a tilt that affected the robot's orientation. The enhanced field strength intensified the robot's motion, and the increased friction force due to the higher flow rate contributed to a greater negative net motion velocity.

Utilizing a hybrid control strategy that combines torque and force, we successfully generated a rotating magnetic field with a gradient counter to the direction of blood flow. This magnetic force augmentation significantly bolstered the robot's ability to navigate against the blood current, as evidenced in Figures 2.11d,e. Upstream traversal against flow velocities of 10, 15, and 20 mL/min (with mean velocities of 5.89, 8.84, and 11.79 mm/s, respectively) was accomplished with magnetic field strengths exceeding 2, 3, and 4 mT. In contrast to scenarios where only torque was applied, the robot maintained its position closer to the tube's boundary, thus eliminating any significant tilt. The efficacy of the hybrid control approach was further substantiated under pulsatile flow conditions. Here, blood flow was

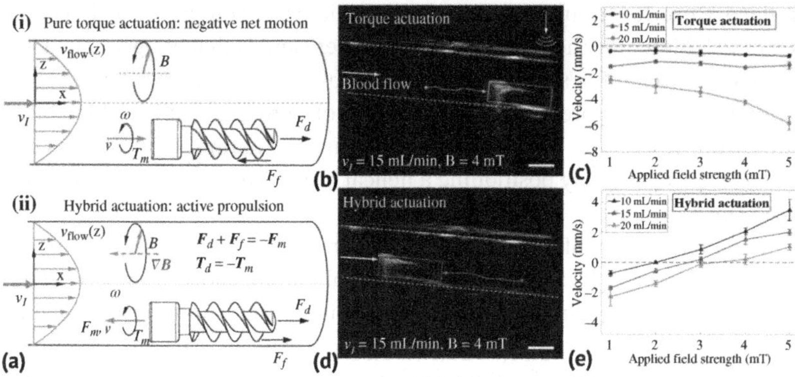

Figure 2.11 Magnetic actuation of a miniature helical robot in blood flow. (a) Schematic of a magnetically actuated helical robot in blood flow. (i) The robot exhibits negative net motion along the flow direction under a uniform rotating magnetic field B (pure torque actuation). (ii) The robot exhibits active propulsion under a rotating field with field gradient ∇B (hybrid actuation). F_m, magnetic force; F_f, friction force; F_d, drag force; T_m, magnetic torque; T_d, drag torque; v_i, input flow rate. (b) Experimental results of magnetic torque-actuated robot locomotion in blood flow. (c) The relationship between the input field strength and locomotion velocity of the robot under the torque actuation. (d) Experimental results of magnetic torque and force-hybrid actuation against blood flow. The solid line in each of (b) and (d) denotes the motion trajectories of the helical robot, and the robot is real-time localized and marked by the rectangles. Scale bars: 3 mm. (e) The relationship between the input field strength and locomotion velocity of the robot under the hybrid actuation. The input frequency is 5 Hz for all trials. Each error bar denotes the standard deviation (SD) from three trials.

modulated by a programmable pump to alternate between a high-velocity phase (12–24 mL/min, with mean velocities of 7.07–14.15 mm/s) over one second and a low-velocity phase (below 0.2 mL/min) over the subsequent second. The findings revealed that the robot could propel itself against the pulsatile flow, achieving a net forward motion that was positively correlated with the strength of the applied magnetic field. These results demonstrated that the torque and force-hybrid control approach could increase the robot's thrust force, indicating the feasibility of the hybrid control approach for navigating a helical robot against blood flow.

2.4.3 Helical Robot-induced Doppler Signals

The ultrasound Doppler signal operates on the principle of the Doppler effect, which involves the detection of changes in frequency of ultrasound waves that are reflected back after interacting with moving objects. In our experiments, the RBCs serve as the ultrasound reflectors responsible for inducing these frequency shifts in the ultrasound waves that are both emitted and reflected. The frequency alteration attributed to the motion of these mobile RBCs can be measured by using the influence exerted by an actively controlled microrobot on the local blood flow. This method allows for the assessment of Doppler signals that are generated as a consequence of the robot's interaction with the blood medium. Simulations have been conducted to explore the impact of the robot's presence on both blood flow dynamics and RBCs. The robot's ability to perturb the natural blood flow is crucial for its role in precisely positioning a helical robot through the use of motion-induced Doppler signals.

To evaluate the effect of the robot's rotation on the standard blood circulation, simulations have been set up under conditions that mimic blood flow (as depicted in Figure 2.12a). These simulations aim to provide insights into how the helical robot's movement can disrupt the typical blood flow patterns, which is a key aspect of utilizing Doppler signals for the robot's navigation and targeting capabilities. Results show that the rotating robot affects the blood flow profile, and the laminar flow profile shows distortion near the robot.

To explore the impact of a rotating robot on the motion of RBCs, a simulation was conducted releasing a set of simulated RBCs (S-RBCs) from the entry of a tube, as depicted in Figure 2.12b. The simulation visually tracks the displacement of these cells. Within the Poincaré maps, all microparticles are categorized into two groups according to their initial *y*-coordinate (*yo*). The phase map illustrates the starting positions of the S-RBCs, with the two groups distinguished according to their initial y-coordinate: one group for $yo < 0$ and the other for $yo > 0$. Until they approach the robot, the S-RBCs maintain a structured arrangement due to the laminar flow condition. Each S-RBC's movement is dictated by the drag force, aligning their velocity with the flow's distribution. Upon interaction with the rotating robot, the flow profile's distortion causes some S-RBCs to alter their initial

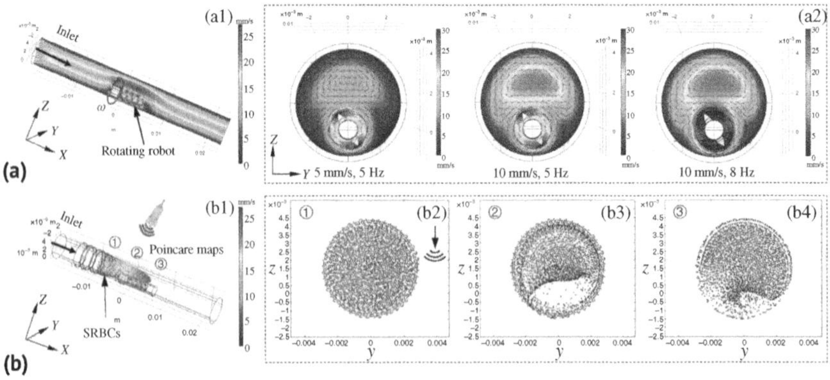

Figure 2.12 (a) Simulation results of a rotating helical robot in blood flow. (b) Simulation results of the motion of simulated RBC flow through the robot.

yz-plane positions. A greater number of S-RBCs shift position and transition to the opposing half-plane as they traverse past the robot. The findings suggest that emitting ultrasound waves from the *z*-axis can provide a Doppler shift indicative of the RBCs' motion near the robot, detectable by an ultrasound imaging system. The simulation further demonstrates that an increased input frequency enhances the robot-induced Doppler signal under flowing conditions, a factor beneficial for real-time localization during vascular system operations.

The experimental examination of the Doppler signal, generated by the robot's motion, was conducted within blood-mimetic environments as shown in Figure 2.13a. Prior to the activation of the rotating magnetic field (at $f = 0$), no Doppler color shift was detected. With the rotation of the robot, which broadens the area of influence, the Doppler color region expanded as the frequency was raised from 1 to 6 Hz. The Doppler display showed opposite flow directions on the two sides of the robot, with the left side indicating flow toward the ultrasound transducer and the right side indicating flow away from it. Simulations revealed that the blood flow induced by the robot was nearly opposite on either side.

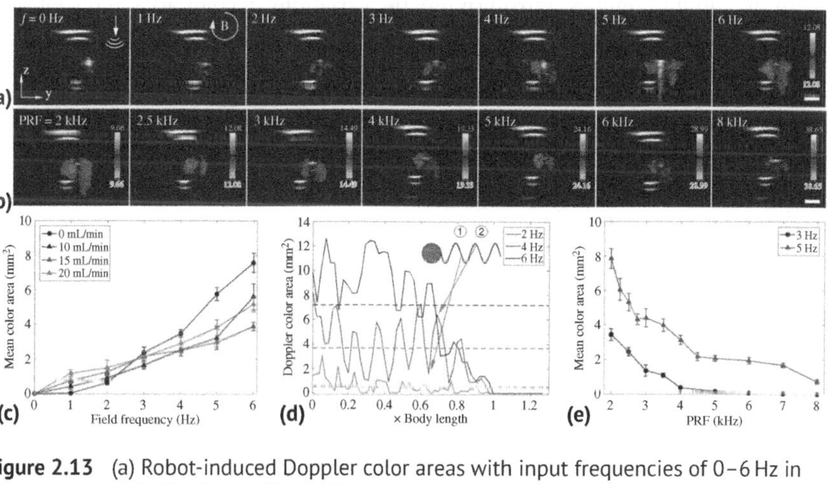

Figure 2.13 (a) Robot-induced Doppler color areas with input frequencies of 0–6 Hz in stagnant blood. PRF: 2.5 kHz. The white arrow and curves denote the ultrasound propagation direction. (b) Robot-induced Doppler color area at PRF of 2–8 kHz. Input frequency: 5 Hz. Scale bars in (a) and (b) are 2 mm. (c) Mean color area under different input frequencies and flow rates. Each mean value is calculated as the mean color area of the robot swimming through the imaging plane from head to tail. (d) Change of Doppler color area when the robot swims through the imaging plane from head to tail. Dashed lines denote the mean values. (e) Mean area of Doppler color under different PRF. Each error bar in (c) and (e) denotes the SD from three trials.

In addition to magnetic field parameters, the pulse repetition frequency (PRF), representing the count of ultrasound pulses emitted within a given time frame, also impacted the Doppler color region. When two successive pulses are directed at moving reflectors (RBCs) over a period of 1/PRF, a phase variance between the returning echoes occurs, indicating the reflectors' movement. The ultrasound system's sensitivity to low-velocity blood flow was enhanced by increasing PRF from 2 to 8 kHz, resulting in a diminished color region as observed in Figure 2.13b. The Doppler signal's mean color areas under flow conditions were also studied. The background blood flow was nearly perpendicular to the ultrasound's propagation path (an insonation angle of approximately 90°), thus exerting minimal effect on the Doppler color region, as depicted in Figure 2.13c.

To quantitatively assess the variations in color areas as the robot navigated through the imaging plane, the mean color area was employed. These areas fluctuated with the robot's progression. Notably, a more extensive color area was noted when the imaging plane intersected the apex of the helical tail (indicated as ② in Figure 2.13d), whereas it shrank when reaching the tail's central axis (indicated as ① in Figure 2.13d). The mean color area was computed, indicating an increment with the input frequency. Furthermore, the variations in color area in response to diverse PRF settings were quantitatively examined by calculating the mean color area, as illustrated in Figure 2.13e.

2.4.4 Ultrasound Doppler Imaging-guided Real-time Navigation in Vascular Environment

Achieving real-time autonomous navigation for the microrobot necessitates the synchronized operation of four integral system modules: the host computer, magnetic actuation system, an ultrasound imaging device integrated with a three-degree-of-freedom (3-DOF) platform, and the helical robot itself.

The robot's real-time navigation was executed under conditions of both laminar and pulsatile blood flow, encompassing maneuvers both upstream and downstream. Employing a color Doppler ultrasound imaging technique, the setup featured an imaging depth of 4 cm and a temporal resolution ranging from 6 to 10 frames per second (FPS). A specially designed 3-DOF robotic manipulator was instrumental in adjusting the ultrasound probe's position and orientation. The magnetic actuation system, comprising three mobile electromagnetic coils, was engineered to facilitate a hybrid torque and force control strategy.

For the experiments, vascular phantoms were created by embedding hand-crafted vascular replicas into a 20% gelatin-aqueous solution. The programmable pump was utilized to modulate blood flow velocity. To mitigate scattering effects due to air or bubbles, the ultrasound probe was in contact with the upper surface of the phantom throughout the testing process.

To achieve upstream navigation within a vascular phantom featuring four distinct routes, as depicted in Figure 2.14a, a hybrid control strategy combining torque and force was implemented to counteract the blood flow, ranging from 10 to 20 mL/min. A rotating magnetic field with a gradient directed against the flow was utilized, with adjustments made to the gradient's direction, as well as the yaw and pitch angles, in accordance with the robot's tracking data and predetermined trajectory. This adaptability is crucial for steering the robot along the intended route while preventing entrapment within the vascular confines.

Critical for precise navigation within intricate vascular networks, the robot's path selection ability is paramount. The pivotal maneuver involves navigating near the convergence points of multiple pathways. For instance, the robot's trajectory was fine-tuned according to the angle of intersection between path 1 and path 2, facilitating on-demand steering (as shown in paths ① to ④ of Figure 2.14c). Upon reaching the junction of paths 2 and 3, a comparable control method was engaged to opt for the desired route (as depicted in paths ④ to ⑦ of Figure 2.14c). Our investigation into the correlation between navigation distance and time confirmed the hybrid control's efficacy in countering blood flow (as illustrated in Figure 2.14b). The mean lateral deviations from the channel's central axis for paths 1 to 4 were measured at 3.92, 3.17, 2.95, and 4.52 mm, equating to 0.54,

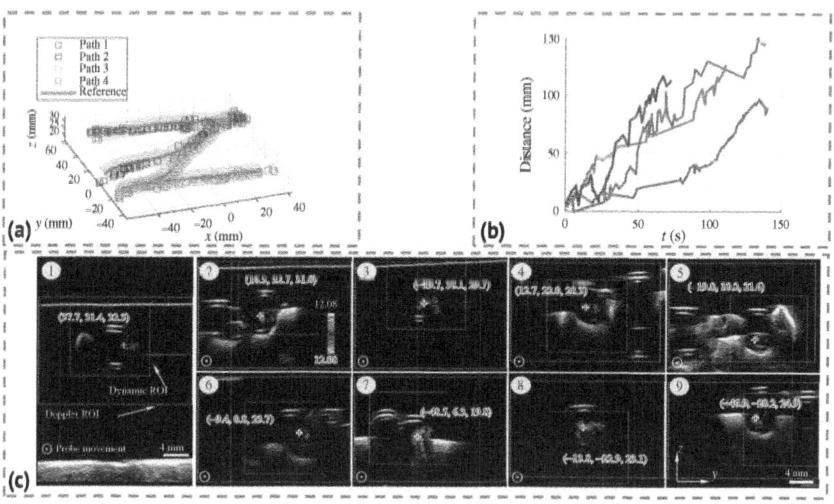

Figure 2.14 (a) The reference lines show the central lines of the vascular channel, and the square points are the real-time tracked positions during the upstream navigation. Input flow rate: 10 mL/min for paths 1 to 3 and 20 mL/min for path 4. (b) Relationship between upstream navigation distance and time in different paths. (c) The ultrasound images show the tracking positions and the dynamic ROIs during navigation, as correspondingly marked in (a). Scale bars: 4 mm.

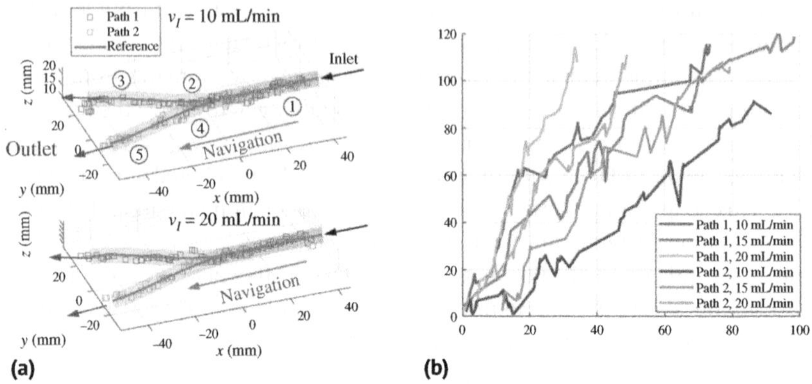

Figure 2.15 (a) Real-time tracked position during downstream navigation with flow rates of 10–20 mL/min. The inlet and two outlets are marked by arrows. (b) Relationship between downstream navigation distance and time in different flowing conditions.

0.43, 0.30, and 0.62 of the body length, respectively. This underscores the robot's capability for real-time tracking and its capacity to navigate effectively within complex vascular structures.

The hybrid control approach, utilizing Doppler imaging for guidance, was confirmed effective during downstream navigation trials. These trials took place within a Y-shaped channel configuration, as illustrated in Figure 2.15a, where a single entry point and dual exit points were plumbed into a system driven by a pump capable of managing flow rates between 10 and 20 mL/min. This setup allowed for the assessment of the Doppler imaging-guided hybrid control strategy in a controlled, downstream navigation scenario.

The strategy's validation was crucial for demonstrating the robot's ability to navigate efficiently through more complex vascular layouts while maintaining precise control over its movement against blood flow. The robot demonstrated mobility along two distinct routes (② to ③ and ④ to ⑤), with the Doppler-induced color areas incorporated into the dynamic regions of interest (ROIs). The average distances between consecutive tracking points were 0.62, 0.65, and 1.01 mm under flow conditions of 10, 15, and 20 mL/min, respectively. The average lateral deviations from the central axis were measured at 2.73, 2.48, and 2.26 mm, equating to 0.37, 0.34, and 0.31 of the body length under flow rates of 10–20 mL/min. The navigation distance's variation during actuation was also examined across a range of flow rates from 10 to 20 mL/min, as depicted in Figure 2.15b. A magnetic field strength ranging from 2 to 3 mT was adjusted according to the input flow rate, with the mean locomotion velocity kept below 3 mm/s. These findings suggest that the proposed strategy is viable for downstream navigation, offering adjustable navigation speed and the ability to choose the path effectively.

2.5 Discussion and Conclusion

2.5.1 Discussion

The prototype RoboMag system, when integrated with a real-time optimization algorithm, has been confirmed to be effective through experimental evidence. This integration significantly enhances the system's flexibility, allowing for the prevention of collisions between the coils and any obstacles present within the operational space. Moreover, the algorithm is instrumental in minimizing the electrical currents; it achieves this by strategically positioning the coils at their most advantageous locations during actuation—a feature not commonly found in prior systems utilizing multiple mobile coils [18, 19]. The balance struck between the swiftness of the optimization process and the precision of its outcomes is notably satisfactory. Furthermore, the operational span of the system has been broadened to encompass a hemispherical region measuring 203 mm in diameter. To bolster performance at elevated frequency bands, a set of more compact coils, characterized by an inductance of 55.5 mH, has been integrated into the system. The experimental assessments indicate that the forces exerted by each coil upon the robots do not surpass 3.5×10^{-4} N [55], suggesting that these forces may diminish further with the deployment of diminutive swimmers.

Despite advancements, there is room for enhancing the current design. The coils' adjustable positioning, while beneficial, poses challenges regarding the precision of magnetic flux density generation when compared with stationary systems [13]. Enhancing system accuracy could be achieved through feedback mechanisms on coil positioning or by implementing closed-loop control strategies for microrobots [10, 56]. Additionally, achieving 5-DOF control for magnetic microrobots is intricate due to modeling complexities [17]. Future research should address more intricate scenarios pertinent to biomedical applications, such as varying flow velocities and the manipulation of microrobots on a smaller scale. Integration with medical imaging systems [57, 58] and the incorporation of closed loop tracking algorithms [10, 56] are critical for the system's ongoing evolution.

2.5.2 Conclusion

In this research, we present a novel magnetic actuation system that leverages the agility of mobile electromagnetic coils, coupled with an innovative control strategy. This strategy facilitates the computation of optimal coil placements in real time, enhancing operational efficiency while also steering clear of potential collisions with workspace obstacles. To showcase this concept, we've developed a prototype named RoboMag. Our system is equipped with a mechanical setup featuring three robotic arms, each with four axes of movement, enabling the manipulation of three mobile coils within a hemispherical region measuring 203 mm in diameter. Through a series of experimental trials, we've validated both the system's capabilities and the efficacy of our control approach. Notably, we've achieved a significant reduction in required currents, with savings of up to 62.7% for rotating-field operations and 61.9% for oscillating-field maneuvers, when compared to traditional methods. Moreover, the system's ability to prevent coil-to-boundary collisions during extended actuation is noteworthy. Additionally, RoboMag has been instrumental in orchestrating the collective motion of microrobotic swarms, showcasing its potential to produce finely tuned dynamic magnetic fields for intricate operational demands.

Appendix Additional Algorithms

Algorithm 1: Collision Detection Algorithm Based on Triangle-ray Intersection

Require: parameter set $\mathbf{p}$; the coordinates of three vertices of J triangular surfaces $\mathrm{v0}_j$, $\mathrm{v1}_j$ and $\mathrm{v2}_j$, $(j = 1, \dots, J)$

Ensure: the result of collision detection $Q(\mathbf{p})$

$Q(\mathbf{p}) \leftarrow 0$
for $i \leftarrow 1$ to n **do**
    calculate oc_i and rc_i from $\mathbf{p}$
    for $j \leftarrow 1$ to J **do**
        calculate t_j, u_j and v_j by solving Eq. (2.6)
        if $t_j \leq 0$ and $u_j \geq 0$ and $v_j \geq 0$ and $u_j + v_j \leq 1$ **then**
            $Q(\mathbf{p}) \leftarrow 1$
        end if
    end for
end for
return $Q(\mathbf{p})$

Algorithm 2: Real-time Optimization on the Positions of Mobile Electromagnetic Coils Based on PSO

Require: location of the robot $\mathbf{d}$, parameters of the demanded magnetic field
Ensure: optimal parameter set $\mathbf{p}_{\text{best}}$ **Initialisation:**

for $s \leftarrow 1$ to S **do**
    if ($\mathbf{p}_{\text{pre}}$ exists) **then**
        if $s = 1$ **then**
            $\mathbf{p}_s^0 \leftarrow \mathbf{p}_{\text{pre}}$
        else
            $\mathbf{p}_s^0 \leftarrow \mathbf{p}_{\text{pre}} + \omega \text{rand}(\mathbf{p}_{\text{min}} - \mathbf{p}_{\text{pre}}, \mathbf{p}_{\text{max}} - \mathbf{p}_{\text{pre}})$
        end if
    else
        $\mathbf{p}_s^0 \leftarrow \text{rand}(\mathbf{p}_{\text{min}}, \mathbf{p}_{\text{max}})$
    end if
    if $C(\mathbf{p}_s^0) < C(\mathbf{p}_{\text{best}}^0)$ **then**
        $\mathbf{p}_{\text{best}}^0 \leftarrow \mathbf{p}_s^0$
    end if
    $\hat{\mathbf{p}}_s \leftarrow \mathbf{p}_s^0$
end for **Main loop:**
$\mathbf{p}_{\text{best}}^0 \leftarrow \mathbf{p}_{\text{best}}^0, \mathbf{v}_s^0 \leftarrow 0, g \leftarrow 1,$
while $(g < G) \wedge (\Delta C(\mathbf{p}_{\text{best}})$ in recent 5 iterations $> \epsilon \vee C(\mathbf{p}_{\text{best}}) > \omega_3)$ **do**
    for $s \leftarrow 1$ to S **do**
        $\mathbf{v}_s^g \leftarrow c_0 \mathbf{v}_s^{g-1} + c_s r_s(\hat{\mathbf{p}}_s - \mathbf{p}_s^{g-1}) + c_g r_g(\mathbf{p}_{\text{best}} - \mathbf{p}_s^{g-1})$
        $\mathbf{p}_s^g \leftarrow \mathbf{p}_s^{g-1} + \mathbf{v}_s^g$
        if $C(\mathbf{p}_s^g) < C(\hat{\mathbf{p}}_s)$ **then**
            $\hat{\mathbf{p}}_s \leftarrow \mathbf{p}_s^g$
        end if
        if $C(\mathbf{p}_s^g) < C(\mathbf{p}_{\text{best}}^g)$ **then**
            $\mathbf{p}_{\text{best}}^g \leftarrow \mathbf{p}_s^g$
        end if
    end for
    $g \leftarrow g + 1$
end while
$\mathbf{p}_{\text{pre}} \leftarrow \mathbf{p}_{\text{best}}$
return $\mathbf{p}_{\text{best}}$

References

1 M. Sitti, H. Ceylan, W. Hu, J. Giltinan, M. Turan, S. Yim, and E. Diller. Biomedical applications of untethered mobile milli/microrobots. *Proceedings of the IEEE*, 103 (2): 205–224, 2015.

2 J. F. Schenck. Safety of strong, static magnetic fields. *Journal of Magnetic Resonance Imaging*, 12 (1): 2–19, 2015.

3 A. W. Mahoney and J. J. Abbott. Five-degree-of-freedom manipulation of an untethered magnetic device in fluid using a single permanent magnet with application in stomach capsule endoscopy. *International Journal of Robotics Research*, 35 (1–3): 129–147, 2016.

4 K. M. Popek, T. Hermans, and J. J. Abbott. First demonstration of simultaneous localization and propulsion of a magnetic capsule in a lumen using a single rotating magnet. In *Proceedings of the IEEE International Conference on Robotics and Automation (ICRA)*, pages 1154–1160, 2017.

5 C. M. Heunis, Y. P. Wotte, J. Sikorski, G. P. Furtado, and S. Misra. The ARMM system–autonomous steering of magnetically-actuated catheters: towards endovascular applications. *IEEE Robotics and Automation Letters*, 5 (2): 705–712, 2020.

6 J. Yu, B. Wang, X. Du, Q. Wang, and L. Zhang, Ultra-extensible ribbon-like magnetic microswarms. *Nature Communications*, 9 (1): 3260, 2018.

7 J. Yu, L. Yang, and L. Zhang. Pattern generation and motion control of a vortex-like paramagnetic nanoparticle swarm. *International Journal of Robotics Research*, 37 (8): 912–930, 2018.

8 Y. Zhang, et al. Real-time tracking of fluorescent magnetic spore–based microrobots for remote detection of *C. diff* toxins. *Science Advances*, 5 (1): eaau9650, 2019.

9 I. S. M. Khalil, et al. Mechanical rubbing of blood clots using helical robots under ultrasound guidance. *IEEE Robotics and Automation Letters*, 3 (2): 1112–1119, 2018.

10 L. Yang, J. Yu, and L. Zhang. Statistics-based automated control for a swarm of paramagnetic nanoparticles in 2-d space. *IEEE Transactions on Robotics*, 36 (1): 254–270, 2020.

11 X. Du, K. F. Chan, X. Xia, P. W. Y. Chiu, and L. Zhang. A magnetically-triggered soft capsule for on-demand mucus collection. In *Proceedings of the IEEE International Conference on Cyborg and Bionic Systems*, pages 250–254, 2018.

12 L. Yang, E. Yu, C.-I. Vong, and L. Zhang. Discrete-time optimal control of electromagnetic coil systems for generation of dynamic magnetic fields with high accuracy. *IEEE/ASME Transactions on Mechatronics*, 24 (3): 1208–1219, 2019.

13 F. Niu, J. Li, W. Ma, J. Yang, and D. Sun. Development of an enhanced electromagnetic actuation system with enlarged workspace. *IEEE/ASME Transactions on Mechatronics*, 22 (5): 2265–2276, 2017.

14 F. Ongaro, S. Pane, S. Scheggi, and S. Misra. Design of an electromagnetic setup for independent three-dimensional control of pairs of identical and nonidentical microrobots. *IEEE Transactions on Robotics*, 35 (1): 174–183, 2019.

15 C. Chautems and B. J. Nelson. The tethered magnet: force and 5-DOF pose control for cardiac ablation. In *Proceedings of the IEEE International Conference on Robotics and Automation (ICRA)*, pages 4837–4842, 2017.

16 S. E. Wright, A. W. Mahoney, K. M. Popek, and J. J. Abbott. The spherical-actuator-magnet manipulator: a permanent-magnet robotic end-effector. *IEEE Transactions on Robotics*, 33 (5): 1013–1024, 2017.

17 P. Ryan and E. Diller. Magnetic actuation for full dexterity microrobotic control using rotating permanent magnets. *IEEE Transactions on Robotics*, 33 (6): 1398–1409, 2017.

18 J. Sikorski, A. Denasi, G. Bucchi, S. Scheggi, and S. Misra. Vision-based 3-D control of magnetically actuated catheter using BigMag—an array of mobile electromagnetic coils. *IEEE/ASME Transactions on Mechatronics*, 24 (2): 505–516, 2019.

19 L. Yang, X. Du, E. Yu, D. Jin, and L. Zhang. DeltaMag: an electromagnetic manipulation system with parallel mobile coils. In *Proceedings of the IEEE International Conference on Robotics and Automation (ICRA)*, pages 9814–9820, 2019.

20 J. Sikorski, I. Dawson, A. Denasi, E. E. Hekman, and S. Misra. Introducing BigMag: a novel system for 3D magnetic actuation of flexible surgical manipulators. In *Proceedings of the IEEE International Conference on Robotics and Automation (ICRA)*, pages 3594–3599, 2017.

21 J. Tuszynski. Triangle/ray intersection. *MATLAB Central File Exchange*, 2020. Accessed December 17, 2024.

22 B. Véron, A. Hubert, J. Abadie, and N. Andreff. Dealing with redundancy of a multiple mobile coil magnetic manipulator: a 3RPR magnetic parallel kinematics manipulator. In *Advances in Robot Kinematics*, pages 201–208, New York, NY, USA, 2018. Springer.

23 T. Whelan, M. Kaess, H. Johannsson, M. Fallon, J. J. Leonard, and J. McDonald. Real-time large-scale dense RGB-D SLAM with volumetric fusion. *International Journal of Robotics Research*, 34 (4/5): 598–626, 2015.

24 Y. Zhang, W. Xu, Y. Tong, and K. Zhou. Online structure analysis for real-time indoor scene reconstruction. *ACM Transactions on Graphics*, 34 (5): 1–13, 2015.

25 D. Jin and L. Zhang. Collective behaviors of magnetic active matter: recent progress toward reconfigurable, adaptive, and multifunctional swarming micro/nanorobots. *Accounts of Chemical Research*, 55 (1): 98–109, 2021.

26 C. K. Schmidt, M. Medina-Sánchez, R. J. Edmondson, and O. G. Schmidt. Engineering microrobots for targeted cancer therapies from a medical perspective. *Nature Communications*, 11: 5618, 2020.

27 H. Chen, H. Zhang, T. Xu, and J. Yu. An overview of micronanoswarms for biomedical applications. *ACS Nano*, 15: 15625–15644, 2021.

28 S. M. Beladi-Mousavi, J. Klein, B. Khezri, L. Walder, and M. Pumera. Active anion delivery by self-propelled microswimmers. *ACS Nano*, 14: 3434–3441, 2020.

29 J. Zhang, Z. Ren, W. Hu, R. H. Soon, I. C. Yasa, Z. Liu, and M. Sitti. Voxelated three-dimensional miniature magnetic soft machines via multimaterial heterogeneous assembly. *Science Robotics*, 6: eabf0112, 2021.

30 D. Jin, K. Yuan, X. Du, Q. Wang, S. Wang, and L. Zhang. Domino reaction encoded heterogeneous colloidal microswarm with on-demand morphological adaptability. *Advanced Materials*, 33: 2100070, 2021.

31 M. Valdez-Garduno, M. Leal-Estrada, E. S. Oliveros-Mata, D. I. Sandoval-Bojorquez, F. Soto, J. Wang, and V. Garcia-Gradilla. Density asymmetry driven propulsion of ultrasound-powered Janus micromotors. *Advanced Functional Materials*, 30: 2004043, 2020.

32 D. Ahmed, A. Sukhov, D. Hauri, D. Rodrigue, G. Maranta, J. Harting, and B. J. Nelson. Bioinspired acousto-magnetic microswarm robots with upstream motility. *Nature Machine Intelligence*, 3: 116–124, 2021.

33 Z. Xiao, S. Duan, P. Xu, J. Cui, H. Zhang, and W. Wang. Synergistic speed enhancement of an electric-photochemical hybrid micromotor by tilt rectification. *ACS Nano*, 14: 8658–8667, 2020.

34 Z. Zheng, et al. Ionic shape-morphing microrobotic end-effectors for environmentally adaptive targeting, releasing, and sampling. *Nature Communications*, 12: 411, 2021.

35 B. Wang, et al. Endoscopy-assisted magnetic navigation of biohybrid soft microrobots with rapid endoluminal delivery and imaging. *Science Robotics*, 6 (52): eabd2813, 2021.

36 Z. Ren, R. Zhang, R. H. Soon, Z. Liu, W. Hu, P. R. Onck, and M. Sitti. Soft-bodied adaptive multimodal locomotion strategies in fluid-filled confined spaces. *Science Advances*, 7 (27): eabh2022, 2021.

37 Q. Wang, J. Yu, K. Yuan, L. Yang, D. Jin, and L. Zhang. Disassembly and spreading of magnetic nanoparticle clusters on uneven surfaces. *Applied Materials Today*, 18: 100489, 2020.

38 L. Wang, et al. Guiding drug through interrupted bloodstream for potentiated thrombolysis by C-shaped magnetic actuation system *in vivo*. *Advanced Materials*, 33: 2105351, 2021.

39 Q. Wang, et al. Reconfigurable magnetic microswarm for accelerating tPA-mediated thrombolysis under ultrasound imaging. *IEEE/ASME Transactions on Mechatronics*, 26 (1), 2267–2277, 2021.

40 H. Ceylan, I. C. Yasa, U. Kilic, W. Hu, and M. Sitti. Translational prospects of untethered medical microrobots. *Progress in Biomedical Engineering*, 1: 012002, 2019.

41 Y. Alapan, U. Bozuyuk, P. Erkoc, A. C. Karacakol, and M. Sitti. Multifunctional surface microrollers for targeted cargo delivery in physiological blood flow. *Science Robotics*, 5 (42): eaba5726, 2020.

42 H. Xu, M. Medina-Sánchez, M. F. Maitz, C. Werner, and O. G. Schmidt. Sperm micromotors for cargo delivery through flowing blood. *ACS Nano*, 14: 2982–2993, 2020.

43 I. S. Khalil. Mechanical rubbing of blood clots using helical robots under ultrasound guidance. *IEEE Robotics and Automation Letters*, 3: 1112–1119, 2018.

44 Q. Wang, L. Yang, J. Yu, P. W. Y. Chiu, Y.-P. Zheng, and L. Zhang. Real-time magnetic navigation of a rotating colloidal microswarm under ultrasound guidance. *IEEE Transactions on Biomedical Engineering*, 67: 3403–3412, 2020.

45 A. V. Singh, et al. Multifunctional magnetic hairbot for untethered osteogenesis, ultrasound contrast imaging and drug delivery. *Biomaterials*, 219: 119394, 2019

46 Q. Chen, F.-W. Liu, Z. Xiao, N. Sharma, S. K. Cho, and K. Kim. Ultrasound tracking of the acoustically actuated microswimmer. *IEEE Transactions on Biomedical Engineering*, 66: 3231–3237, 2019.

47 Q. Wang, et al. Magnetic navigation of collective cell microrobots in blood under ultrasound Doppler imaging. *IEEE/ASME Transactions on Mechatronics*, 27: 3174–3185, 2021.

48 S. Mutlu, O. Yasa, O. Erin, and M. Sitti. Magnetic resonance imaging-compatible optically powered miniature wireless modular Lorentz force actuators. *Advanced Science*, 8: 2002948, 2021.

49 L. Xie, et al. Photoacoustic imaging-trackable magnetic microswimmers for pathogenic bacterial infection treatment. *ACS Nano*, 14: 2880–2893, 2020.

50 D. Li, C. Liu, Y. Yang, L. Wang, and Y. Shen. Micro-rocket robot with all-optic actuating and tracking in blood. *Light: Science & Applications*, 9: 84, 2020.

51 A. Aziz, J. Holthof, S. Meyer, O. G. Schmidt, and M. Medina-Sánchez. Dual ultrasound and photoacoustic tracking of magnetically driven micromotors: From in vitro to *in vivo*. *Advanced Healthcare Materials*, 10: 2101077, 2021.

52 A. Aziz, et al. Medical imaging of microrobots: toward *in vivo* applications. *ACS Nano*, 14: 10865–10893, 2020.

53 Q. Wang and L. Zhang. External power-driven microrobotic swarm: from fundamental understanding to imaging-guided delivery. *ACS Nano*, 15: 149–174, 2021.

54 S. Pané, et al. Imaging technologies for biomedical micro-and nanoswimmers. *Advanced Materials Technologies*, 4: 1800575, 2019.

55 L. Leon, F. M. Warren, and J. J. Abbott. Optimizing the magnetic dipole-field source for magnetically guided cochlear-implant electrode array insertions. *Journal of Medical Robotics Research*, 3 (1): 1850004, 2018.

56 L. Yang, Y. Zhang, Q. Wang, K.-F. Chan, and L. Zhang. Automated control of magnetic spore-based microrobot using fluorescence imaging for targeted delivery with cellular resolution. *IEEE Transactions on Automation Science and Engineering*, 17 (1): 490–501, 2020.

57 D. Li, D. Dong, W. Lam, L. Xing, T. Wei, and D. Sun. Automated *in vivo* navigation of magnetic-driven microrobots using OCT imaging feedback. *IEEE Transactions on Biomedical Engineering*, 67 (8): 2349–2358, 2020.

58 Q. Wang, L. Yang, J. Yu, P. W. Chiu, Y.-P. Zheng, and L. Zhang. Real-time magnetic navigation of a rotating colloidal microswarm under ultrasound guidance. *IEEE Transactions on Biomedical Engineering*, 67: 3403–3412, 2020.

3

Deep Reinforcement Learning Framework-based System Control for Soft Microrobots

This chapter proposes a flow rate rejection control strategy based on a deep reinforcement learning (DRL) framework to further optimize the control algorithm of the system based on the hardware conditions of the previous chapter, enabling it to drive the soft magnetic miniature robot (SMMR) in an unstructured environment and improve control accuracy and repeatability.

3.1 Introduction

In the contemporary landscape of technology, magnetically actuated robots have garnered considerable attention within biomedical domains, owing to their diverse and promising applications, spanning from disease diagnosis and minimally invasive surgical interventions to targeted drug delivery [1–3]. Nam and his team introduced a unique magnetic helical robot, capable of helical movement, precise drug delivery to designated areas, and mechanical drilling to clear obstructions in human tubular structures [4]. Furthermore, a magnetically steered, self-folding microrobot [5] has been devised for autonomous navigation and targeted drug delivery to cancer cells, enabling swift drug release at the intended site. Wang [6] and colleagues presented a novel magnetic robot designed for sustained drug release, which exhibits stable and continuous motion while facilitating quantitative drug administration. The robot's motion capabilities were assessed through in vitro experiments utilizing a sophisticated motion control approach. Despite their remarkable potential, magnetically actuated robots confront significant challenges in dynamically evolving environments and achieving precise control, particularly in adverse biofluid conditions.

To tackle the potential application of a SMMR in environments that undergo dynamic changes, we report a control framework based on DRL, mapping magnetic field parameters to robot states, to precisely control a physical SMMR to reach the goal and finish hovering tasks in the presence of flow rates. Autonomous manipulation of underwater robots [7, 8] and grasping of aerial robots [9] in research, goal-reaching, and hovering tasks are researched and implemented.

Firstly, an SMMR with head-tail morphology, driven by a mobile electromagnetic actuation system equipped with three electromagnetic coils [10] is designed. Thereafter, a simulator based on neural networks (NNs) was constructed, which was used to train the control policy. Then, the soft actor-critic (SAC) [11] algorithm based on a gated recurrent unit (GRU) [12] is exploited and implemented using history state action and the estimated flow rates as a part of inputs. Due to the uncertainties of the system and the inaccuracy of the magnetic field, the randomization during training is applied to bridge the mismatch between simulation model and real-world dynamics. Lastly, the trained control policy is transferred to a physical SMMR for goal-reaching and hovering tasks, while rejecting flow rates.

3.2 System Design and Modeling

As for electromagnetic coil-based mobile systems, the magnetic field is generated by coils. The coils are movable, which is able to enlarge the working area. This system has good scalability and control bandwidth. However, among the previous systems, single-coil-based systems face the inertia problem [13]; for systems with multiple coils [14, 15], the coupled movements of the coils will make them difficult to reach optimal positions for actuation, especially when obstructions exist within the workspace during clinical applications. Meanwhile, further improvement in energy efficiency is also a challenge.

Compared with these two types of systems, as shown in Figure 2.1, proposing a magnetic actuation system with decoupled mobile coils will solve the problems of single-coil-based and multiple-coil-based systems, and has higher energy efficiency due to the enhanced flexibility. Magnetic-torque propulsion and magnetic-force actuation are two techniques employed for remote control of magnetic robots. For a robot situated at a spatial location denoted as **d**, the magnetic torque τ_d and force $\mathbf{F}_d$ that are applied to it can be calculated by following equations:

$$\tau_d = \mathbf{m}_d \times \mathbf{B} \tag{3.1}$$

$$\mathbf{F}_d = (\mathbf{m}_d \cdot \nabla)\mathbf{B} \tag{3.2}$$

where **B** represents the magnetic flux density at the robot's position, and $\mathbf{m}_d$ denotes the magnetic moment of the magnet or magnetized material within the robot, a model is proposed to generate the required magnetic torque and force at a specific location **d**.

The modeling process is based on coordinate transformation and linear superposition. As shown in Figure 2.2, we define three types of coordinate systems: world frame (with no superscript), robotic arm frame (R), and coil frame (C). As for the world frame, the center of the working space is assigned as the origin ($\mathbf{O}_\omega$ in Figure 2.2). The origin for the frame of the ith robotic arm is designated as the base of each arm (${}^R\mathbf{O}_{r,i}$ in Figure 2.2). In the case of the ith coil's frame, the origin is positioned at the center of the coil (${}^C\mathbf{O}_{c,i}$ in Figure 2.2), and z-axis is along the axial direction of the coil. The right-hand rule is used to define the axes of coordinate system.

For a magnetic medical robot that is located at position $\mathbf{d} = (x_d, y_d, z_d)$ under the world frame, its relative position with the ith (i = 1, ... , n) coil can be calculated by transforming $\mathbf{d}$ into ith coil's frame, by using rotating matrix ${}^c_w\mathbf{T}$ and translational vector ${}^c_w\mathbf{t}$. The rotational movement for transformation from world frame to ith coil's frame is represented by matrix ${}^c_w\mathbf{T}$, and ${}^c_w\mathbf{t}$ stands for translational movement. Their values are based on the real-time location $\mathbf{O}_{c,i}$ and orientation $\hat{\mathbf{r}}_{c,i}$ of the coil. The coordinate of $\mathbf{d}$ under coil's frame is denoted as ${}^c\mathbf{d}_i$.

To get the magnetic field at ${}^c\mathbf{d}_i$, we build and calibrate a data-driven magnetic field model of a single coil. The model is based on finite element method data and calibrated according to the measured magnetic flux density [16]. From the model, the magnetic flux density at ${}^c\mathbf{d}_i$ with a current of I_i = 1A can be calculated, and denoted as ${}^c\hat{\mathbf{B}}_i$. The computed magnetic flux density, denoted as ${}^c\hat{\mathbf{B}}_i$, will be converted to the world frame and referred to as $\hat{\mathbf{B}}_i$. Consequently, the required currents in the coils $\mathbf{I} = [I_1, I_2, \dots, I_n]^T$ for generating magnetic flux density $\mathbf{B}$ at point $\mathbf{d}$ can be calculated as:

$$\mathbf{I} = \hat{\mathbf{M}}^{-1} \cdot \mathbf{B} \tag{3.3}$$

where $\hat{\mathbf{M}}$ is composed of three linearly independent base vectors: $\hat{\mathbf{M}} = \left[\widehat{\mathbf{B}_i}, \widehat{\mathbf{B}_j}, \widehat{\mathbf{B}_k}\right]$, $(i, j, k = 1, \dots, n, i \neq j \neq k)$.

The base vectors for magnetic force actuation can be calculated in a method which is similar to calculate magnetic flux density as:

$$\hat{\mathbf{F}}_i = {}^c_w\mathbf{T}^{-1} \cdot \hat{\mathbf{F}}_i = {}^c_w\mathbf{T}^{-1} \cdot \left({}^c\mathbf{m}_d \cdot \nabla\right)^c \hat{\mathbf{B}}_i = {}^c_w\mathbf{T}^{-1} \cdot {}^c\hat{\mathbf{G}}_i \cdot {}^c_w\mathbf{T} \cdot \mathbf{m}_d \tag{3.4}$$

where $\mathbf{m}_d$ is the magnetic moment of the robot under world frame, ${}^c\hat{\mathbf{G}}_i$ is a matrix representing the gradient of magnetic flux density [17] at ${}^c\mathbf{d}$ when I_i=1A. The currents required for generating force $\mathbf{F}$ at $\mathbf{d}$ can be addressed by:

$$\mathbf{I} = \hat{\mathbf{H}}^{-1} \cdot \mathbf{F} \tag{3.5}$$

where $\hat{\mathbf{H}}$ consists of three linearly independent base vectors: $\hat{\mathbf{H}} = \left[\hat{\mathbf{F}}_i, \hat{\mathbf{F}}_j, \hat{\mathbf{F}}_k\right]$, $(i, j, k = 1, \dots, n, i \neq j \neq k)$.

Based on the magnetic field model, a real-time optimization algorithm based on PSO can be applied to optimize the position of the electromagnetic coil, which is able to calculate the optimal position of the moving coils. The purpose of the algorithm is to meet the following conditions:

1. avoid collision between coils and surrounding obstacles;
2. generate the accurate magnetic field at a specified spatial location;
3. reduce the applied currents to minimize heat generation;
4. minimize mechanical movements during operation.

To enhance system efficiency [18] and expedite the convergence time of the optimization process, a strategy is adopted wherein the mobile coils maintain alignment with the robot's axial direction and follow its movements. Consequently, the parameters governing each coil's position can be streamlined to two variables: the distance ($d_{\text{coil},i}$) between the coil's tip and the robot, and the angle of inclination ($\alpha_{\text{coil},i}$) between the coil and the horizontal plane (as illustrated in Figure 2.2). The proposed optimization algorithm aims to identify the optimal set of parameters $\mathbf{p} = \{d_{\text{coil},i}, \alpha_{\text{coil},i}\}$ $(i = 1, \ldots, n)$ that fulfills the aforementioned objectives, which are formulated through the cost function of the optimization process. For the purpose of collision avoidance, an algorithm based on triangle-ray intersection is implemented to detect potential collisions between the system and irregular environmental barriers, contingent upon the application of a parameter set $\mathbf{p}$. The outcome of this collision detection serves as a component within the objective function utilized in the optimization process.

Under specific operational scenarios, RGB-D imaging techniques can be employed to acquire representations of the barriers, which are then subsequently preserved as three-dimensional (3D) models [19, 20] in the STereoLithography (STL) format, facilitated by the application of 3D reconstruction algorithms. In the STL files, the spatial configuration of the barrier's 3D model is delineated by J triangles in space, where each triangle is characterized by the coordinates of its three vertices, denoted as $\mathbf{v0}_j, \mathbf{v1}_j, \mathbf{v2}_j$, $(j = 1, \ldots, J)$, (as shown in Figure 2.3). Given that the ith mobile coil occupies a position denoted as $\mathbf{o}_{c,i}$, and $\mathbf{q}_i$ represents a point situated on the exterior surface of this coil, we can define $\overrightarrow{\mathbf{q}_i\mathbf{o}_{c,i}}$ as the vector that directs from $\mathbf{q}_i$ to $\mathbf{o}_{c,i}$. Here $\mathbf{q}_i$ and $\mathbf{o}_{c,i}$ are the functions of parameter set $\mathbf{p}$. The $\mathbf{q}_i$ can be selected on the edge of the coil, where collisions are more likely to happen, however, in this chapter, to be calculated quickly for real-time optimization, we choose the center of the surface on coil's tip as $\mathbf{q}_i$. Consequently, we extend the perimeter of the obstruction radially to create an updated boundary for the point $\mathbf{q}_i$. It is essential that the discrepancy between the initial boundary and this new boundary exceeds the coil's radius to mitigate the risk of potential collisions occurring at the coil's extremities. Vectors $\overrightarrow{\mathbf{v0}_j\mathbf{v1}_j}$ and $\overrightarrow{\mathbf{v0}_j\mathbf{v2}_j}$ are defined similarly, and they determine a plane $\{\mathbf{v0}_j\mathbf{v1}_j\mathbf{v2}_j\}$ on which the jth triangle locates. The vector $\mathbf{e}(\overrightarrow{\mathbf{q}_i\mathbf{o}_{c,i}})$, represents a unit vector emanating from $\mathbf{q}_i$ and aligned in the same direction as $\overrightarrow{\mathbf{q}_i\mathbf{o}_{c,i}}$. For each triangular facet comprising the 3D boundary model, the variables t_j , u_j, and v_j must be determined through a solution process.

$$\left[-\mathbf{e}\left(\overrightarrow{\mathbf{q}_i\mathbf{o}_{c,i}}\right), \overrightarrow{\mathbf{v0}_j\mathbf{v1}_j}, \overrightarrow{\mathbf{v0}_j\mathbf{v2}_j}\right]\begin{bmatrix} t_j \\ u_j \\ v_j \end{bmatrix} = \mathbf{q}_i - \mathbf{v0}_j \tag{3.6}$$

where t_j is the distance between $\mathbf{q}_i$ and the intersection point on plane $\{\mathbf{v0}_j, \mathbf{v1}_j, \mathbf{v2}_j\}$ are the barycentric coordinates of the intersection point. If $t_j \leq 0$, $u_j \geq 0, v_j \geq 0$, and $u_j + v_j \leq 1$, we can say that the intersection point is in the same direction as the vector $\overrightarrow{\mathbf{q}_i\mathbf{o}_{c,i}}$, and that it lies in the interior of triangle j. Consequently, a ray $-t_j\mathbf{e}(\overrightarrow{\mathbf{q}_i\mathbf{o}_{c,i}})$ originates from $\mathbf{q}_i$ and in the direction of $\overrightarrow{\mathbf{q}_i\mathbf{o}_{c,i}}$ intersects with triangle j. Considering the 3D barrier that consists of J triangles, if there exists a triangle j' that has intersection with ray $-t_{j'}\mathbf{e}(\overrightarrow{\mathbf{q}_i\mathbf{o}_{c,i}})$, $t_{j'} \geq 0, t_{j'} \in R$, then collision will occur between the electromagnetic coil and the barrier. In summary, the entire sequence of events can be formulated as follows:

$$Q(\mathbf{p}) = \begin{cases} 1(\exists j \in [1, j], t_j < 0 \wedge u_j \geq 0 \wedge v_j \geq 0 \wedge u_j + v_j \leq 1) \\ 0(\text{Others}) \end{cases} \tag{3.7}$$

The function $Q(\mathbf{p})$ determines the outcome of collision detection when the parameter set p is utilized. A value of 1 signifies that a collision will occur, whereas a value of 0 indicates no collision will take place. The process is described in Algorithm 1 in Appendix.

As shown in Algorithms 2 and 3 in Appendix, it outlines the procedure for the real-time coil's position optimization algorithm. This algorithm iterates the optimization process over G generations, with each generation comprising S parameter sets. The notation $\mathbf{p}_s^g$ represents the sth parameter set within the gth generation, $(s = 1, \ldots, S, g = 1, \ldots, G)$. The parameters are constrained within specified boundaries, denoted as $\mathbf{p}_{\min}$ and $\mathbf{p}_{\max}$, which represent the lower and upper limits, respectively.

Usually, where a periodic alternative magnetic field is implemented, in order to minimize the current during the driving of the magnetic torque, we need to calculate the demanded amount of current required during each period of the magnetic field. Define $I_i\left(\frac{k}{mf}, d_{\text{coil},i}, \alpha_{\text{coil},i}, \mathbf{d}\right)$ to be the demanded current in the ith coil at time $t = \frac{k}{mf}$, which can be calculated through Eq. (3.3). f is the demanded frequency of the periodic dynamic field. Consequently, if we apply the parameter set $\mathbf{p}_s^g = \{d_{\text{coil},i}, \alpha_{\text{coil},i}\}$ $(i = 1, \ldots, n)$, the total demanded currents for all the n coils of the system, within a single cycle of the dynamic field, can be quantified through calculation:

$$I(\mathbf{p}_s^g) = \sum_{i=1}^{n}\sum_{k=1}^{m}\left(I_i\left(\frac{k}{mf}, d_{\mathrm{coil},i}, \alpha_{\mathrm{coil},i}, \mathbf{d}\right)\right)^2 \tag{3.8}$$

Subsequently, we proceed to compute the optimization cost function, which is defined as

$$\ell(\mathbf{p}_s^g) = \omega_1 I(\mathbf{p}_s^g) + \omega_2 \parallel \mathbf{p}_s^g - \mathbf{p}_{\mathrm{pre}} \parallel + \omega_3 Q(\mathbf{p}_s^g) \tag{3.9}$$

In this equation, $\parallel \mathbf{p}_s^g - \mathbf{p}_{\mathrm{pre}} \parallel$ represents the variation between the current set of parameters and the previously optimal set, which can be interpreted as the magnitude of the amount of mechanical movements. The expression $Q(\mathbf{p}_s^g)$ indicates the outcome of collision detection when parameter set s is implemented, as defined by Eq. (3.7). This equation comprises three components, each responsible for managing the required electrical currents, mechanical displacements, and collision avoidance, respectively. The coefficients ω_1, ω_2, and ω_3 represent the respective weights assigned to these aspects, usually with $\omega_3 \to \infty$ to emphasize collision avoidance, while ω_1 and ω_2 can be adjusted according to specific requirements. Larger ω_1 will lead to a lower current to achieve higher efficiency, increasing the value of ω_2 diminishes the mechanical movements arising from minimal parameter alterations, thereby broadening the control bandwidth. In this context, assign ω_1 a value of 1 and ω_2 a value of 2 ($\omega_1 = 1, \omega_2 = 2$). The optimization process operates in real-time concurrent with the system's functionality. The inputs for this optimization algorithm consist of the real-time positional data of the magnetic medical robot, denoted as $\mathbf{d}$, along with the configuration and specific attributes of the magnetic field, such as rotating field or oscillating field, pitch angle, direction angle. The algorithm outputs the most optimal set of parameters. Whenever the input data undergoes a change, the optimization routine is invoked anew, and the resultant optimal parameters are stored as $\mathbf{p}_{\mathrm{pre}}$. This stored parameter set is subsequently utilized in subsequent optimization iterations to expedite the convergence process. The optimization process can be divided into initialization and searching, as in Algorithm 2. Upon initiation of the actuation process and the first execution of the optimization, the initial values of the parameter sets are established through:

$$\mathbf{p}_s^0 = \mathrm{rand}(\mathbf{p}_{\min}, \mathbf{p}_{\max})(s = 1, \dots, S) \tag{3.10}$$

Where $\mathrm{rand}(a, b)$ is utilized for generating a random vector within the specified range from the lower bound a to the upper bound b, during the execution of an optimization process in the context of actuation, the initialization of the parameter sets for the current optimization iteration begins with the outcome of the preceding optimization process, denoted as $\mathbf{p}_{\text{pre}}$. This concept is mathematically represented by the following equation:

$$\begin{cases} \mathbf{p}_s^0 = \mathbf{p}_{\text{pre}} \quad (s = 1) \\ \mathbf{p}_s^0 = \mathbf{p}_{\text{pre}} + \omega_{\text{rand}}(\mathbf{p}_{\min} - \mathbf{p}_{\text{pre}}, \mathbf{p}_{\max} - \mathbf{p}_{\text{pre}}) \\ (s = 2, \ldots, S) \end{cases} \tag{3.11}$$

where ω is a manually assigned constant between 0 and 1. Once the initial generation of S parameter sets has been established, the magnetic field and collision detection functions are calculated using these sets, subsequently leading to the evaluation of cost functions. The parameter set with the best cost is designated as the initial global optimum, denoted as $\mathbf{p}_{\text{best}}$. Building upon these initial parameter sets, the subsequent step involves searching for the optimum over G generations of iterative processes. The determination of the S parameter sets for gth generation $g(1, \ldots, G)$ can be formulated through the following equations:

$$\begin{cases} \mathbf{v}_s^g = c_0\mathbf{v}_s^{g-1} + c_s r_s(\hat{\mathbf{p}}_s - \mathbf{p}_s^{g-1} + c_g r_g(\mathbf{p}_{\text{best}} - \mathbf{p}_s^{g-1})) \\ \mathbf{p}_s^g = \mathbf{p}_s^{g-1} + \mathbf{v}_s^g \end{cases} \tag{3.12}$$

where $\mathbf{v}_s^g$ is the speed of the value change of the sth parameter set at the gth generation. It comprises three components: inertia, the influence of its own best till gth generation $\hat{\mathbf{p}}_s$, and the influence of global best till gth generation $\mathbf{p}_{\text{best}}$. The inertia weight is designated as c_0, c_s and c_g are coefficients to determine the amount of influence that is made by the personal best and the global best, respectively. These coefficient values are specified through constraint parameters, with c_0 set to 0.730 and both c_s and c_g assigned values of 1.496. Additionally, r_s and r_g represent random values within the range of 0 to 1. Upon completion of G generations, the current global best solution ($\mathbf{p}_{\text{best}}$) is transmitted to the magnetic field generation and mechanical actuation systems for implementation. Subsequently, the previous global best solution ($\mathbf{p}_{\text{pre}}$) is updated to the current $\mathbf{p}_{\text{best}}$ for subsequent optimization iterations. The detailed optimization iteration process is described in Algorithm 2 in Appendix.

3.3 Construction of the Control Framework

In recent years, DRL has led to very successful agents that are able to manage more challenging and complicated tasks with limited prior knowledge. In this chapter, our objective is to train a robust control policy that can perform goal-reaching and hovering task for an SMMR subjecting to flow rates. In this section, a GRU-based SAC framework is introduced in detail (Figure 3.1a,b).

In the realm of DRL, the core mechanism revolves around executing actions that optimize a certain reward. At each time step t, the system perceives a state St within a state space S, upon which the control policy determines an action $A_t \in A$. Subsequent to executing this action, a new measurement of the state S_{t+1} is obtained and the agent calculates a scalar reward $R_t \in R$. As noted earlier, our objective is to precisely navigate the robot from its current pose $P = [x, y, \psi]$ to a desired goal pose $Pg = [xg, yg, \psi_g]$ as swiftly as possible, followed by keeping hovering for a while.

1. *State*: The state space S is defined by a tuple with six elements $O = [\rho, \theta, \phi, \upsilon, \omega, \gamma]$, which satisfies

$$\begin{cases} \rho = \sqrt{(x_g - x)^2 + (y_g - y)^2} \\ \sigma = \arctan 2((y_g - y), (x_g - x)) - \psi \\ \phi = \psi_g - \psi - \sigma \end{cases} \tag{3.13}$$

where ρ represents the real-time distance from the robot position to the target position, as shown in Figure 3.1c. σ denotes the yaw angle deviation between the target yaw angle and the real-time yaw angle of the robot. ϕ denotes the

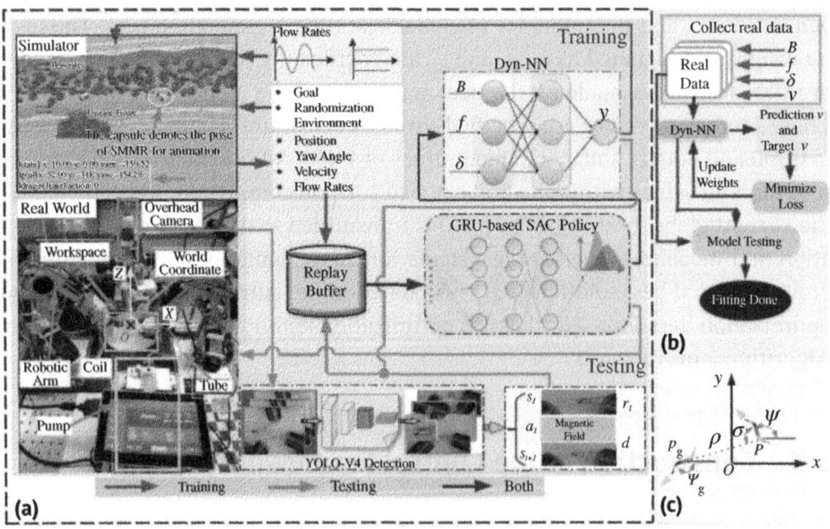

Figure 3.1 (a) Pipeline for the DRL-based control framework. (b) Dyn-NN training process. (c) Definition of a coordinate system.

angle relationship among target angle, current yaw angle, and σ. Transform the range of σ and ϕ to $[-\pi, \pi]$. γ represents whether the robot hovers on the target pose for the desired time.

2. *Action*: Based on the parameters of the conical rotating magnetic field, the action space is defined by $A = [B, f, \delta, \omega_B]$, set these parameters as $B \in [1, 7]$, $f \in [0, 3]$,$\delta \in [0, 0.3\pi]$, and $\omega_B \in [-\pi/4, \pi/4]$. During the training, all actions are rescaled to the range $[-1,\ 1]$.
3. *Reward*: The reward function used to train and evaluate agent performance is defined as:

$$\mathbf{R} = \mathbf{R}_1 + \mathbf{R}_2 + \mathbf{R}_3 \tag{3.14}$$

where

$$\begin{cases} \mathbf{R}_1 = -k_1 \parallel \rho \parallel -k_2 \parallel \sigma \parallel -k_3 \parallel \phi \parallel \\ \mathbf{R}_2 = -k_4\sqrt{\upsilon^2 + w^2} \\ \mathbf{R}_3 = k_5\gamma \end{cases} \tag{3.15}$$

$k_i(i = 1, \ldots, 5)$ represents the corresponding weights, which are values determined by experience to facilitate the balance of different indicators. The first assessment meets the requirements of the goal, and the second measures energy consumption. The third is the bonus, which describes the hovering demand. It is worth mentioning that we remap the scope of the reward function to the range of $[0, 1]$, because the normalized reward function maximizes the convergence.
4. *Goal*: The goal of this task is to navigate the robot to a user-defined target location (i.e., the location of the tissue used for drug delivery) and remain hovering (drug release) for a period of time. In the training phase, after the current goal is reached within the design tolerance, a new target is randomly sampled within the boundary of the predefined area.
5. *Episode*: The episode terminates when the policy reaches 30 consecutive cycles, or when the strategy fails to complete the task within 400 time steps.

Then utilize the Dynamic Neural Network (Dyn-NN) to fit the dynamic model of the system and construct the simulator. As for the experimental data used to construct simulator, due to the noise in the measurement and the changes in the experimental process, the physical data have inherent uncertainty, it is still a rough approximation of the physical setting. Because of the real gap between the simulator and the real world, the strategy of training purely on the simulated data often can not be extended to the real world. In order to include these uncertainties in the model, an attractive way to improve the robustness of control strategies is randomization, which has been applied in many applications [21]. Therefore, instead of just training strategies under a specific dynamic model, we train the policy under a variety of distributions of dynamic models that generalize to reality. The manual design of randomized parameters can carry out successful policy training according to some expertise in the experiment.

1. State Noise: The quantization error caused by camera calibration vibration and eye-hand calibration can not be ignored. In order to better simulate the types of noise we expect to experience in reality, we simulate these errors by adding a small amount of uncorrelated Gaussian noise to the current robot attitude $P = [x, y, \psi]$.
2. Action Noise: Physical robots will experience many effects that our simulator does not model. Therefore, we add the relevant Gaussian noise to all the actions to explain the imperfections of the magnetic actuation system. During data collection using this strategy, we add noise with a specific standard deviation (Std) to each action and then apply it to the simulation.
3. Position Freezing: The visual detection method sometimes temporarily loses the trajectory of the robot, and we model by freezing the position of the robot in the simulator with a small probability of γ for a short time. This strategy can enhance the policy learning how to interact with the environment without affecting the convergence speed of training.

For the above reasons, we choose to randomize the following parameters, as shown in Table 3.1, where the mean of noise is zero and there is a specific Std [22], where the Std of the action space represents the percentage of the action range.

After further preparation, the algorithm is trained to work. The detailed training procedure for our actor-critic framework is summarized in Algorithm 4 in the Appendix. The training process of our task of using behavior-criticism structure is summarized in algorithm. In our implementation, a random goal and a random starting position, as well as a set of randomization parameters, will be sampled from their corresponding ranges at the beginning of every timestep. Initialize the policy network and value network of SAC, including the temperature α of entropy. In addition, we add traffic information to the simulator, which will affect the line speed of SMMR. Specifically, during each episode, if the number of timesteps has

Table 3.1 Parameters randomization.

Parameter	Value	Parameter	Value
P	Standard deviation: 1.5 mm	σ	Standard deviation: 0.1 rad
ϕ	Standard deviation: 0.1 rad	B	Standard deviation: 16%
f	Standard deviation: 10%	δ	Standard deviation: 12%
ω_B	Standard deviation: 5%	γ	0.1/s

not yet reached the threshold denoted as T_L, actions will be randomly drawn from the predefined action space. When the number of time steps exceeds T_L, the action will be determined by the policy network, which will generate the current action from the historical state-action buffer, current state, last action, and estimated traffic. The value network computes the state value function by utilizing the historical state-action buffer, the present state, the chosen action, the anticipated traffic sequence, and a set of randomized parameters. When the number of transitions in D exceeds the number of small batches of N, the weight parameters of policy network, value network, and temperature α will be updated. The training process will be terminated when the maximum number of sets $M(1000)$ is reached. Each episode consists of $T(400)$ time steps, corresponding to about 40 s per episode. Finally, a learning strategy network with parameter θ_π is obtained and used as a control strategy.

A crucial parameter in the training procedure that warrants discussion is the history time span, denoted as T_h, which represents the length of time of the historical state-action and is used to infer dynamics. From Figure 3.2, we can observe that using different historical time spans will lead to different rewards and convergence performance. However, the relationship between the history time span and the reward is not monotonic. Roughly speaking, long time history span is used for the modeling of slow change system, and short time history span is used for the modeling of fast change system. In addition, we performed a 100 episode task to evaluate the success rate of training control strategies for each historical time span, as shown in Figure 3.3. We can see that when the historical time span is $1s$, a higher success rate can be achieved. Combined with the reward curve in Figure 3.2, we choose the historical time span of $1s$ in the next simulation and experiment.

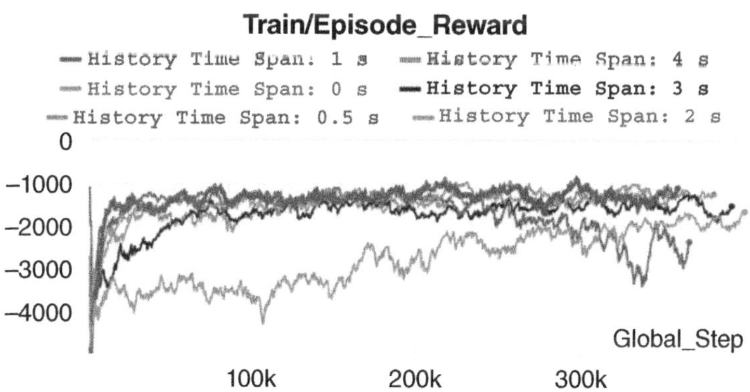

Figure 3.2 Comparison of average reward under different history time spans during training in the presence of flow rates.

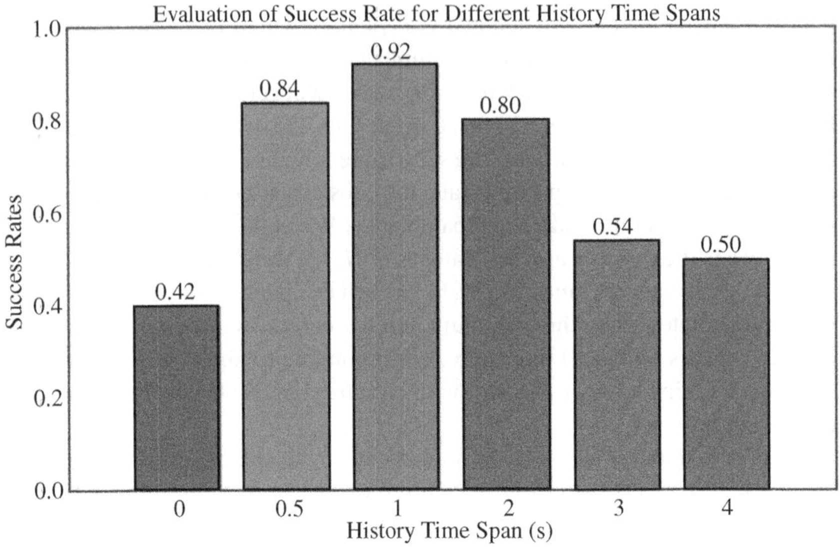

Figure 3.3 Comparison of success rates for evaluation of control policies under different history time spans.

3.4 Experimental Validation

In this section, real-world experiments are performed directly to evaluate the method on a physical robot without additional fine-tuning. First, the experimental setup and procedure for the real-world environment are described. Second, the relevant experimental scenarios are designed and implemented. Finally, the experimental results are analyzed to verify the validity of the method.

Firstly, the SMMR is placed in a silicone tube filled with a viscous silicone fluid to simulate a biological fluid. A 35 ml syringe is mounted on a programmable fluid pump driven by a computer-controlled linear motor that generates flow through the tubes. The magnetic actuation system can generate a magnetic flux density in excess of 10 mT; however, as the magnetic flux density increases, the end-effector of the three robotic arms will jitter violently due to the mutual inductance of the three electromagnetic coils. Therefore, the maximum flux density was set to 7 mT to avoid this phenomenon as much as possible. The design specifications of the mobile magnetic actuation system are listed in [10].

To account for the deformation of the SMMR during swimming, the YOLO-V4 framework is utilized for detecting and tracking its position. Initially, capture images of the robot in its swimming mode under various magnetic field configurations using an overhead camera. Following this, construct a dataset comprising 400 such images. Subsequently, proceed to train the YOLO-V4 model weights based on this newly created dataset. Once these weights are well trained, they will be saved and the location of the machine will be predicted through the bounding box. It is worth noting that our experiments take place within the X–Y plane, under the assumption that $Z = 0$. The robot's instantaneous position relative to the global coordinate system can be determined through a process of image-to-world calibration. At the commencement of each experimental session, the SMMR is positioned within the tubes with an identical initial posture, while the target posture remains constant along the negative x-axis of the global coordinate system. It is crucial to highlight that, throughout each session, the SMMR operates without knowledge of the flow rates. The control algorithm is trained using a desktop computer featuring an Intel $i9 - 11900k$ processor, 32 GB of RAM, and an $NvidiaGeForceRTX3080$ graphics card with 10 GB of dedicated memory.

1. **Scenario I: Varied Flow Rates:** In this section, we assess and contrast the efficacy of our approach with the SAC and ADRC methods under identical experimental conditions. The SAC baseline is sourced from $OpenAI$.

   Initially, the SMMR is positioned within silicone tubes measuring 5 mm in inner diameter, with the tube arrangement aligned predominantly along the x-axis. We adopt the identical start and end positions as those stipulated in the simulation illustrated in Figure 3.4. Furthermore, the parameters for the various flow rates are consistent with the settings depicted in Figure 3.4a. For the sake of generality, each method undergoes three repetitive trials. To pro vide a more thorough evaluation of the experimental outcomes, we introduce four assessment metrics: convergence time (the initial duration to reach the predefined convergence zone for successful trials), the mean of steady error (MSE: upon reaching the convergence zone, the recorded distance to the target), the Std of the steady error, and the count of successful trials.

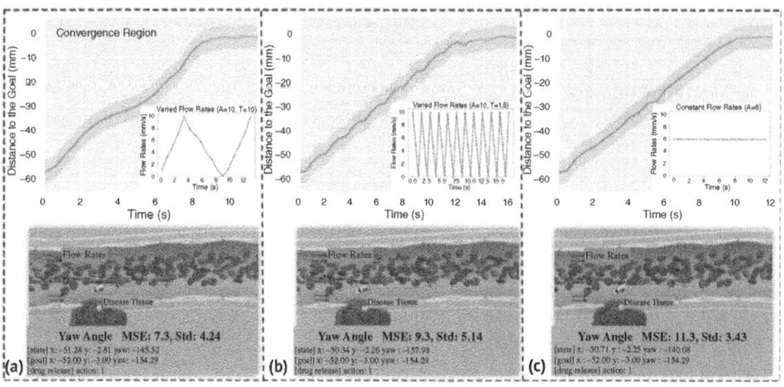

Figure 3.4 Simulation results of goal-reaching and hovering task. (a) Varied flow rates in a sine wave with A 10 and T 10. (b) Varied flow rates in a sine wave with A 10 and T 1.5. (c) Constant flow rates with A 6. For each scenario, the shaded area indicates the desired convergence region. The evolution of the yaw angle is depicted using MSE and the Std.

The comparative performance outcomes of the three methodologies are presented in Figure 3.5, encompassing statistical evaluation metrics and the temporal evolution of linear velocity. During the plotting of the curves, negative values for the distance to the goal indicate that the SMMR's position along the x-axis has not surpassed the target position (i.e., −52 mm), otherwise, the value is positive. It is evident that all three methods are capable of guiding the SMMR toward the goal in the presence of flow rates. However, they show different performances in terms of the designed evaluation indicators. Specifically, the ADRC method successfully accomplishes the goal-reaching and hovering tasks in two out of three trials (i.e., ADRC − trials − 1 and ADRC − trials − 2), with an overshoot occurring at approximately 25 s in the third trial, potentially attributed to magnetic field inaccuracies stemming from cumulative robotic arm errors. The SAC method demonstrates a suboptimal flow rate rejection performance, characterized by high-amplitude oscillations in linear velocity, and only completes the task in the third trial, primarily due to the SAC baseline's inability to learn an effective policy under flow rate conditions. Notably, our method surpasses the other two in terms of success rates, with all three trials being successfully completed. Additionally, our method exhibits a faster convergence speed, approximately 11.7 s (ADRC : 19.3s, SAC : 37.3s), which is crucial for real-time control in practical applications. In terms of MSE, our method achieves superior results (1.513 ± 0.712) compared to ADRC (1.902 ± 0.634) and SAC (2.096 ± 0.388 in the third trial). Snapshots from our first trial (ours − trials − 1) are depicted in Figure 3.6.

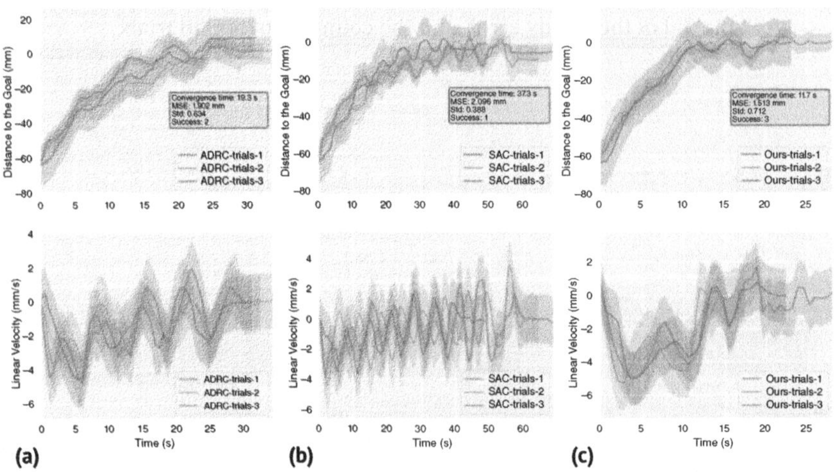

Figure 3.5 Performance comparison of the ADRC method, SAC method, and our method, including the designed evaluation indicators and linear velocity from experiments. (a) ADRC method. (b) SAC method. (c) Our method.

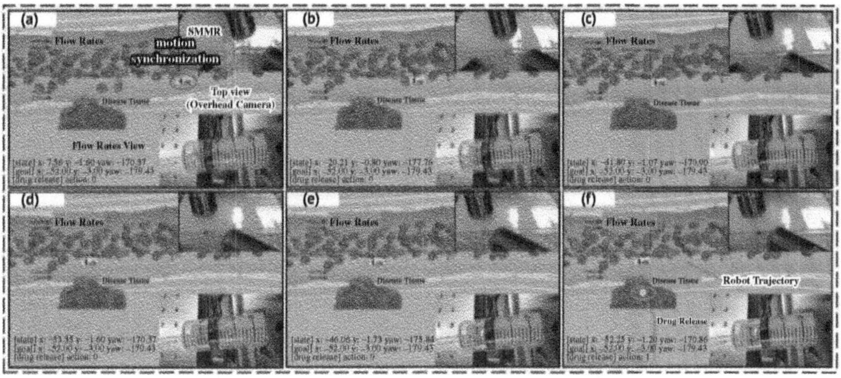

Figure 3.6 Image sequences of goal-reaching and hovering control in varied flow rate Scenario I under our method. (a–c) Goal-reaching process. (d–f) Hovering control process and finally complete the task. (a) $t = 0$, (b) $t = 5$, (c) $t = 10$, (d) $t = 12$, (e) $t = 17$, (f) $t = 21$.

When compared to the ADRC and SAC methods using the designed assessment metrics, our method demonstrates superior performance over these traditional methods, providing valuable insights into potential biomedical applications. Furthermore, the results offer a feasibility study for the practical application of DRL in miniature robots.

2. **Scenario II: Constant Flow Rates:** To further validate our approach, we implemented a scenario involving constant flow rates, which represents a common case in practical applications, using the same control policy parameters as those employed in Scenario I. We conducted three trials with a flow rate of 6 mm/s and presented the results in Figure 3.7. We can observe that our method still successfully navigates the SMMR converge to its desired goal and maintain around it within small errors. However, during the third trial, a slightly slower convergence rate was observed compared to the other two trials. This may be attributed to inaccuracies in the magnetic field resulting from imprecise coil postures, akin to the issues encountered in Scenario I.
3. **Scenario III: Navigation in Two Silicone Tubes with Different Diameters:** The SMMR's flexible and adaptable body exhibits numerous advantageous traits in confined spaces, including its ability to conform to various tube shapes, as exemplified in Figure 3.8. To demonstrate the motion capabilities of the SMMR, a phantom was crafted utilizing two distinct silicone tube diameters, interconnected through a specialized connector. These silicone tubes, designated as D5 − 3 (indicating an outer diameter of 5 mm and an inner diameter of 3 mm), were securely mounted within the workspace of the magnetic actuation system, as illustrated in Figure 3.9a. The navigation pathway in this scenario is reconstructed by tracing the SMMR's position within the tubes, which represents the tubes' center line, as shown

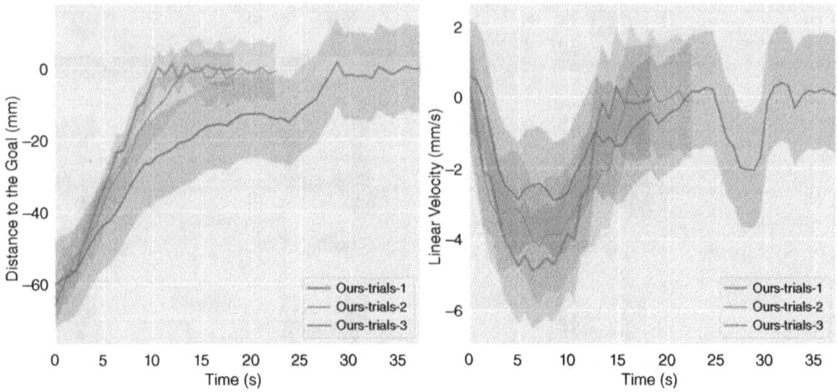

Figure 3.7 Performance of our method under the constant flow rate scenario.

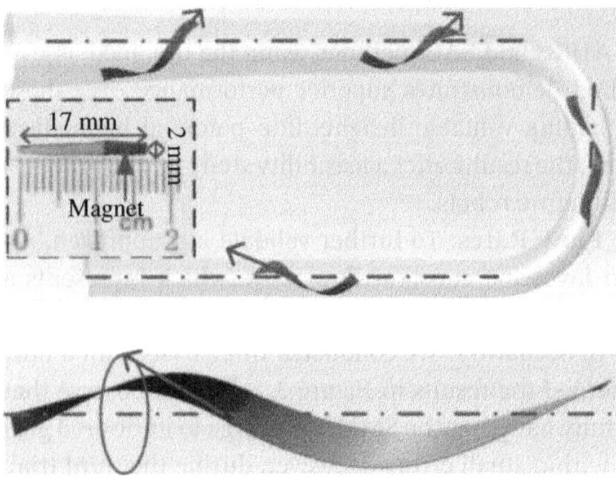

Figure 3.8 Dimension of SMMR and its locomotion under its ability of adaptive deformation.

in Figure 3.9a. Given its complexity, featuring multiple curvatures, the objective of this scenario is formulated as following a specified path on the horizontal plane. We consider this scenario as an extension of the goal-reaching task. The path is segmented into a sequence of predefined points, where upon reaching the current goal, the next target is designated. The desired magnetic field orientation is determined based on these predefined points. During the navigation of the SMMR within the silicone tube D5 − 3, depicted in Figure 3.9b, the implemented control strategy adjusts by narrowing the conical angle range to accommodate the tube's smaller dimensions. Subsequently, the SMMR transitions through the connector, as illustrated in Figure 3.9c. It is worth noting that visual tracking is temporarily lost during

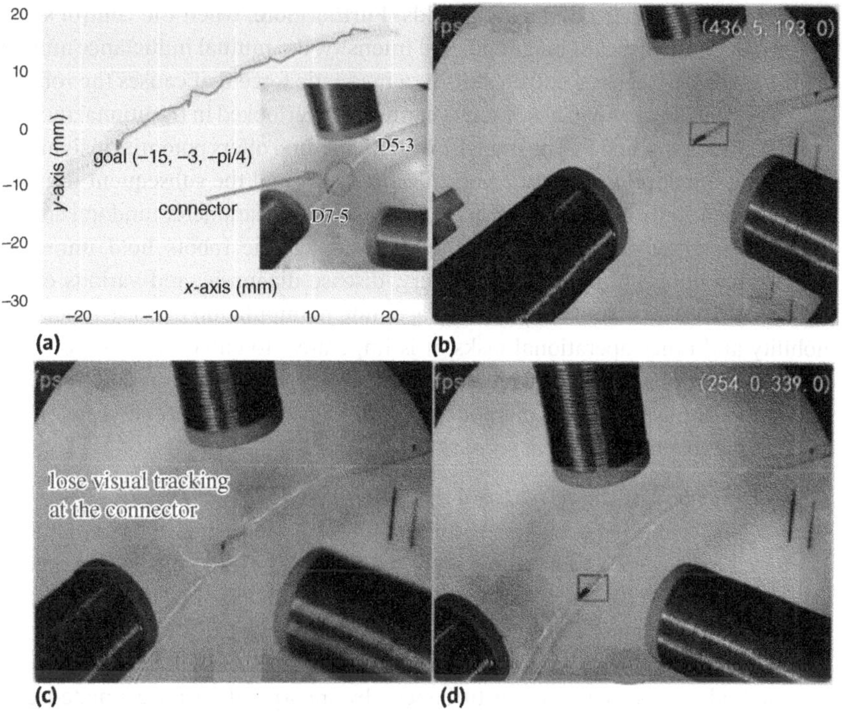

Figure 3.9 Demonstration of the passing through two kinds of silicone tubes with different diameters under flow rates in Scenario I. (a) Layout of tubes and the results of trajectory tracking. (b–d) Image sequences of passing through the tubes and achieving goal-reaching and hovering at the final waypoint.

this passage due to the occlusion. To address this challenge, the position freezing technique, incorporated within the randomization process, is employed. Following this, the SMMR continues its journey into the larger silicone tube D7 − 5, where the conical angle range is expanded, as shown in Figure 3.9d. Ultimately, the SMMR achieves hovering at the designated goal, showcasing its adaptive motion capabilities attributed to its soft body characteristics.

3.5 Discussion and Conclusion

Creating an optimal magnetic field in practical settings is crucial for the SMMR's ability to accurately reject flow rates. As discussed before, deviations in the magnetic field produced by the mobile magnetic actuation system can compromise the SMMR's performance. Specifically, our observations indicate that after multiple attempts, the robotic arms experience diminished performance in positioning the electromagnetic coils due to joint clearance, leading to discrepancies between the intended and realized magnetic fields. Furthermore, when the control strategy produces a significant magnetic field intensity, the mutual inductance among the three electromagnetic coils results in a magnetic force that causes the robotic arms to vibrate intensely. These factors cannot be overlooked in traditional model-based control approaches. The model-free DRL method offers potential insights by suggesting the development of a versatile simulator and the subsequent transfer of the trained control policy to the actual robot through parameter randomization.

The design and production techniques for magnetic robots hold immense promise for applications in drug delivery, disease diagnosis, and various other domains. However, to ensure swift and accurate manipulation of these robots for mobility and other operational tasks, it is imperative to enhance the efficacy of the control methodologies. Despite the advancements, there remains a substantial scope for refining the control methods to fully unlock the transformative potential of magnetic robots in practical scenarios.

1. **Comparison of the ADRC and the GRU-based SAC** In Scenario II, Section 3.4, we presented a comparative analysis of the performance of three methods based on the evaluation metrics that we designed. The results of our experiments revealed that the SAC method exhibited suboptimal performance and lower success rates. Consequently, this section delves deeper into the characteristics of the ADRC and GRU-based SAC methods. Specifically, the ADRC method compensates for disturbances by incorporating an estimated disturbance as a feed forward component in the control input. Although this approach has appropriate performance, the output of the controller with feed forward structure may not be optimal [23]. Moreover, the constraints on the control action may not be satisfied for actual values.

   When implementing the ADRC method, our focus is exclusively on adjusting the magnitude and direction of the magnetic field. Additionally, we maintain the maximum values for the magnetic field's frequency (i.e., $f = 3$) and conical angle (i.e., $\delta = 0.3\pi$) to ensure that SMMR attains sufficient speed to counter the flow rates. To demonstrate the effectiveness of our approach, we conduct a comparison of the magnetic field magnitudes generated by the two methods.

   Figure 3.10 illustrates that, in some instances, the ADRC method exceeds the predefined maximum magnitude of the magnetic field due to the absence of control action constraints. However, when the control action is implemented on the actual robot, its value is restricted to ensure it does not surpass this threshold. While both methods are capable of completing the task, the ADRC's output tends to be more aggressive. Regarding our proposed method, it is worth noting that the flow rate information is encoded within the historical state-action pairs. Consequently, during the policy training process, the embedding of flow rates is also acquired. The advantage of our method lies in its utilization of the historical state-action mechanism.

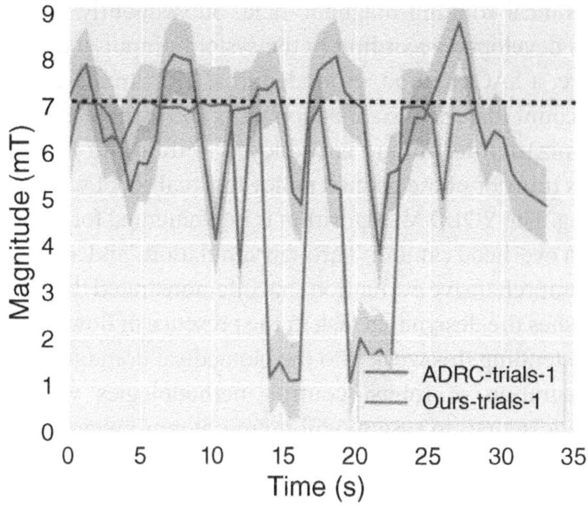

Figure 3.10 Comparison of the magnitude of magnetic field for the ADRC method and our method in experiment Scenario I.

2. **Limitations and Potential Applications:** In our experimental setup, we utilized silicone tubes filled with dynamic viscous silicon fluid as phantoms to emulate biofluids. The SMMR navigates within this phantom and can be tracked using an overhead camera system. However, it is important to acknowledge that there exist inherent discrepancies between laboratory conditions and real-world applications. The current camera-based vision tracking system encounters numerous limitations when applied in biomedical contexts. Alternatively, biomedical imaging techniques, such as C-arm for optional intraoperative intermittent X-rays, could be employed to locate magnetic robots. Additionally, within the magnetic actuation system, the coil postures are not optimized during the tracking process, potentially resulting in reduced actuation efficiency. When considering flow rates in biomedical scenarios, accurately estimating their true values is highly complex. The RWD laser speckle imaging system [18] provides a solution by offering quantitative velocity information of blood flow with high spatial resolution. This system analyzes scattered light signals from particles within biological tissues (such as red blood cells) that are illuminated by a coherent beam from a pulsed ruby laser.

This chapter explores the goal-reaching and hovering control of a SMMR utilizing a DRL framework to counteract flow rates. The overall control strategy encompasses a simulator, a DRL algorithm, and a real-time visual tracking technique. Initially, an SMMR with a head-tail configuration is designed and actuated in a helical manner through a conical rotating magnetic field. Subsequently, based on a Dyn-NN, a simulator is developed according to the system's input and output parameters. Furthermore, a SAC method, grounded in a GRU architecture, is introduced, taking into account historical state-action pairs and estimated flow rates. Additionally, a randomization technique is employed to train the control policy, enabling the seamless transfer of the trained policy to a real robot without the need for parameter tuning. The YOLO-V4 algorithm is implemented for tracking the robot's position via an overhead camera. Through simulations and a series of experimental scenarios, comprehensive evaluations have demonstrated that our method effectively accomplishes the designated task in the presence of flow rates, hinting at the potential for extending this work into the biomedical domain.

We believe that the integration of robust control methodologies with a highly maneuverable magnetic actuation system will represent a promising area of research for the application of magnetic robots in biomedical contexts.

Appendix Additional Algorithms

Algorithm 1: Collision Detection Algorithm Based on Triangle-ray Intersection

Input: parameter set $\mathbf{p}$; the coordinates of three vertices of J triangular surfaces $\mathbf{v0}_j$, $\mathbf{v1}_j$ and $\mathbf{v2}_j$, $(j = 1, \ldots, J)$
Output: the result of collision detection $Q(\mathbf{p})$

```
1:   Q(p) ← 0
2:   for i ← 1 to n do
3:      calculate o_{c,i} and r_{c,i} from p
4:      for j ← 1 to J
5:         calculate t_j, u_j and v_j by solving Eq. (3.6)
6:         if (t_{j≤0} and u_{j≥0} and v_{j≥0} and
        u_j + v_j ≤ 1) then
7:            Q(p) ← 1
8:         end if
9:      end for
10:  end for
11:  return Q(p)
```

Algorithm 2: Real-time Optimization on the Positions of Mobile Electromagnetic Coils Based on PSO

Input: location of the robot **d**, parameters of the demanded magnetic field (e.g., pitch angle and direction angle of the rotating field)

Output: optimal parameter set $\mathbf{p}_{\text{best}}$ initialization:

for $s \leftarrow 1$ to S **do**
  if ($\mathbf{p}_{\text{pre}}$ exists) **then**
    $\mathbf{p}_s^0 \leftarrow \mathbf{p}_{\text{pre}}$ for $s = 1$
    $\mathbf{p}_s^0 \leftarrow \mathbf{p}_{\text{pre}} + \omega_{\text{rand}}(\mathbf{p}_{\min} - \mathbf{p}_{\text{pre}}, \mathbf{p}_{\max} - \mathbf{p}_{\text{pre}})$ for $s > 1$
  else
    $\mathbf{p}_s^0 \leftarrow \text{rand}(\mathbf{p}_{\min}, \mathbf{p}_{\max})$
  end if
  if ($\ell(\mathbf{p}_s^0 < \ell(\mathbf{p}_{\text{best}}^0))$) **then**
    $\mathbf{p}_{\text{best}}^0 \leftarrow \mathbf{p}_s^0$
  end if
  $\hat{\mathbf{p}}_s \leftarrow \mathbf{p}_s^0$
end for

Main loop :

$\mathbf{p}_{\text{best}} \leftarrow \mathbf{p}_{\text{best}}^0, \mathbf{v}_s^0 \leftarrow 0, g \leftarrow 1$
while $(g < G) \wedge (\triangle\ell(\mathbf{p}_{\text{best}}))$ in recent five iterations $> \epsilon \vee \ell(\mathbf{p}_{\text{best}}) > \omega_3)$
  for $s \leftarrow 1$ to S **do**
    $\mathbf{v}_s^g \leftarrow c_0\mathbf{v}_s^{g-1} + c_s r_s(\hat{\mathbf{p}}_s - \mathbf{p}_s^{g-1}) + c_g r_g(\mathbf{p}_{\text{best}} - \mathbf{p}_s^{g-1})$
    $\mathbf{p}_s^g \leftarrow \mathbf{p}_s^{g-1} + \mathbf{v}_s^g$
   if ($\ell(\mathbf{p}_s^g) < \ell(\hat{\mathbf{p}}_s)$) **then**
    $\hat{\mathbf{p}}_s \leftarrow \mathbf{p}_s^g$
   end if
   if ($\ell(\mathbf{p}_s^g) < \ell(\mathbf{p}_{\text{best}})$) **then**
    $\mathbf{p}_{\text{best}} \leftarrow \mathbf{p}_s^g$
   end if

Algorithm 3: Real-time Optimization on the Positions of Mobile Electromagnetic Coils Based on PSO

Input: location of the robot **d**, parameters of the demanded magnetic field (e.g., pitch angle and direction angle of the rotating field)

Output: optimal parameter set $\mathbf{p}_{\text{best}}$ initialization:

    end for
    $g \leftarrow g+1$
end while
$\mathbf{p}_{\text{pre}} \leftarrow \mathbf{p}_{\text{best}}$
return $\mathbf{p}_{\text{best}}$

Algorithm 4: GRU-based SAC

Data: Initialize value networks Q_1, Q_2 and policy network π with weights $\theta^{Q_1}, \theta^{Q_2}, \theta^{\pi}$, respectively. Initialize target network $\bar{Q}_1, \bar{Q}_2$ with weights $\bar{\theta}^{Q_1} \leftarrow \theta^{Q_1}, \bar{\theta}^{Q_2} \leftarrow \theta^{Q_2}$. Initialize replay buffer D, temperature α, Randomization space U, Simulator Ξ, history time span T_h, control period $\triangle t$, length of history state-action buffer H is $T_L = T_h/\triangle t$

for *episode* $m = 1$ to M **do**
    Initialize a random goal and receive initial observation state s_1;
    for *timestep* $t = 1$ to T **do**
        Uniformly sample parameters $\mu - U$, generate randomized environment $\Xi \leftarrow U(\mu)$;
        Sample flow rates ξ
        if $t \leq T_L$ **then**
            Sample an action from A randomly;
            Execute action a_t, receive reward r_t and new state s_{t+1} in the Ξ;

Algorithm 4: GRU-based SAC

Data: Initialize value networks Q_1, Q_2, and policy network π with weights θ^{Q_1}, θ^{Q_2}, θ^{π}, respectively. Initialize target network $\bar{Q}_1$, $\bar{Q}_2$ with weights $\bar{\theta}^{Q_1} \leftarrow \theta^{Q_1}$, $\bar{\theta}^{Q_2} \leftarrow \theta^{Q_2}$. Initialize replay buffer D, temperature α, Randomization space U, Simulator Ξ, history time span T_h, control period $\triangle t$, length of history state-action buffer H is $T_L = T_h / \triangle t$

Update $H_{t+1} \leftarrow (H_t, a_t, s_{t+1})$;
end
if $t > T_L$ **then**
Select an action a_t from policy
a_t - $\pi(s_t, H_t, a_{t-1}, \xi \mid \theta^{\pi}, \Xi)$;
Execute action at, receive reward r_t and new state s_{t+1} in the Ξ;
Update $H_{t+1} \leftarrow (H_t, a_t, s_{t+1})$;
Store transition tuple $(H_t, a_t, a_{t-1}, r_t, H_{t+1}, \xi, \mu)$ into D;
if *training* then
Sample a random mini-batch of N transitions from D;
for *each gradient step* **do**
$\theta^{Q_i} \leftarrow \theta^{Q_i} - \lambda_Q \nabla_{\theta Q_i} J_Q(\theta^{Q_i})$, $i \in 1, 2$;
$\theta^{\pi} \leftarrow \theta^{\pi} - \lambda_{\pi} \nabla_{\theta^{\pi}} J_{\pi}(\theta^{\pi})$;
$\alpha \leftarrow \alpha - \lambda_{\pi} \nabla_{\alpha} J(\alpha)$;
$\bar{\theta}^{Q_i} \leftarrow \tau \theta^{Q_i} + (1 - \tau) \bar{\theta}^{Q_i}$;
end
end
end
Update $s_t \leftarrow s_{t+1}$, $H_t \leftarrow H_{t+1}$, $a_{t-1} \leftarrow a_t$;
end
end

References

1 P. E. Dupont, et al. A decade retrospective of medical robotics research from 2010 to 2020. *Science Robotics*, 6 (60): eabi8017, 2021.

2 W. Hu, G. Z. Lum, M. Mastrangeli, and M. Sitti. Small-scale soft-bodied robot with multimodal locomotion. *Nature*, 554 (7690): 81–85, 2018.

3 T. Xu, C. Huang, Z. Lai, and X. Wu. Independent control strategy of multiple magnetic flexible millirobots for position control and path following. *IEEE Transactions on Robotics*, 38 (5): 2875–2887, 2022.

4 J. Nam, W. Lee, J. Kim, and G. Jang. Magnetic helical robot for targeted drug-delivery in tubular environments. *IEEE/ASME Transactions on Mechatronics*, 22 (6): 2461–2468, 2017.

5 K. T. Nguyen, et al. A magnetically guided self-rolled microrobot for targeted drug delivery, real-time x-ray imaging, and microrobot retrieval. *Advanced Healthcare Materials*, 10 (6): 2001681, 2021.

6 Z. Wang, S. Guo, J. Guo, Q. Fu, L. Zheng, and T. Tamiya. Selective motion control of a novel magnetic-driven minirobot with targeted drug sustained-release function. *IEEE/ASME Transactions on Mechatronics*, 27 (1): 336–347, 2021.

7 R. Wang, S. Wang, Y. Wang, L. Cheng, and M. Tan. Development and motion control of biomimetic underwater robots: a survey. *IEEE Transactions on Systems, Man, and Cybernetics: Systems*, 52 (2): 833–844, 2020.

8 Q. Cao, R. Wang, T. Zhang, Y. Wang, and S. Wang. Hydrodynamic modeling and parameter identification of a bionic underwater vehicle: Robdact. *Cyborg and Bionic Systems*, 2022: 9806328, 2022.

9 Y. C. Liu and C. Y. Huang. DDPG-based adaptive robust tracking control for aerial manipulators with decoupling approach. *IEEE Transactions on Cybernetics*, 52 (8): 8258–8271, 2021.

10 X. Du, M. Zhang, J. Yu, L. Yang, P. W. Y. Chiu, and L. Zhang. Design and real-time optimization for a magnetic actuation system with enhanced flexibility. *IEEE/ASME Transactions on Mechatronics*, 26 (3): 1524–1535, 2020.

11 T. Haarnoja, A. Zhou, P. Abbeel, and S. Levine. Soft actor-critic: off-policy maximum entropy deep reinforcement learning with a stochastic actor. In *International Conference on Machine Learning*, pages 1861–1870, 2018. PMLR.

12 K. Cho. On the properties of neural machine translation: encoder-decoder approaches. *arXiv preprint arXiv:1409.1259*, 2014.

13 C. M. Heunis, Y. P. Wotte, J. Sikorski, G. P. Furtado, and S. Misra. The ARMM system-autonomous steering of magnetically-actuated catheters: towards endovascular applications. *IEEE Robotics and Automation Letters*, 5 (2): 705–712, 2020.

14 J. Sikorski, A. Denasi, G. Bucchi, S. Scheggi, and S. Misra. Vision-based 3-D control of magnetically actuated catheter using bigmag—an array of mobile electromagnetic coils. *IEEE/ASME Transactions on Mechatronics*, 24 (2): 505–516, 2019.

15 L. Yang, X. Du, E. Yu, D. Jin, and L. Zhang. Deltamag: an electromagnetic manipulation system with parallel mobile coils. In *2019 International Conference on Robotics and Automation (ICRA)*, pages 9814–9820, 2019, IEEE.

16 J. Sikorski, I. Dawson, A. Denasi, E. E. Hekman, and S. Misra. Introducing bigmag—a novel system for 3D magnetic actuation of flexible surgical manipulators. In *2017 IEEE International Conference on Robotics and Automation (ICRA)*, pages 3594–3599, 2017, IEEE.

17 F. Niu, J. Li, W. Ma, J. Yang, and D. Sun. Development of an enhanced electromagnetic actuation system with enlarged workspace. *IEEE/ASME Transactions on Mechatronics*, 22 (5): 2265–2276, 2017.

18 B. Véron, A. Hubert, J. Abadie, and N. Andreff. Dealing with redundancy of a multiple mobile coil magnetic manipulator: a 3RPR magnetic parallel kinematics manipulator. In *Advances in Robot Kinematics 2016*, pages 201–208, 2017, Springer.

19 T. Whelan, M. Kaess, H. Johannsson, M. Fallon, J. J. Leonard, and J. McDonald. Real-time large-scale dense RGB-D slam with volumetric fusion. *The International Journal of Robotics Research*, 34 (4–5): 598–626, 2015.

20 Y. Zhang, W. Xu, Y. Tong, and K. Zhou. Online structure analysis for real-time indoor scene reconstruction. *ACM Transactions on Graphics (TOG)*, 34 (5): 1–13, 2015.

21 X. B. Peng, M. Andrychowicz, W. Zaremba, and P. Abbeel. Sim-to-real transfer of robotic control with dynamics randomization. In *2018 IEEE International Conference on Robotics and Automation (ICRA)*, pages 3803–3810, 2018, IEEE.

22 K. Cho, B. Van Merriënboer, C. Gulcehre, D. Bahdanau, F. Bougares, H. Schwenk, and Y. Bengio. Learning phrase representations using RNN encoder-decoder for statistical machine translation. *arXiv preprint arXiv:1406.1078*, 2014.

23 B. Vatankhah and M. Farrokhi. Nonlinear model-predictive control with disturbance rejection property using adaptive neural networks. *Journal of the Franklin Institute*, 354 (13): 5201–5220, 2017.

4

Reconfigurable Electromagnetic Actuation System and Robust Control in Fluid Environments

This chapter reports a reconfigurable electromagnetic actuation system (REMA) and a multi-objective optimization (MOO) method for generating the Pareto-optimized coil configuration to achieve field and force control in large workspaces. Furthermore, a robust three-dimensional (3D) path following control framework is constructed for achieving accurate microrobotic navigation under fluid flow conditions.

4.1 Introduction

Remote magnetic actuation represents an emerging technology that harnesses magnetic torque and force to manipulate magnetic robots for minimally invasive diagnostics and therapeutic interventions within the human body [1, 2]. To effectively control magnetic robots for the reliable execution of biomedical applications, the magnetic actuation system constitutes a pivotal component [3, 4]. Nonetheless, the autonomous manipulation of a variety of magnetic robots with distinct actuating fields under a specific magnetic actuation system in expansive workspaces is often challenging due to the scarcity of field dexterity and the limited accessibility of the workspace [5, 6].

Developments in magnetic actuation systems are categorized by magnetic source types (electromagnetic coils and permanent magnets), field characteristics (uniform or nonuniform), and source mobility (moving or stationary) [7, 8]. Each platform has unique advantages, and the choice depends on the application scenario. Stationary electromagnetic actuation (EMA) systems like OctoMag, MiniMag, and others [9, 10], generate complex fields by rapidly changing electric currents. Mobile EMA platforms offer larger workspaces, enhancing flexibility and compatibility with external devices. This discussion focuses on EMA systems designed for manipulating and controlling magnetic robots in expansive workspaces.

Multiple mobile EMA systems with varying coil configurations have been developed for managing magnetic torque and force across large workspaces [11, 12]. These systems manipulate magnetic fields by adjusting the position and orientation of a coil or coil array mounted on a robotic arm with multiple degrees of freedom (DOFs). Sikorski et al.'s system, with a mobile coil on a UR10 arm, extends the range of controllable magnetic fields [13], but it is limited in generating complex fields and forces. Hoang et al.'s dual EMA (DEMA) system offers rapid actuation and high-frequency rotating fields for a magnetic capsule endoscope [14]. Another system with a pair of movable electromagnets controls torque or force by adjusting coil distances [15], though its range is limited. TrinityMag, with a three-coil array on a UR-10 robot, optimizes 3D magnetic fields for rotation but not for gradient forces [16]. A four-coil system on a 6-DOF arm uses a model-based simulator for current computation, but it faces challenges with current sensitivity at specific angles [17].

While mobile EMA systems offer a satisfactory workspace, they encounter limitations in generating dexterous fields and gradients due to the restricted number of coils or fixed coil arrays. It has been established in [11] that the performance of EMA systems is influenced not only by the quantity of coils but also by their configurations. Several studies have investigated the use of multiple independent mobile coils for magnetic robot actuation. For example, the DeltaMag [18] and RoboMag [19] systems were designed with three mobile coils actuated by parallel mechanisms and robotic arms, respectively. However, these systems were primarily concerned with generating rotating fields or torque-based manipulation. In conclusion, achieving simultaneous control of magnetic torque and force within a larger workspace remains a formidable challenge.

This study presents a REMA designed for broad workspaces and automated control of various magnetic robots. The system's advancement stems from the independent 3D motion of three coils on 6-DOF robotic arms, paired with an optimized coil configuration for autonomous operations. The tri-coil design achieves two key functions: it generates rotating fields in any direction by adjusting currents, ensuring field isotropy in a small region [11], and it produces a variety of magnetic fields and gradients, as two sources can independently control eight components [6]. Moreover, a 3D hand position approach was used to model the kinematics of a helical millirobot. An adaptive finite-time sliding mode control (AFTSMC) method, combined with a finite-time disturbance observer (FTDOB), ensures convergence under disturbances. The control framework's effectiveness was validated through extensive experiments. The major contributions of this work can be summarized as follows:

1. We detail the development of a REMA system using three robotic arms for precise control of magnetic fields and gradients, including design, calibration, and modeling.

2. We introduce a MOO for coil configuration to enable rotating isotropic field control and integrated field-force control.
3. Experiments with various magnetic robots validate the system's performance, matching or exceeding existing mobile EMA systems.
4. We propose a robust control law combining AFTSMC and FTDOB to ensure the helical millirobot's convergence to a target path. An adaptive mechanism based on FTDOB estimates mitigates chattering and addresses input saturation. Stability analysis using a Lyapunov function confirms finite-time convergence.
5. The control framework's precision and convergence in complex fluids are confirmed through experiments with a robotic magnetic manipulation system and a helical millirobot prototype.

4.2 System Design and Modeling

This section outlines the design process and core components of our REMA, which enables versatile magnetic manipulation. Our system's flexibility and adaptability hinge on its capacity to produce tailored magnetic fields that align with the patient's anatomy and medical instruments, while preventing coil-environment collisions.

4.2.1 Design of System

In this section, we introduce the design workflow of our REMA and its main composition, which are the key design components that allow the platform to perform dexterous magnetic manipulation. The dexterity and scalability of our system primarily depend on its ability to generate desired magnetic fields based on the patient's body shape and curvature, as well as the surrounding medical instruments, while simultaneously avoiding collisions between coils and the environment.

4.2.1.1 REMA Platform Overview

Figure 4.1 delineates the comprehensive layout of our REMA platform, which is strategically positioned within expansive workspaces to facilitate image-guided magnetic manipulations. Table 4.1 offers a detailed synopsis of the technical specifications inherent to the REMA system. This system is comprised of three portable coils that are articulated on three autonomous 6-Degrees of Freedom (6-DOF) robotic limbs, enabling the generation of magnetic fields across vast operational spaces. The attainment of a versatile three-coil arrangement, or coil configuration, is feasible through the dexterous manipulation of the trio of robotic limbs

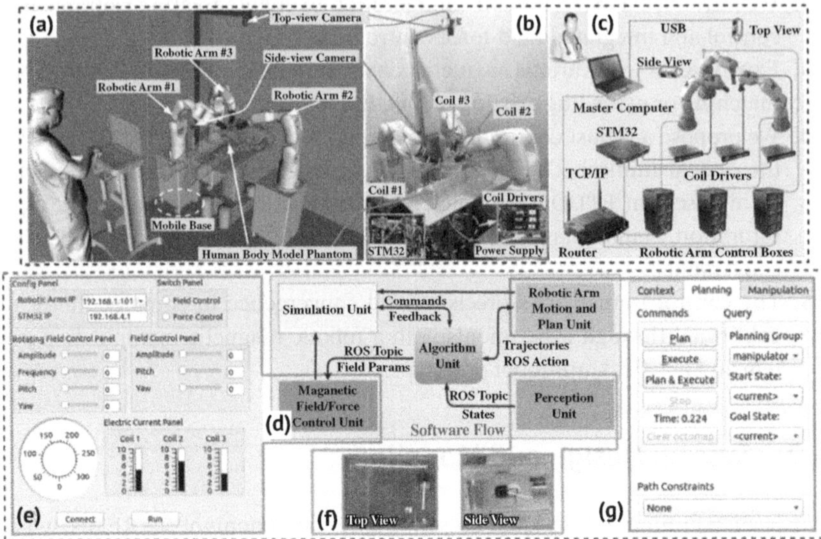

Figure 4.1 Overview framework of REMA. REMA is a magnetic actuation system designed to implement autonomous magnetic manipulations in human-scale workspaces. (a) Developed simulation environment based on ROS architecture. (b) Developed real-world REMA system. It comprises three mobile coils that are actuated by three independent 6-DOF robotic arms. The low-level controller STM32 and three coil drivers are deployed to manipulate the electric currents flowing into coils. (c) Schematic depiction of hardware flow. (d) Architecture of software flow. (e) GUI for magnetic field/force control unit. (f) Image processing results under perception unit. (g) GUI for robotic arm motion and plan unit.

across diverse application contexts. The robotic limb in question, specifically the SJ-603-A model from Anno Robot, boasts a spherical operational envelope with a radius spanning 540 mm, alongside a payload capacity accommodating up to 3 kg, complemented by a precision repeatability of $\pm$ 0.1 mm. Owing to the installation of each robotic limb on a mobile platform, as depicted in Figure 4.1a, their mobility is enhanced, allowing for relocation to various operational theaters and seamless assimilation with medical imaging apparatuses, including X-ray fluoroscopic devices.

4.2.1.2 Hardware Design

The following describes the general hardware design workflow of REMA. In designing our electromagnetic coil system, we've opted for a different approach compared to those that use the largest possible coils to achieve stronger magnetic fields. Given that our robotic arm has a payload capacity of 3 kg, this sets a limit on

Table 4.1 General specifications of components in our system.

	Parameters	Value	Parameters	Value
	DOFs	6	Working radius	540 mm
	Payload	3 kg	Position repeatability	±0.1 mm
	Maximum velocity	0.5 m/s	Weight	23 kg
	Wire diameter	0.9 mm	Number of turns	960
	Overall length	115 mm	Outer diameter	50 mm
	Core length	110 mm	Core diameter	16 mm
	Maximum current	10 A	Weight	1.2 kg
	Workspace size(L*W*H)	450*300*350 mm^3	Number of coils	3
	Maximum field(ROI center)	17 mT	Maximum gradient (ROI center)	120 mT/m
	Camera resolution	1280*1024	Frame rates	50 Hz
	Communication interface	ROS	Communication frequency	20 Hz

the weight of the coil. Additionally, for larger coils, thermal management becomes a significant factor, which poses a challenge in mobile EMA systems. Therefore, we focused on factors such as weight, diameter, and field generation capability when designing the coil, guided by our finite-element method (FEM) simulation outcomes. Consequently, we've constructed a coil with a ferromagnetic core to enhance its magnetic field strength, weighing in at 1.2 kg. The specifics of this coil design are outlined in Table 4.1.

Our system employs a pair of perpendicular digital cameras, models MER2-160-227U3C from DaHeng, to image the workspace from both top and side angles. Operating at a frame rate of 50 frames per second, these cameras produce images with a resolution of 1280 × 1024 pixels after downsampling. We've developed image processing techniques to reconstruct the environment and determine the spatial location of the magnetic robots, which is crucial for both experimental measurements and closed-loop feedback control.

For managing real-time control of the electric currents into the three coils, we utilize an STM32 control board functioning as a low-level embedded controller. This board interfaces with the high-level master computer to receive the required current values. Upon reception, these currents are converted into pulse-width modulation (PWM) signals and sent to servo drivers (ESCON 70/10, Maxon Inc.) with a maximum output current of 10 A. As depicted in Figure 4.1b, a 1000-W power supply is used to power the system. The control commands are transmitted from the high-level master computer to the STM32 and robotic arm control units through TCP/IP socket communication. Figure 4.1c displays the complete hardware operation flow.

The comprehensive software architecture is illustrated in Figure 4.1d, demonstrating how five key components interact via ROS communication on a Linux system powered by Ubuntu 18.04, featuring an Intel 8GHz CPU, 32GB RAM, and an Nvidia RTX-3080 GPU. The magnetic field/force control unit, as detailed in Figure 4.1e, takes in the specified parameters to direct the magnetic robots' actions. Meanwhile, the perception unit processes live video from two cameras to monitor the magnetic robots' movements, as shown in Figure 4.1f. The robotic arm motion and planning unit manages the coordinated control of the three arms using the ROS-Moveit interface, depicted in Figure 4.1g. The algorithm unit generates control commands using feedback from the robots' status and environmental data. Our software is designed to seamlessly incorporate user-friendly controls and optimization libraries, enabling the development of algorithms alongside real-time physics engines, all facilitated by ROS communication, thus enhancing realistic physical interaction capabilities.

To showcase magnetic manipulation and assess our proposed controller, we've created a simulation environment modeled after real-world conditions, as shown in Figure 4.1a. This environment includes three key parts: robotic arm kinematics, magnetic field modeling, and dynamics of the magnetic robot. For a realistic simulation, the magnetic fields are based on an actual coil model. We've also implemented a dynamic model of the magnetic robot to evaluate its movement and to accurately reflect real-world operations. Simulations can help us identify and resolve potential joint constraints and singularity issues before the system is put into use.

4.2.1.3 Calibrations

Accurate calibration of the REMA is crucial for precise actuation of magnetic robots. An eye-on-base calibration was conducted to ascertain the relative poses of the three robotic arms and two cameras. The *easy_handeye* package [20] provided a straightforward software solution for implementing this calibration.

Field calibration is of particular importance. Previous research [21] has explored field calibration for individual coils. Field data obtained from a magnetometer and calculated via FEM simulation were used to derive a continuous function representation of the field, which heavily relies on FEM field data. However, FEM modeling typically does not account for manufacturing imperfections, inhomogeneous materials, or unmodeled disturbances.

Leveraging the high positioning accuracy of the robotic arm in this study, an automated field data collection software was developed to record coil position and field data pairs. As shown in Figure 4.2a, the coil traversed all grid cells with a 3-mm spacing via the robotic arm. The grid-cell size can be adjusted based on actual situations. In Figure 4.2b, a commercial 3D magnetometer (TLE493D W2B6, Infineon Inc.) was placed at a fixed position for actual field collection. Next, a 3D vector-field interpolation function based on the regular grid interpolator was established using the sampled field data pairs to compute field values over the 3D space.

To assess the effectiveness of our field calibration procedure, we selected four distinct points within the workspace. At these points, we utilized a magnetic field

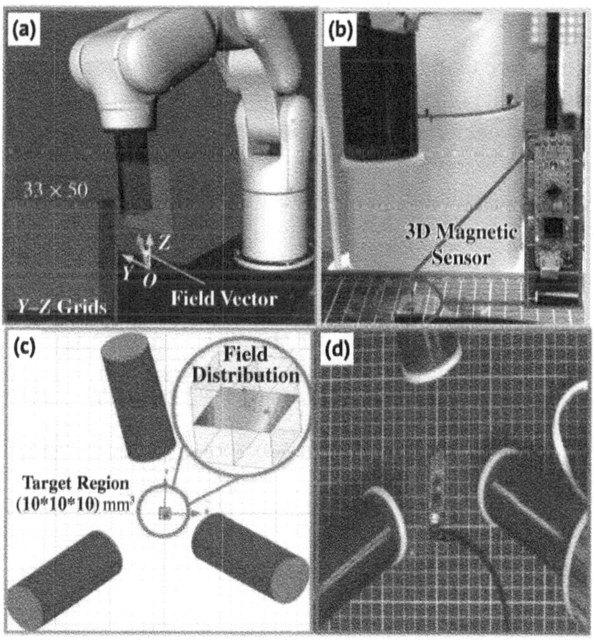

Figure 4.2 Automated collection of field data reading from magnetic sensor via robotic arm and three-coils field calibration. (a) Developed GUI for the collection of field data. (b) Real-world field collection for one coil. (c) Three-coil field calibration via FEM analysis. (d) Real-world field collection of three coils using a magnetometer.

sensor to measure the actual magnetic fields. Concurrently, we derived the calibrated fields using our proposed interpolation function. For comparative purposes, we also employed the calibration technique outlined in [21], which combines FEM simulations and empirical measurements to construct a look-up table. Table 4.2 presents a summary of the measured field $\mathbf{B}_{meas}$, the calibrated field $\mathbf{B}_{FEM}$ from [21], and the calibrated field $\mathbf{B}_{ours}$ obtained through our approach. It is important to note that the coordinates of the sampled points are referenced from the center tip of the coil, and the electric current applied was set at 1 A. To more thoroughly examine how the calibration performances of the FEM method and our method differ, we calculated both the magnitude ratios of $\mathbf{B}_{FEM}$ to $\mathbf{B}_{meas}$ and $\mathbf{B}_{ours}$ to $\mathbf{B}_{meas}$, as well as the angles between $\mathbf{B}_{FEM}$ and $\mathbf{B}_{meas}$ and between $\mathbf{B}_{ours}$ and $\mathbf{B}_{meas}$. The findings indicated that in our method, the ratio of the field amplitude is notably closer to 1. This suggests that our calibrated outcomes are in excellent agreement with the measured data. In addition, the angular error in our method is considerably lower than that in [21]. Typically, the main causes of error are probably the tolerances in coil manufacturing and any misalignments during the system's assembly process.

For an EMA system with more than one or two coils, especially when they are not sufficiently far from each other or generate high amplitude fields, the relationship between coil currents and generated fields is nonlinear, and the superposition principle may not hold due to these nonlinear effects and interactions of ferromagnetic material saturation. Several calibration methods for multi-coil systems have

Table 4.2 Comparison of the field calibration results for our method and FEM method.

	[0,0,60] (mm)	[0,−40,60]	[30,30,70]	[−10,33,56]
Measurements (mT) $\mathbf{B}_{meas}$	[0.01,0.02,0.98]	[0.01,−0.41, 0.62]	[0.34,0.31,0.40]	[−0.20,0.43,0.80]
Methods in [21] $\mathbf{B}_{FEM}$	[0.0.1.17]	[0.00,−0.45, 0.75]	[0.25,0.25,0.61]	[−0.15,0.48,0.92]
Our Method $\mathbf{B}_{ours}$	[0,0,1.03]	[0.00,−0.40, 0.60]	[0.31,0.31,0.45]	[−0.23,0.42,0.73]
$\frac{\|\mathbf{B}_{FEM}\|}{\|\mathbf{B}_{meas}\|}$	1.19	1.17	1.15	1.13
$\frac{\|\mathbf{B}_{ours}\|}{\|\mathbf{B}_{meas}\|}$	1.05	0.97	1.03	0.93
$\angle(\mathbf{B}_{meas}, \mathbf{B}_{FEM})$	1.31°	2.62°	18.97°	4.25°
$\angle(\mathbf{B}_{meas}, \mathbf{B}_{ours})$	1.31°	0.80°	5.12°	3.22°

been introduced by employing a static 3D array of magnetic sensors placed at the center of the workspace [3, 22].

To investigate accurate field modeling for REMA, the field characteristics of the three coils were modeled via FEM analysis, as shown in Figure 4.2c. Subsequently, Figure 4.2d displays the measurements of the fields using the magnetometer placed in the target position for a given coil configuration similar to that in Figure 4.2c. Then, four different field directions with an amplitude of 1 mT were generated at the target position using the superposition principle according to the calculated three-coil currents I_c respectively. Finally, I_c was increased with increasing multiples to observe the linearity of the generated field.

Figure 4.3 presents the comparative results of the FEM simulation and the measurements obtained from the magnetometer. An approximately linear field-to-current relationship was observed at low currents, where magnetic saturation is low and the linear assumption can be applied. The linearity performance decreased significantly in the high-current region due to strong saturation with an increase in current. However, an applicable linear field range satisfies the requirements of this study.

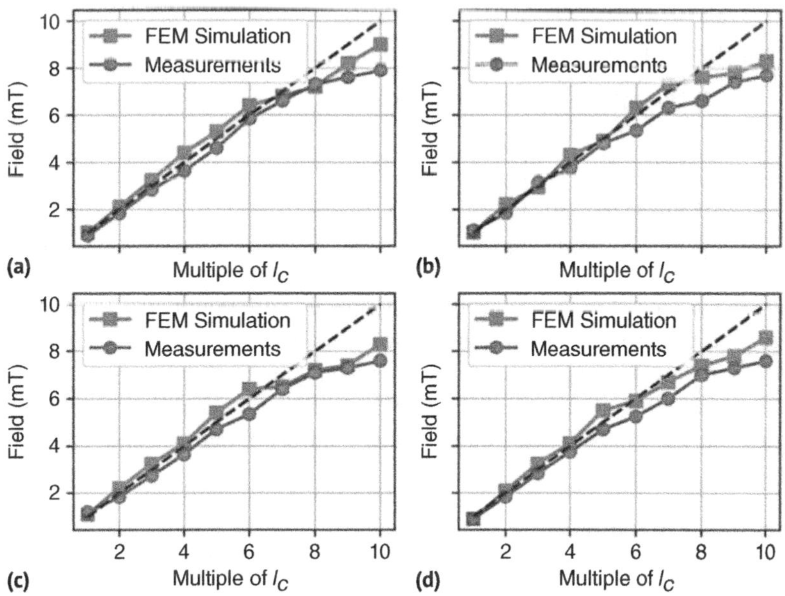

Figure 4.3 Performance of field generation using the FEM simulation and REMA. (a) Field direction (1, 1, 1). (b) Field direction (−1, 1, 1). (c) Field direction (−1, −1, −1). (d) Field direction (1, −1, −1).

4.2.2 Modeling of System

For REMA, the fields and forces generated are computed relative to the position $\mathbf{p}_w \in R^3$ within the global coordinate system, denoted as w. Consequently, it's necessary for each coil to convert its magnetic fields into this global frame. Given that the calibrated unit-current fields are initially defined in the local coil frame c, specific equations are utilized to determine the unit-current field for the kth coil ($k = 1, 2$, and 3) in relation to the global frame.

$$\begin{cases} \mathbf{p}_{c,k} = ({}^w\mathrm{H}_r \cdot {}^r\mathrm{H}_c)^{-1} \cdot k \oplus (\mathbf{p}_w), \\ \mathbf{b}_{w,k}(\mathbf{p}_w) = ({}^w\mathrm{H}_r \cdot {}^r\mathrm{H}_c)_k \oplus \mathbf{b}_{c,k}(\mathbf{p}_{c,k}) \end{cases} \tag{4.1}$$

where ${}^w\mathrm{H}_r \in R^{4\times4}$ and ${}^r\mathrm{H}_c \in R^{4\times4}$ denote the homogeneous transformation matrices from the robotic arm base frame r to the world frame, and that from the coil frame to the robotic arm base frame, respectively. Note that the ${}^w\mathrm{H}_r$ is derived using eye-on-base calibration. The notation $\oplus$ represents a homogeneous transformation that describes transforming a vector expressed in frame i to frame j.

Following this, the magnetic field $\mathbf{b}(\mathbf{p}_w)$ produced by the three coils can be approximated as being linearly dependent on the electric currents $\mathbf{i} = [i_1 \quad i_2 \quad i_3]^T$ [23].

$$\begin{aligned} \mathbf{b}(\mathbf{p}_w) &= \mathrm{A}_b(\mathbf{p}_w)\mathbf{i} \\ &= \left[\mathbf{b}_{1,w}(\mathbf{p}_w) \quad \mathbf{b}_{2,w}(\mathbf{p}_w) \quad \mathbf{b}_{3,w}(\mathbf{p}_w)\right] \\ &= \begin{bmatrix} b_{1x} & b_{2x} & b_{3x} \\ b_{1y} & b_{2y} & b_{3y} \\ b_{1z} & b_{2z} & b_{3z} \end{bmatrix} \begin{bmatrix} i_1 \\ i_2 \\ i_3 \end{bmatrix} \end{aligned} \tag{4.2}$$

where the $\mathrm{A}_b(\mathbf{p}_w) \in R^{3\times3}$ denotes the unit-current actuation matrix for a given point $\mathbf{p}_w$ in the workspace, similar to a Jacobian matrix.

Fundamentally, any magnetized robot experiences magnetic torques and forces when situated within a magnetic field. The magnetic torque and force exerted on a magnetized robot positioned in a magnetic field $\mathbf{b}(\mathbf{p}_w)$ can be represented as described in [24, 25].

$$\tau_m = \int_{V_m} \mathbf{m} \times \mathbf{b}(\mathbf{p}_w)\, \mathrm{d}V_m \tag{4.3}$$

$$\mathbf{f}_m = \int_{V_m} (\mathbf{m} \cdot \nabla)\mathbf{b}(\mathbf{p}_w)\, \mathrm{d}V_m \tag{4.4}$$

where V_m is the total volume of the magnet embedded in the head of the robot.

The primary goal of an EMA system is to offer the essential DOFs needed to control a variety of magnetic robots. This largely hinges on the arrangement of the coils and the quantity of the electromagnets involved. To achieve complete independent control over both torque and force within a uniformly magnetic workspace, a minimum of eight stationary coils is required, with adjustments made solely through electric currents. Nevertheless, if the coil setup can be dynamically altered via a mobile platform, the number of inputs necessary for automated control can be significantly reduced. As confirmed in [11], the effectiveness of EMA systems is influenced not just by the number of coils but also by their specific configurations. Thus, the REMA platform can deliver varying levels of performance by simply adjusting the positions and orientations of the coils. Moreover, the selection of the coil configuration can be guided by the specific manipulation tasks intended.

4.3 Multi-objective Optimization Method for Coil Configuration Generation

4.3.1 Problem Formulation

For different application scenarios, the required actuating fields should be evaluated by designing relevant evaluation metrics to generate a satisfactory coil configuration. Consider the case of rotating field control: a uniform rotating field induces rotational torque, facilitating the dependable movement of a helical robot without incurring extraneous magnetic gradient forces that could hinder its motion. Furthermore, the system must ensure the generation of rotating fields at any position and orientation to circumvent control singularities. However, crafting a singular evaluation metric to characterize the desired rotating fields proves to be a formidable task.

Consequently, devising multiple evaluation metrics becomes essential to quantify and appraise the intended rotating field. It should be noted that each coil has a total of 6-DOFs for control: a 5-DOF pose (neglecting the coil rotation along its principal axis) and a 1-DOF electric current. The nonlinear correlation between the multiple evaluation metrics and the DOFs of the three coils is intricate. The MOO technique [26] may offer an effective solution for handling multiple objective functions (i.e., multiple evaluation metrics) within a search space (i.e., the poses of the three coils). The subsequent subsection elucidates the application of the MOO technique to resolve our problem.

4.3.2 Definition of Coil Configuration Space

To execute the MOO for devising the optimal coil configuration, it's necessary to define a search space—specifically, a coil configuration space (CCS) that integrates the design constraints pertaining to the coil's position and orientation.

Accordingly, we establish a global reference frame $O_{w\text{-}xyz}$ and a local coil frame $O_{c\text{-}xyz}$ to characterize the coil's configuration, as illustrated in Figure 4.4. The generalized coordinate system for each coil frame is delineated by five elements $\{\mathbf{x}^i = (L^i, \theta^i, \phi^i, \alpha^i, \beta^i) \in R^5, i = 1, 2, 3\}$ with respect to the world frame. Certain constraints must be designed for these parameters to ensure a collision-free coil pose (see Tables 4.3 and 4.4).

We regard the collective pose parameters of the three coils as the CCS. For the optimization problem at hand, the CCS, denoted as $\mathcal{X}$, is formally defined as follows:

$$\mathcal{X} := \{\mathbf{x} = (L^1, \theta^1, \phi^1, \alpha^1, \beta^1, \dots, L^3, \theta^3, \phi^3, \alpha^3, \beta^3) \in R^{15}\} \tag{4.5}$$

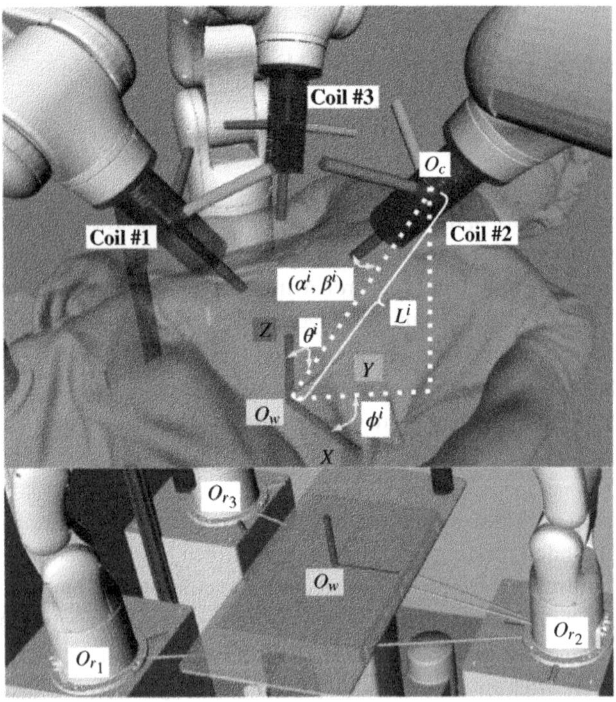

Figure 4.4 Definition of optimization parameters to describe a coil pose and relative positions of three robotic arms under the frame O_w. The optimization parameters are described using five elements $(L^i, \theta^i, \phi^i, \alpha^i, \beta^i)$ expressed in the frame O_w. Note that elements of (L^i, θ^i, ϕ^i) denote the spherical coordinates to determine the position of the frame O_c. Elements of (α^i, β^i) denote angles that the Z-axis of the frame O_c rotates around the X-axis and Y-axis of the frame O_w, respectively.

Table 4.3 Position and orientation of each robotic arm's base relative to world frame.

	X(m)	Y(m)	Z(m)	Angle around Z (°)
Robotic arm #1	−0.470	−0.310	−0.115	45
Robotic arm #2	0.580	0	−0.115	180
Robotic arm #3	−0.470	0.310	−0.115	−45

Table 4.4 Constraints of optimization parameters (i = 1, 2, 3).

	L^i(m)	θ^i (°)	ϕ^1 (°)	ϕ^2 (°)	ϕ^3 (°)	α^i (°)	β^i (°)
χ_0	0.15	45	225	0	130	0	0
$\chi_{\min}$	0.11	30	195	−45	90	−18	−18
$\chi_{\max}$	0.25	60	270	45	165	18	18

4.3.3 Definition of Evaluation Metrics

In the motion modes of magnetic robots, we primarily consider helical propulsion helical propulsion (rotating field control) and gradient pulling (field and force control). Therefore, we consider the definition of evaluation metrics for rotating field control and combined field and force control, respectively.

1. Rotating Field Control: The preferred coil configuration should ensure that the magnetic fields produced are aligned with the target field patterns. Isotropy can be assessed by analyzing field samples taken across various orientations within the region of interest. To simplify the calculation without loss of generality, we set the desired field as a unit field $\hat{\mathbf{b}}(\mathbf{p}_w)$ at position $(\mathbf{p}_w)$. Within the designated area, the magnitude and orientation of all actual fields $\mathbf{b}(\mathbf{p}_i)$ must correspond precisely to those of the desired field $\hat{\mathbf{b}}(\mathbf{p}_w)$. Therefore, we designed the two evaluation metrics to represent the isotropy of the field:
   ① $\|\mathbf{b}\|_{\text{iso}}$ measures the amplitude similarity between the actual and the desired field.

$$\|\mathbf{b}\|_{\text{iso}} = 1 - \frac{\sum_{i=0}^{i=N-1} \left| \|\mathbf{b}(\mathbf{p}_i)\| - \left\|\hat{\mathbf{b}}(\mathbf{p}_w)\right\| \right|}{N} \tag{4.6}$$

   ② $(\mathbf{b}_{\text{cosine}})_{\min}$ assesses the minimum degree of alignment between the target and measured fields, aiming to maximize this value to ensure optimal performance under the most conservative conditions.

$$(\mathbf{b}_{\text{cosine}})_{\min} = \min\left(\frac{\mathbf{b}(\mathbf{p}_i)\cdot\hat{\mathbf{b}}(\mathbf{p}_w)}{\|\mathbf{b}(\mathbf{p}_w)\|\left\|\hat{\mathbf{b}}(\mathbf{p}_w)\right\|}\right), i = 0, \dots, N-1 \tag{4.7}$$

Executing precise movements across various positions and orientations without singularities is essential. For REMA, this can be analyzed by examining the field actuation matrix A_b. The singular values $\sigma_i (i=1,2,3)$ of matrix A_b are pivotal for assessing the efficiency of coil current to field mapping [11]. A higher minimum singular value σ_3 minimizes the peak coil current necessary for extreme control situations near singularities. Additionally, approximating the manipulability ellipsoid to a spherical shape enhances directional control capabilities. This is realized by maximizing the reciprocal of the condition number κ to be close to 1.

Based on the analysis mentioned above, we utilized the field isotropy metrics $||\mathbf{b}||_{\text{iso}}$ and $(\mathbf{b}_{\text{cosine}})_{\min}$ along with the actuation matrix A_b isotropy indexes σ_3 and $1/\kappa$, as objective functions for optimizing coil configurations via the MOO approach.

$$\mathbf{x}^* = \arg \max_{\mathbf{x} \in \mathcal{X}} \{||\mathbf{b}||_{\text{iso}}, (\mathbf{b}_{\text{cosine}})_{\min}, \sigma_3, 1/\kappa\} \text{ s.t. } \mathcal{X}_{\min} \leq \mathcal{X} \leq \mathcal{X}_{\max} \tag{4.8}$$

2. Combined Field and Force Control: Force control is more complex to analyze than field control because it requires a five-component gradient vector to ascertain the resulting three-component force vector. When applying a magnetic force on a robot, it is imperative to account for field control, as a force cannot be applied without an underlying field. In scenarios where a magnetic robot operates freely and its orientation must be managed—such as with capsule robots—it is logical to presume that the robot will naturally align with the applied field. Orchestrating fields and forces within the operational space of REMA is intricate due to difficulties in explicitly mapping field and force variables to the positioning and current of the three coils. Nonetheless, field control is achievable through adjustments in the electric currents of the coils alone. Furthermore, by employing a variable coil setup, the workspace can be surveyed for gradient distributions that meet the desired force criteria. It is also essential to factor in the manipulability measure for each robotic arm to circumvent singularities during the optimization process. Thus, with a given field and force, field control is attainable by modifying the currents in the three coils, fine-tuning the actual force to a target value within tolerable margins, and improving the Yoshikawa manipulability measure for each arm. Consequently, the MOO problem can be articulated as follows:

$$\mathbf{x}^* = \arg \max_{\mathbf{x} \in \mathcal{X}} \{e_{\mathbf{f}}, m_1, m_2, m_3\} \text{ s.t. } \mathcal{X}_{\min} \leq \mathcal{X} \leq \mathcal{X}_{\max} \tag{4.9}$$

where $e_{\mathbf{f}} = -|\mathbf{f}_d - \mathbf{f}(\mathbf{x})||$, $\mathbf{f}_d$ denotes the desired force, and $\mathbf{f}(\mathbf{x})$ represents the actual force for a specific coil configuration $\mathbf{x}$. The Yoshikawa manipulability $m_i = \sqrt{\det(\mathbf{J}_i(\mathbf{q})\mathbf{J}_i^T(\mathbf{q}))}$, $i = 1, 2, 3$, which describes how well-conditioned the robotic arm is to achieve an arbitrary velocity in the Cartesian space. $\mathbf{J}(\mathbf{q})$ represents the Jacobian matrix of the robotic arm, which is a function of the joint position vector $\mathbf{q}$.

4.3.4 Multi-objective Optimization Method

When optimizing coil configurations, improving one performance metric could adversely affect another, necessitating a balanced approach aligned with application requirements. The MOO technique is adept at identifying optimal solutions amidst competing objectives.

With a defined CCS and performance metrics in place, MOO can extract the Pareto set. We utilized the nondominated sorting genetic algorithm (NSGA-II), a widely-adopted MOO algorithm, renowned for its efficient exploration of extensive search spaces, to swiftly and effectively address the MOO challenge. Based on the above analysis, by combining the designed objective functions and the defined CCS $\mathcal{X}$, the process of determining the optimal coil configuration is formulated as an MOO problem to maximize the performance criteria.

$$\mathbf{x}^* = \arg\max_{\mathbf{x}\in\mathcal{X}} \{f_1, f_2, f_3, f_4\} \text{ s.t. } \mathcal{X}_{\min} \leq \mathcal{X} \leq \mathcal{X}_{\max} \tag{4.10}$$

where f_l denotes the objective function. The steps for obtaining a Pareto-optimized coil configuration via the MOO method are outlined in Algorithm 1. This approach accommodates two distinct control modes—rotating field control and combined field-force control—to produce varied coil setups. The NSGA-II algorithm, which we utilize, commonly employs the hypervolume metric to assess performance. This metric evaluates the space between a set reference point and the solution set provided [27]. Essentially, a larger hypervolume of the solution set derived from the MOO algorithm signifies superior algorithmic efficacy.

4.4 System Validation and Analysis

To assess the performance of the developed REMA platform and the proposed magnetic manipulation techniques, a series of experiments were executed across various scenarios. These encompassed directing helical swimmers via rotating field control, guiding catheters using field (torque) control, and driving a prototype capsule with a combination of field and force control. Additionally, the use of a simulator environment allowed for debugging and tuning of the control methods and relevant parameters in advance. Subsequently, real-world tests were conducted to confirm the platform's performance.

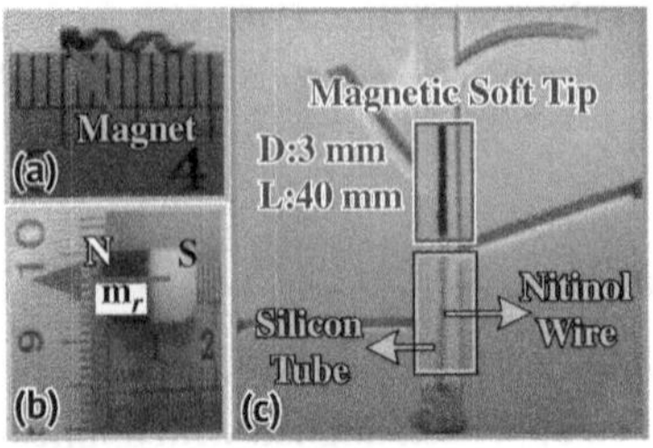

Figure 4.5 Overview of the different magnetic robots for the real-world experiments. (a) Magnetic helical swimmer. (b) Mockup capsule. (c) Magnetically steerable catheter in a phantom with different branches.

4.4.1 Experimental Setup

To ascertain the efficacy of our rotating field control approach, we constructed a helical swimmer equipped with a Grade-N52 NdFeB magnet (1 mm in diameter, 2 mm in height) positioned at its head, oriented such that its dipole moment is perpendicular to the axis of rotation, as illustrated in Figure 4.5a.

Furthermore, to showcase the combined field and force control, we engineered a mockup capsule using 3D-printed resin, which is depicted in Figure 4.5b. This capsule incorporates a magnet (4 mm in diameter, 6 mm in height) with its dipole moment aligned parallel to the axis. Alongside these untethered robots, we tested the system's ability to navigate a catheter through a phantom with multiple branches, as presented in Figure 4.5c. The catheter was assembled from three primary components: a magnetic soft tip, a nitinol guidewire, and an outer silicone tube.

4.4.2 Validation of Rotating Field Control

To validate the performance of generated rotating fields, the actuation of a helical swimmer in a simulation environment and a real-world 3D free space were implemented, respectively.

1. Actuation of a Virtual Helical Swimmer in Simulation Environments: Figure 4.6a-1–a-4 illustrates the trajectory of a virtual helical swimmer in the simulation environment, guided by the generated rotating fields. Before this simulation, an optimal coil configuration was determined following Algorithm 1 (refer to Appendix). The path-following capability of the virtual swimmer was demonstrated within a life-sized operational space. At each step, the electric currents were adjusted to maintain alignment of the rotational field with the predefined trajectory. Consequently, the virtual swimmer adhered to the reference path, steered by the rotational fields.

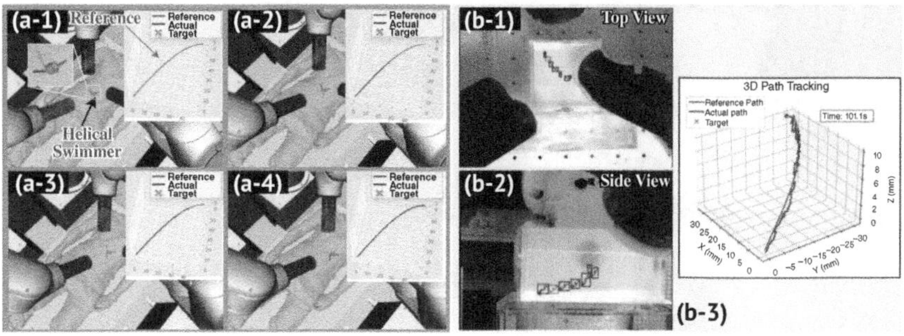

Figure 4.6 Magnetically actuating a helical swimmer under the guidance of the rotating field generated by a Pareto-optimized coil configuration. (a-1–a-4): The actuation of a helical swimmer in the proposed simulation environment. (b-1–b-3): The results of 3D path tracking of a helical swimmer under the rotating field control.

2. 3D Path Tracking of a Helical Swimmer in a Cubic Container Phantom: In 3D space, the control performance of a helical swimmer is influenced by its own weight and external disturbances. Consequently, a pure-pursuit strategy was utilized to regulate the swimmer's pitch and yaw, facilitating precise 3D trajectory tracking.

   Top and side views of the helical swimmer's trajectory (as shown in Figure 4.6b-1 and b-2) were captured to illustrate its positioning throughout the path-following process. Figure 4.6b-3 depicts the swimmer's ability to adhere closely to a 3D path, maintaining minimal deviation.

4.4.3 Navigation of Magnetic Catheters Under Controlled Field

To further validate the feasibility of the REMA platform, the catheter depicted in Figure 4.5c was steered magnetically through various branches of a phantom, monitored in real time via imaging. The navigation of the catheter involves applying a magnetic field (torque) to the magnet at its tip, causing the catheter's flexible body to bend.

The catheter was propelled by a human's manipulation, with the three coils monitoring its tip's real-time location. Magnetic fields were generated to guide the catheter's tip into the desired branch by leveraging magnetic torque, causing it to bend and align with the field. Figure 4.7a–d demonstrate this navigation sequence. The results confirm the catheter's successful traversal through the phantom's distinct branches, highlighting the method's adaptability.

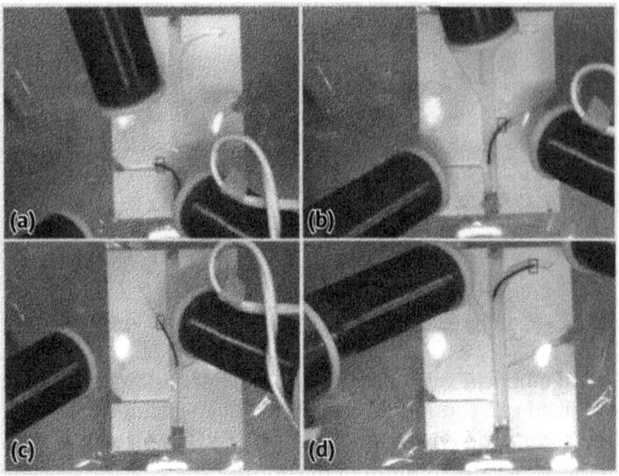

Figure 4.7 Results of magnetically actuating a catheter under the guidance of the field control. (a–d) Display the navigation of a magnetic catheter into different phantom branches under field control.

4.4.4 Actuation of a Capsule Robot Using Combined Field and Force Control

To validate the effectiveness of the combined field-force control, an experiment was undertaken where a capsule was directed to follow a predefined trajectory with varying orientations in 3D space. The capsule was placed in a cube filled with 80% glycerol, aligning its direction with the magnetic field as it moved.

Figure 4.8 illustrates the experimental results of a capsule robot tracking a prearranged path in a cubic container. The experiment was divided into three distinct motion stages. Notably, during Motion Stages I and III, the direction of the desired force vector did not coincide with that of the field vector, indicating a misalignment between the heading and the direction of motion. For each stage, the coil configuration was determined to generate the requisite field and force as per the planned trajectory. The velocity of the capsule was primarily influenced by the magnitude of the applied force. Specifically, when the directions of the field and force remained constant throughout each stage, changes in their magnitudes, scaled uniformly, impacted only the electric currents, not altering the coil configuration. We utilized a proportional-integral-derivative (PID) controller to compute the scaling factor from the capsule's real-time position, which was then applied to adjust the amplitudes of the field and force. Guided by this strategy, the capsule successfully followed the planned path across the three stages. Furthermore, the field-priority force control method was found to effectively balance the generation of field and force with the manipulability of the robotic arms. The experiments detailed herein underscore the efficacy of the REMA platform and its control strategy in steering a capsule through 3D spaces.

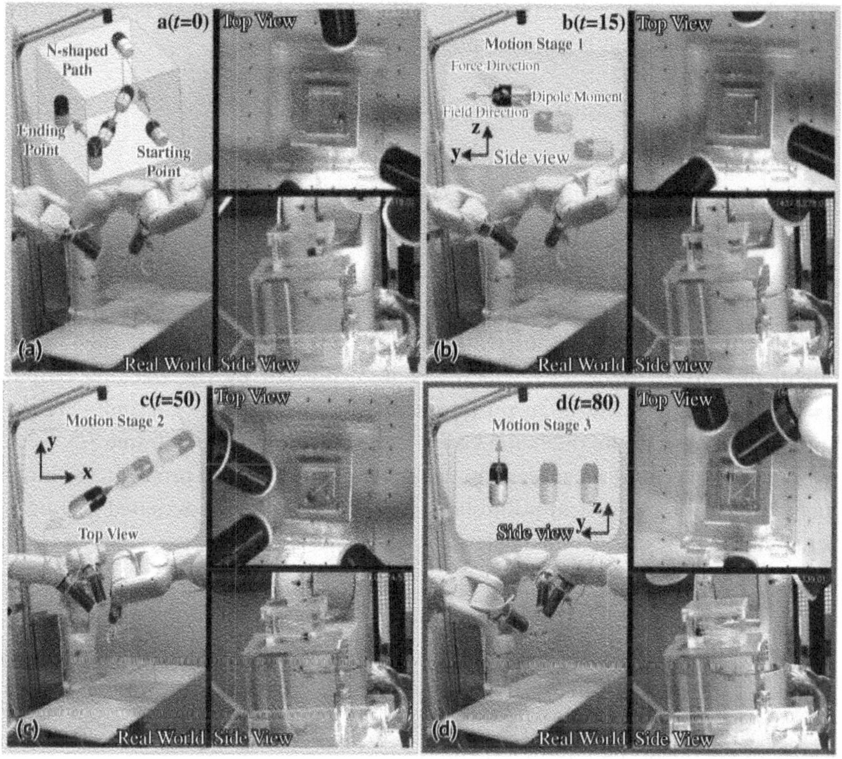

Figure 4.8 Actuation of a mockup capsule under the guidance of the combined field and force control method to track a planned path in the 3D space. The overall process is divided into three motion stages. Motion Stage 1: (a–b) show the motion stage 1 that the capsule's heading maintains horizontal (along the y-axis) and the force along the y-axis. Motion Stage 2: (b–c) show the motion stage 2 that the angle between the capsule's heading and the applied force direction is 180°. Motion Stage 3: (c–d) show the motion stage 3 that the capsule's heading maintains vertical (z-axis) and the force along the y-axis.

4.5 Construction of Path Control Framework

4.5.1 Kinematic Model of Helical Millirobot

In this section, the kinematic model of the helical millirobot is constructed based on the 3D hand position approach. Furthermore, designing appropriate guidance and control laws enables the helical millirobot to track a 3D reference path, ensuring all signals in the closed-loop system remain bounded and tracking errors converge to an arbitrarily small vicinity of the sliding surface within finite time.

The kinematic equation of the helical millirobot in the 3D space is given by the following five-DOF mathematical model:

$$\begin{cases} \dot{x} = u\cos(\psi)\cos(\theta) - \upsilon\sin(\psi) + w\cos(\psi)\sin(\theta) \\ \dot{y} = u\sin(\psi)\cos(\theta) + \upsilon\cos(\psi) + w\sin(\psi)\sin(\theta) \\ \dot{z} = -u\sin(\theta) + w\cos(\theta) \\ \dot{\theta} = q \\ \dot{\psi} = r/\cos(\theta) \end{cases} \tag{4.11}$$

where $[x, y, z]$ represents the position. θ and ψ are the pitch and yaw angles (all in the world frame $\mathcal{W}$), which can be expressed by the tracking output $\eta = [x, y, z, \theta, \psi]^T$; while $[u, \upsilon, w]$ represents the linear velocity. q and r are the angular velocities in the pitch and yaw directions (all in the body frame $\mathcal{H}$).

Taking into account the underactuated nature of this control problem, the kinematic model for the helical millirobot can be separated into actuated and non-actuated dynamics, as detailed below:

$$\dot{\eta} = \mathbb{R}(\eta)\boldsymbol{u}_a + \delta(\eta, \boldsymbol{w}) \tag{4.12}$$

where $\boldsymbol{u}_a = [u, q, r]^T$ and $\boldsymbol{w} = [\upsilon, w]^T$ denote the velocity vectors of the helical millirobot in actuated and unactuated directions, respectively. The functions $\mathbb{R}(\boldsymbol{\eta})$ and $\delta(\eta, \boldsymbol{w})$ can be considered as the kinematic matrix and the kinematic disturbance, respectively, which can be expressed as:

$$\mathbb{R}(\eta) = \begin{bmatrix} \cos(\psi)\cos(\theta) & 0 & 0 \\ \sin(\psi)\cos(\theta) & 0 & 0 \\ -\sin(\theta) & 0 & 0 \\ 0 & 1 & 0 \\ 0 & 0 & \dfrac{1}{\cos(\theta)} \end{bmatrix}$$

$$\delta(\eta, \boldsymbol{w}) = \begin{bmatrix} -\upsilon\sin(\psi) + w\sin(\theta)\cos(\psi) \\ \upsilon\cos(\psi) + w\sin(\theta)\sin(\psi) \\ w\cos(\theta) \\ 0 \end{bmatrix} \tag{4.13}$$

For the tracking output η of the control system described in Eq. (4.12), the geometric center of the controlled object is a common choice. However, a significant issue arises as this geometric center is unaffected by control inputs in the yaw and pitch directions. This is particularly problematic for underactuated systems, where the kinematic matrix's (4.13) singularity introduces unstable dynamics. To resolve this, we looked to reference [28] to expand the concept of the hand position point to the 3D positioning method used in helical millirobots. Consequently, the tracking output is now defined as the position of a virtual point on the helical millirobot's longitudinal symmetry axis, located a fixed distance l from the principal axis's midpoint (refer to Figure 4.9), which is now rewritten as follows:

$$\boldsymbol{h}(\eta) = \begin{bmatrix} x + l\cos(\theta)\cos(\psi) \\ y + l\cos(\theta)\sin(\psi) \\ z - l\sin(\theta) \end{bmatrix} \tag{4.14}$$

By differentiating (4.14), the kinematic model of the helical millirobot can be rearranged into 3D hand position dynamics as follows:

$$\dot{\boldsymbol{h}}(\eta) = \mathbb{J}(\eta)\boldsymbol{u}_a + \boldsymbol{J}_\delta(\eta, \boldsymbol{w}) \tag{4.15}$$

where $\boldsymbol{u}_a = [u, q, r]^T$ denotes the velocity vector of a helical millirobot in the actuated direction in the frame $\mathcal{H}$. $\mathbb{J}(\eta)$ is regarded as the transformation matrix denoting the effect of internal dynamics, and $\boldsymbol{J}_\delta(\eta, \boldsymbol{w})$ can be regarded as the

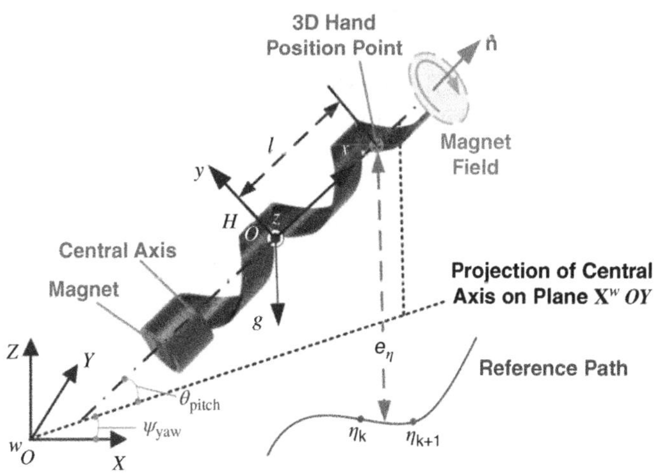

Figure 4.9 Definitions of world frame $\mathcal{W}$ and body frame $\mathcal{H}$. The helical millirobot is actuated by rotating magnetic fields to follow a reference path using a selected 3D hand position point.

input–output map of the kinematic lumped disturbances in the world frame $\mathcal{W}$, including external disturbances and system uncertainties, which can be expressed by:

$$\mathbb{J}(\eta) = \begin{bmatrix} \cos(\psi)\cos(\theta) & -l\sin(\theta)\cos(\psi) & -l\sin(\psi) \\ \sin(\psi)\cos(\theta) & -l\sin(\theta)\sin(\psi) & l\cos(\psi) \\ -\sin(\theta) & -l\cos(\theta) & 0 \end{bmatrix}$$
$$\boldsymbol{J}_{\delta}(\eta, \boldsymbol{w}) = \begin{bmatrix} -v\sin(\psi) + w\sin(\theta)\cos(\psi) \\ v\cos(\psi) + w\sin(\theta)\sin(\psi) \\ w\cos(\theta) \end{bmatrix} \tag{4.16}$$

In (4.14), when $l > 0$, the yaw angle ψ and pitch angle θ play a significant role in influencing the tracking output. Hence, effective control of the orientation angle of the millirobot becomes crucial in maintaining the stability of the 3D hand position point. In contrast to the kinematic matrix (4.13), no singularity exists in the matrix $\mathbb{J}(\boldsymbol{\eta})$ for all $\theta, \psi \in \mathbb{R}$.

4.5.2 Estimation of Lumped Disturbance

The following assumption and lemma will be utilized in the observer design.

Assumption 1: The lumped disturbance $\boldsymbol{J}_{\delta}$ and its derivative are assumed to be unpredictable, varying, and limited in energy, so the unknown disturbance and its derivative satisfy $\boldsymbol{J}_{\delta} \leq \alpha < \infty$ and $\dot{\boldsymbol{J}}_{\delta} \leq \beta < \infty$, where α and β are the bounded.

Lemma 1: For the differentiator taking on the form as:

$$\begin{cases} \dot{\xi}_0(t) = v_0(t) + u(t) \\ v_0(t) = -\lambda_0|\xi_0 - f_0(t)|^{\frac{n}{n+1}}\operatorname{sign}(\xi_0 - f_0(t)) + \xi_1(t) \\ \dot{\xi}_1(t) = v_1(t) \\ v_1(t) = -\lambda_1|\xi_1 - v_0|^{\frac{n-1}{n}}\operatorname{sign}(\xi_1 - v_0) + \xi_2(t) \\ \vdots \\ \xi_{n-1}(t) = v_{n-1}(t) \\ v_{n-1}(t) = -\lambda_{n-1}|\xi_{n-1} - v_{n-2}|^{\frac{1}{2}}\operatorname{sign}(\xi_{n-1} - v_{n-2}) + \xi_n(t) \\ \dot{\xi}_n(t) = -\lambda_n\operatorname{sign}(\xi_n - v_{n-1}). \end{cases} \tag{4.17}$$

The following equalities could hold in the absence of input noises after a finite time of transient process:

$$\xi_0 = f_0(t); \quad \xi_i = v_{i-1} = f_0^{(i)}(t), \quad (i = 1, \dots, n). \tag{4.18}$$

In this study, a finite-time observer to estimate the lumped disturbance can be designed as follows:

$$\begin{cases} \dot{\hat{h}}_\eta = V_\eta + \hat{J}_\delta \\ \hat{J}_\delta = -\lambda_{\beta_1} \left|\hat{h}_\eta - h_\eta\right|^{1/2} \text{sign}\left(\hat{h}_\eta - h_\eta\right) + \hat{J}^* \\ \dot{\hat{J}}^* = -\lambda_{\beta_2} \text{sign}\left(\hat{J}^* - J_\delta\right) \end{cases} \tag{4.19}$$

where $\dot{\hat{h}}_\eta$ and $\hat{J}_\delta$ denote the estimated motion velocity and lumped disturbance within the frame $\mathcal{W}$, respectively. V_η represents the velocity vector transformed into frame $\mathcal{W}$, and J^* is an intermediate variable. The observer gains λ_{β_1} and λ_{β_2} dictate the convergence rate. With estimation errors defined as $\tilde{h}_\eta = \hat{h}_\eta - h_\eta$ and $\tilde{J}^* = \hat{J}^* - J_\delta$, and by differentiating $\tilde{h}_\eta$ and $\tilde{J}^*$ over time, the error equations can be formulated as follows:

$$\begin{cases} \dot{\tilde{h}}_\eta = -\lambda_{\beta_1} \tilde{h}_\eta^{1/2} \text{sign}\left(\tilde{h}_\eta\right) + \tilde{J}^* \\ \dot{\tilde{J}}^* = -\lambda_{\beta_2} \text{sign}\left(\hat{h} - J_\delta\right) - \dot{J}_\delta \end{cases} \tag{4.20}$$

where $\dot{J}_\delta$ represents the derivative of the disturbance, which is presumed to be bounded for all $t \geq 0$ as per Assumption 1. As per Lemma 1, the state estimation of the lumped disturbance can be achieved within a finite time period.

$$\hat{J}^* = J_\delta \tag{4.21}$$

with the error dynamics converging to the bounded region. Following the similar manners, the finite time of the transient process T_1 is given by:

$$T_1 \leq \frac{2}{\zeta} V_0^{1/2} \tag{4.22}$$

where V_0 is a positive definite quadratic function denoted by $V_0 = s^T \Gamma s$ with the vector $s = \left[\left|\tilde{h}_\eta\right|^{(1/2)} \text{sign}\left(\tilde{h}_\eta\right), \text{sign}\left(\tilde{J}^*\right)\right]^T$. The parameter Γ is a constant, symmetric, and positive definite matrix, while ζ denotes the constant related to the reaching time.

The following assumption and lemma will be utilized in the controller design:

Assumption 2 [29]: In practice, the control law (velocity vector) u_a determined by the saturated function of the nominal control command u_{a_0} can be described by:

$$u_a = \text{sat}(u_{a_0}) = \text{sign}(u_{a_0}) \min\{|u_{a_0}|, u_{a,\max}\} \tag{4.23}$$

where $u_{a,\max}$ denotes the positive magnitude constraint.

Lemma 2 [30]: The nth-order smooth nonlinear switching function $F(X) : \mathcal{R} \rightarrow \mathcal{R}$ can be represented as:

$$F(X) = \begin{cases} 0, & \text{if } |X| \leq X_a \\ 1 - \cos^n\left(\frac{\pi}{2}\sin^n\left(\pi\frac{|X|^2 - X_a^2}{2X_b^2 - 2X_a^2}\right)\right), & \text{otherwise} \\ 1, & \text{if } |X| \geq X_b. \end{cases} \tag{4.24}$$

The requirement is $0 < X_a < X_b$, with X_a and X_b being small positive constants that can be freely chosen to establish the transient behavior within the interval (X_a, X_b) and define the robust control region beyond (X_a, X_b).

4.5.3 Design of Adaptive Finite-time Sliding Mode Control

This section presents the proposed Anti-slip Model Control (ASMC) framework for the 3D path tracking of the helical millirobot. Initially, a manifold must be designed to ensure the system trajectory aligns with the desired properties, including error convergence. Drawing on the helical model for the path tracking task, the sliding surface is defined as follows:

$$\boldsymbol{S}_r = \boldsymbol{e}_\eta + \int_0^t \left(\alpha_e \boldsymbol{e}_\eta(\tau) + \beta \boldsymbol{e}_\eta^{m/n}(\tau)\text{sign}(\boldsymbol{e}_\eta(\tau))\right) d\tau \tag{4.25}$$

Where $\boldsymbol{e}_\eta = \boldsymbol{h}_\eta - \boldsymbol{\ell}_\eta$ represents the tracking error between the actual state vectors $\boldsymbol{h}_\eta = [x, y, z]^T \in \mathbb{R}^3$ and the reference path $\boldsymbol{\ell}\eta = [x_d, y_d, z_d]^T \in \mathbb{R}^3$. m and n are odd integers satisfying $0 < m/n < 1$. α_e and β_e are strictly positive constants that influence the convergence rate of the tracking error. With the 3D hand position approach, the error dynamics of the closed-loop system can be derived by taking the time derivative of the sliding-mode surface $\boldsymbol{S}_r$ as follows:

$$\dot{\boldsymbol{S}}_r = \mathbb{J}(\boldsymbol{\eta})\boldsymbol{u}_a + \boldsymbol{J}_\delta(\boldsymbol{\eta}, \boldsymbol{w}) - \dot{\boldsymbol{\ell}}_\eta + \alpha_e \boldsymbol{e}_\eta + \beta_e \boldsymbol{e}_n^{m/n}\text{sign}(\boldsymbol{e}_\eta). \tag{4.26}$$

In dynamic fluid environments, the millirobot must increase its rotational frequency to counteract the impact of flow rate. However, exceeding the upper limits on rotational frequency can lead to step-out and compromise control accuracy. To mitigate this adverse effect, it is advantageous to eliminate the frequency constraint. Therefore, we propose an auxiliary system that provides feedback on saturation values to ensure finite-time stability [31]. The states of this auxiliary system are employed to develop an adaptive control strategy as follows:

$$\dot{\chi}_r = -k_1\chi_r - F(\chi_r)\frac{\left|\boldsymbol{S}_r^T \Delta\boldsymbol{u}_a\right| + 0.5\rho_r^2 \Delta\boldsymbol{u}_a^T \Delta\boldsymbol{u}_a}{\chi_r} - k_2\text{sign}^\rho(\chi_r) + \rho_r \Delta\boldsymbol{u}_a \tag{4.27}$$

Where $\chi_r \in \mathbb{R}^3$ represents the auxiliary state variable. k_1, k_2, and ρ_r denote the positive gain constants. ρ can take any value within (0, 1) and $F(\chi_r)$ is defined as in Lemma 2. $\Delta \boldsymbol{u}_a = \boldsymbol{u}_a - \boldsymbol{u}_{a_0} = \text{sat}(\boldsymbol{u}_{a_0} - \boldsymbol{u}_{a_0})$ is the input saturation value, representing the difference associated with practical perspective and nominal control command. Considering the dynamics of the helical millirobot in the presence of lumped disturbances and input saturation, the nominal AFTSMC control command is derived.

$$\boldsymbol{u}_{a_0} = -\mathbb{J}^{+}(\eta)\left[K_{\text{SMC}} - \dot{\boldsymbol{e}}_\eta + \hat{\boldsymbol{J}}_\delta(\eta, \boldsymbol{w}) + \lambda_r \chi_r + \left(\alpha_e \boldsymbol{e}_\eta + \beta_e \boldsymbol{e}_\eta^{m/n}\text{sign}(\boldsymbol{e}_\eta)\right)\right] \tag{4.28}$$

The parameter λ_r is a positive constant. The term $\hat{\boldsymbol{J}}_\delta(\eta, \boldsymbol{w})$ represents the vector of lumped disturbances, which can be estimated as per equation (4.29). In the design of the reaching law for SMC K_{SMC}, a common conservative approach is to incorporate a high gain into the controller to ensure system stability. However, this method can lead to overestimation of the gain. To address this, an adaptive mechanism is introduced, which significantly enhances performance by adjusting the controller gains in response to additive perturbations with uncertain boundaries. The proposed reaching law is as follows:

$$K_{\text{SMC}} = K_1 \boldsymbol{S}_r + K_2 |\boldsymbol{S}_r|^p \text{sign}(\boldsymbol{S}_r) \tag{4.29}$$

with the adaptive structure as:

$$\dot{K}_2(t) = \begin{cases} k_\alpha \text{sign}(|\boldsymbol{S}_r| - \mu), & K_2(t) > K_{\min} \\ K_{\min}, & K_2(t) \le K_{\min} \end{cases} \tag{4.30}$$

Where K_1, $K_{\min}$, and k_α are positive real numbers, and $0 < p < 1$. The parameter μ, also a positive real number, serves as the threshold for detecting the loss of sliding mode. The adaptive mechanism adjusts $K_2(t)$ at a rate determined by k_α to counteract disturbances, ensuring that the sliding vector approaches the reference within a finite time [32]. Once the sliding vector touches the sliding surface, $K_2(t)$ is set to the constant minimum value $K_{\min}$. This controller is designed to adapt the gains to achieve minimal effort, ensuring system stability and preventing chattering.

The following lemmas are used for the stability analysis of the sliding phase.

Lemma 3: Suppose $(\alpha, \beta) \in \mathbb{R}$, the following Young's inequality satisfies: $\alpha\beta \le \varepsilon^p \frac{|\alpha|^p}{\rho} + \frac{|\beta|^q}{\varepsilon^q q}$, once $\varepsilon > 0$, $\rho > 1$, and $q > 1$ meet $(\rho - 1)(q - 1) = 1$.

Lemma 4 [33]: For any real number $\alpha > 0$, $\beta > 0$, and $0 < \rho < 1$, an extended Lyapunov condition of finite-time stability can be given:

$$\dot{V}(\mathbf{x}) + \alpha V(\mathbf{x}) + \beta V^\rho(\mathbf{x}) \le 0 \tag{4.31}$$

where the settling time can be estimated by:

$$T_f \leq t_0 + \frac{1}{\alpha(1-\rho)} \ln \frac{\alpha V^{1-\rho}(t_0) + \beta}{\beta} \tag{4.32}$$

Lemma 5 [33]: Considering the system $\mathbf{R} = f(\mathbf{x}(t))$, an extended Lyapunov condition of practical finite-time stability can be guaranteed with the form as:

$$\dot{V}(\mathbf{x}) \leq -\alpha V(\mathbf{x}) - \beta V^{\rho}(\mathbf{x}) + \varepsilon \tag{4.33}$$

with $\alpha, \beta \in \mathbb{R}_+$, $0 < \rho < 1$, and $0 < \varepsilon < \infty$. Then, the system trajectory $\dot{\mathbf{x}} = \mathbf{x}(t)$ is practical finite-time stable and converges into the bounded region. The residual set of the solution of this system is given by $Q_{\text{region}} = \left\{\lim_{t\to T_f} \mid V(\mathbf{x}) \leq \min\{\varepsilon_0/\alpha, (\varepsilon_0/\beta)^{1/\rho}\}\right\}$ with $\varepsilon_0 = \varepsilon/(1-\theta_0)$ and $\theta_0 \in (0, 1)$. The finite settling time is estimated by:

$$T_f \leq \max \begin{bmatrix} t_0 + \dfrac{1}{\theta_0 \alpha(1-\rho)} \ln \dfrac{\theta_0 \alpha V^{1-\rho}(t_0) + \beta}{\beta} \\ t_0 + \dfrac{1}{\alpha(1-\rho)} \ln \dfrac{\alpha V^{1-\rho}(t_0) + \theta_0 \beta}{\theta_0 \beta} \end{bmatrix} \tag{4.34}$$

4.5.4 Stability Analysis

Considering the dynamic model of the helical millirobot, which includes lumped disturbances and input saturation, and adhering to Assumptions 1 and 2, the AFTSMC along with the supplementary system ensures that the tracking errors will converge to a neighborhood close to the origin within a finite time frame. The candidate for the composite Lyapunov function is defined as follows:

$$V = \frac{1}{2}S_r^2 + \frac{1}{2}\chi_r^2 + \frac{1}{2}\tilde{K}_2^2 \tag{4.35}$$

which is composed of $V_r = \frac{1}{2}S_r^2 + \frac{1}{2}\chi_r^2$ to verify the global stability and $V_k = \frac{1}{2}\tilde{K}_2^2$ to describe the internal convergence of the adaptive gain. $\tilde{K}_2 = \hat{K}_2 - \bar{K}_2$ is the parameter adaptation error. Taking the first-order derivative of the Lyapunov function V_r yields:

$$\dot{V}_r = S_r\dot{S}_r + \chi_r\dot{\chi}_r \tag{4.36}$$

In consideration of the nominal control law and the auxiliary system, substituting (4.28) and $u_a = u_{a_0} + \Delta u_a$ into (4.26), and setting $\Delta u = J(\eta)\Delta u_a$ will lead to:

$$\dot{S}_r = -K_1 S_r - \lambda_r \chi_r - \hat{K}_2 S_r^p + \Delta u \tag{4.37}$$

Based on Lemma 3, the inequalities $-\lambda_r\chi_r S_r \le \left(\frac{\lambda_r^2 S_r^2}{2}\right) + \left(\frac{\chi_r^2}{2}\right)$, $\rho_r\chi_r\Delta u \le \left(\frac{\rho_r^2\Delta u^2}{2}\right) + \left(\frac{\chi_r^2}{2}\right)$ and $S_r\Delta u \le \left(\frac{S_r^2}{2}\right) + \left(\frac{\Delta u^2}{2}\right)$ can be satisfied. In view of (1.27), (1.30), and (1.37), the derivation of the Lyapunov function V_r leads to:

$$
\begin{aligned}
\dot{V}_r \le & -S_r\left(K_1 S_r + \lambda_r\chi_r + \hat{K}_2 S_r^p\right) + S_r\Delta u \\
& -k_1\chi_r^2 - F(\chi_r)\chi_r\frac{|S_r\Delta u| + 0.5\rho_r^2\Delta u^2}{\chi_r} \\
& -k_2\chi_r \mathrm{sig}^\rho(\chi_r) + \rho_r\chi_r\Delta u \\
\le & -\left(K_1 - \frac{\lambda_r^2}{2}\right)S_r^2 - (k_1 - 1)\chi_r^2 - K_{\min}\left(S_r^2\right)^{\frac{(p+1)}{2}} \\
& -k_2\left(\chi_r^2\right)^{\frac{(p+1)}{2}} + \frac{\Delta u^2\rho_r^2(1 - F(\chi_r))}{2} \\
& -|S_r\Delta u|(F(\chi_r) - \mathrm{sign}(S_r\Delta u)).
\end{aligned} \tag{4.38}
$$

Then, we will prove that the trajectory of tracking errors eventually converge to the small compact set from the two cases.

Case 1: For $F(\chi_r) < 1$, (4.38) can be rearranged as follows:

$$
\begin{aligned}
\dot{V}_r \le & -\left(K_1 - \frac{\lambda_r^2}{2}\right)S_r^2 - (k_1 - 1)\chi_r^2 - K_{\min}\left(S_r^2\right)^{\frac{(p+1)}{2}} \\
& -k_2\left(\chi_r^2\right)^{\frac{(p+1)}{2}} + \frac{\Delta u^2\rho_r^2}{2} + S_r\Delta u \\
\le & -\kappa_a V_r - \kappa_b V_r^{\frac{(p+1)}{2}} + \varepsilon_r
\end{aligned} \tag{4.39}
$$

where $\kappa_a = \min\{2K_1 - \lambda_r^2 - 1, 2k_1 - 2\}$, $\kappa_b = \min\{2^{(p+1)/2}K_{\min}, 2^{(p+1)/2}k_2\}$, and $\varepsilon_r = \left[\Delta u^2(\rho_r^2 + 1)/2\right]$.

The tracking signals are bounded in the small neighborhoods around the origin $Q_{\text{region}} = \left\{\lim_{t\to T_f} |V(\mathbf{x})| \le \min\left\{\frac{\varepsilon_r}{(1-\theta_0)k_a}, \left(\frac{\varepsilon_r}{(1-\theta_0)k_b}\right)^{2/(p+1)}\right\}\right\}$ with $\theta_0 \in (0, 1)$ in finite time T_2.

$$
T_2 \le \max\begin{bmatrix} T_0 + \dfrac{2}{\theta_0 k_a(1-\rho)}\ln\dfrac{\theta_0 k_a V^{\frac{1-\rho}{2}}(t_0) + k_b}{k_b} \\ T_0 + \dfrac{2}{k_a(1-\rho)}\ln\dfrac{k_a V^{\frac{1-\rho}{2}}(t_0) + \theta_0 k_b}{\theta_0 k_b} \end{bmatrix} \tag{4.40}
$$

Since, $S_r^2/2 \leq V_r$, it follows that the integral sliding mode surface S_r can reach the bounded region in finite time:

$$|S_r| \leq \sqrt{2} * \min\left\{\frac{\varepsilon_r}{(1-\theta_0)k_a}, \left(\frac{\varepsilon_r}{(1-\theta_0)k_b}\right)^{2/(p+1)}\right\} \tag{4.41}$$

Therefore, it means that S_r is bounded in the small set around zero, which subsequently implies that χ_r is uniformly bounded similarly.

Case 2: For $F(\chi_r) = 1$, it implies that $\varepsilon_r = 0$, (4.38) can be rearranged as follows: $\dot{V}_r \leq -(K_1 - \lambda_r^2/2)\,S_r^2 - (k_1 - 1)\chi_r^2 - K_{\min}\left(S_r^2\right)^{(p+1)/2} - k_2\left(\chi_r^2\right)^{(\rho+1)/2}$. Therefore, $\dot{V}_r + \kappa_a V_r + \kappa_b V_r^{(\rho+1)/2} \leq 0$ with $\kappa_a = \min\{2K_1 - \lambda_r^2, 2k_1 - 2\}$, $\kappa_b = \min\left\{2^{(p+1)/2}K_{\min}, 2^{(\rho+1)/2}k_2\right\}$. Based on Lemma 4, the settling time can be estimated by:

$$T_2 \leq T_0 + \frac{2}{k_a(1-\rho)} \ln \frac{k_a V^{\frac{1-\rho}{2}}(t_0) + k_b}{k_b} \tag{4.42}$$

During the reaching phase from the initial state to the sliding state, it is reasonable to assume that there exists an upper bound $\bar{K}_2$ of the adaptive gain $\hat{K}_2$ until the sliding manifold can be reached ($|S_r| < u$). To analyze the internal convergence of the adaptive gain, deriving the function V_k

$$\dot{V}_k = \tilde{K}_2\dot{\hat{K}}_2 = \left(\hat{K}_2 - \bar{K}\right) k_\alpha \mathrm{sign}(S_r - u) \tag{4.43}$$

During the reaching phase, $|S_r| > u$ and $\hat{K}_2 < \bar{K}_2$ are established. So, the condition $\dot{V}_k < 0$ can be satisfied, indicating the existence of a finite time during which the adaptive gain converges. Furthermore, during the sliding phase, $\dot{V}_k = (\hat{K}_2 - \bar{K})K_{\min} < 0$ can also be guaranteed.

To further demonstrate that the tracking errors would drive into small compact sets, we need to consider the Lyapunov function as:

$$V_e = \frac{1}{2}e_\eta^2 \tag{4.44}$$

As the derivative of e_η can be expressed as: $\dot{e}_\eta = \dot{S}_r - \alpha_e e_\eta - \beta_e e_\eta^{m/n}\mathrm{sign}(e_\eta)$, the derivative of V_e will lead to:

$$\dot{V}_e = e_\eta\left(\dot{S}_r - \alpha_e e_\eta - \beta_e e_\eta^{m/n}\mathrm{sign}(e_\eta)\right) \tag{4.45}$$

Since, S_r, χ_r, and K_2 will converge to the small region around origin, (1.45) can be derived as $\dot{V}_e \leq -2\alpha_e V_e - 2^{(m+n)/2n}\beta_e V_e^{(m+n/2n)}$. According to Lemma 4, the tracking error will converge to a bounded region in finite time.

4.6 Experimental Validation and Results

4.6.1 Experimental Setup and Procedure

This section demonstrates the performance of the proposed control framework through the physical experiment containing the comparative analysis.

The 3D printed millirobot prototype, measuring 1.8 mm in diameter and 10 mm in length, features an NdFeB magnet (1 mm diameter, 2 mm length) at its head, oriented perpendicular to the central axis. An EMA system is used to control the closed-loop motion of the millirobot, as illustrated in Figure 4.10a.

The system includes three custom electromagnetic coils on three independently controlled manipulators (SJ-603-A and Anno Robot), capable of tracking the millirobot's position over a broad area. These coils can generate rotating fields in any direction within the workspace by combining three independent magnetic vectors. Technical specifications are detailed in Table 4.1. For real-time tracking, two cameras (MER2-160-227U3C and DaHeng) are integrated with a YOLO deep learning architecture. The software architecture, built on ROS, manages communication between the electromagnetic system and the control framework.

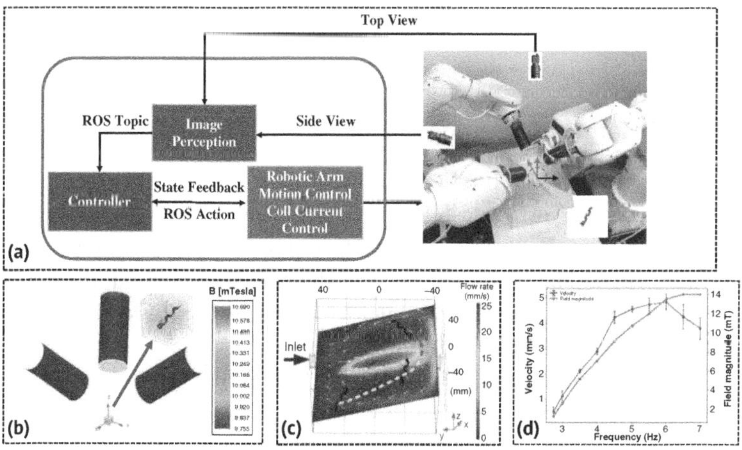

Figure 4.10 Overview framework of the experimental setup. (a) EMA and its closed-loop control framework. The magnetic actuation system can adjust the position of the electromagnetic coils to keep pace with the helical millirobot. (b) Magnetic field distribution generated by the EMA in a cubic space with a side length of 10 mm, centered at the helical millirobot's position. (c) Flow rate distribution around the given reference path. The lower and upper dashed paths represent the upstream and downstream navigation, respectively. (d) The curve labeled "Velocity" illustrates the frequency dependency of the velocity, showing the evolution of the locomotion performance of the helical millirobot, while the curve labeled "Field magnitude" provides the recommended field magnitude required to achieve the corresponding frequency.

Due to the electromagnetic system's flexible mobility, we only need to consider the magnetic field within a region of interest, specifically a 10 mm cubic space centered at the millirobot's location. To further explore the local uniformity of the generated magnetic field, we use ANSYS Electromagnets to model the field characteristics of the target area. When the desired magnetic field value is [0, 0, 10 mT], the spatial distribution in the cubic space is depicted in Figure 4.10b, indicating the uniformity error is less than 10%. Considering the small scale and mobility of the helical millirobot, along with the coils' flexibility, it is reasonable to assume a uniform magnetic field model around the millirobot.

The helical millirobot's 3D navigation is performed in an open tank measuring $80 \times 80 \times 40$ mm^3, filled with 85 wt% glycerol to simulate a low Reynolds number environment. A tunable peristaltic pump is connected to generate fluid flow, emphasizing the complexity of the flow rate within the tank. Figure 4.10c shows the instantaneous flow rate distribution on a reference plane, calculated by COMSOL multiphysics at an inlet flow rate of 50 mm/s. As the millirobot tracks the reference path, it undergoes downstream and upstream navigation in two stages. Notably, the millirobot's locomotion performance deteriorates rapidly when exceeding the step-out characteristic frequency of 6 Hz, as depicted in Figure 4.10d. Thus, the proposed control framework incorporates the saturation constraint of the rotating field frequency to maintain tracking performance in fluid flow conditions.

4.6.2 Scenario I: Targeted Hovering

In the context of a dynamic biofluid environment, where the goal is to achieve the targeted drug delivery at the desired lesion site using a magnetic robot, our proposed method enables the millirobot to perform precise drug release through targeted hovering. To validate the robustness of the proposed method, three sets of flow rates (0, 75, and 200 mm/s) at the inlet were applied to assess the targeted hovering performance of the helical millirobot.

In Figure 4.11a, the initial position of the helical millirobot (10 mm, 0, 0) was 10 mm from the target hovering position (0, 0, 0). The robot entered the hovering state P_{hover} when its position was within ± 1 mm of the target. At $t = 20$ s, it reached the target and maintained this state until $t = 45$ s. Introducing a 75 mm/s flow rate at $t = 45$ s caused a 4.5 mm displacement (Figure 4.11b). Using flow disturbance estimation and control, the robot returned to P_{hover} at $t = 85$ s. Increasing the flow rate to 200 mm/s at $t = 120$ s resulted in a 26 mm displacement. The method regulated the rotation frequency and heading, enabling the robot to resist disturbances and return to P_{hover} at $t = 225$ s. These results confirm the method's effectiveness in controlling the robot's hovering under variable flow rates.

4.6.3 Scenario II: Comparisons of 3D Path Tracking Performance Under Different Control Methods

To evaluate the proposed method's ability to track a reference path under fluid flow conditions, we introduced a preset 3D path shown in Figure 4.12a. The fluid flow at the inlet was set to 50 mm/s. Experiments. Experiments compared this method with IBVS-PID and conventional SMC methods. Evaluations used three indicators: (1) Euclidean norm of tracking errors (deviation from the reference path); (2) standard deviation (SD) of tracking error; and (3) tracking completion time. Parameters for all methods are listed in Table 4.5.

The tracking results for three controllers are shown in Figure 4.12. From Figure 4.12a, all methods successfully controlled the helical millirobot to track the reference path, but with varying performance.

The proposed method enabled smoother tracking against fluid flow compared to SMC and IBVS-PID, which exhibited fluctuations and greater path deviation. Table 4.6 shows the proposed method's superior performance (0.367 ± 0.204 mm) over SMC (0.680 ± 0.344 mm) and IBVS-PID (0.937 ± 0.502 mm). This indicates robustness against disturbances, surpassing SMC and IBVS-PID. The proposed method also demonstrated faster convergence and tracking speed, approximately 76.2 seconds, compared to 82.5 seconds for SMC and 96.9 seconds for IBVS-PID.

The proposed controller's anti-saturation capability, as shown in Figure 4.12c, maintains rotation frequency below the 6 Hz limit. In contrast, SMC exceeds this

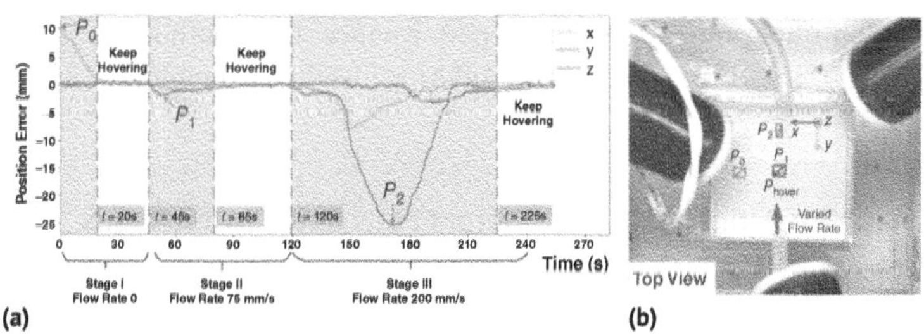

Figure 4.11 Results of targeted hovering of the helical millirobot under varied fluid flow. (a) Evolution of position errors of the helical millirobot. (b) Top-view key snapshots of the experiment process.

Table 4.5 Parameters of three different controllers.

Controllers	Parameters
Proposed	$K_1 = 0.7, \lambda_{\beta_1} = 0.8, \lambda_{\beta_2} = 0.8, \lambda_r = 0.3$ $\alpha_e = 0.1, \beta_e = 0.2, k_1 = 1.5, k_2 = 0.6, \rho_r = 0.3$
SMC	$M_1 = 0.7, M_2 = 0.7$
IBVS-PID	$K_p = 0.05, K_d = 0.03, K_i = 0.02$

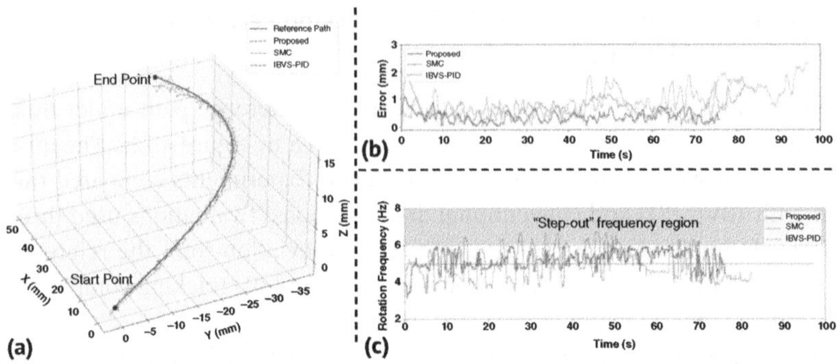

Figure 4.12 Comparison results of path following under the proposed control, SMC, and IBVS-PID method. (a) Traces of the helical millirobot's trajectory. (b) Evolution of path following errors. (c) Comparison of the rotating frequency of the magnetic field generated by three different control methods.

Table 4.6 Comparison of evaluation indicators between controllers.

Controllers	Euclidean norm (mm)	Standard deviation (mm)	Completion time (s)
Proposed	0.367	0.204	76.2
SMC	0.680	0.344	82.0
IBVS-PID	0.937	0.502	96.9

frequency, which undermined the control performance. Our method surpasses SMC and IBVS-PID in performance metrics, offering insights for biomedical applications.

4.6.4 Scenario III: Path Tracking in Different Flow Rate Environments

We explore a scenario where fluid flow and boundary effects impact the tracking process. Figure 4.13 captures the top and side views of the millirobot during this process. Initially, the robot lies on the substrate. Motion initiates boundary forces. During tracking, it faces varying flow rates and fluid effects. From T_0 to T_1, the fluid imposed a drag on the robot's movement; from T_2 to T_3, the fluid flow acted as a push for the robot's movement. This simulates upstream and downstream tracking. The robot detaches, resists disturbances, and converges to the path, as shown in Figure 4.13.

Figure 4.14 shows tracking results with flow rates of 75, 100, and 125 mm/s. Despite disturbances, the proposed control successfully navigates the helical millirobot to the reference path with minor errors. The control framework regulates rotation frequency and orientation in real time, resisting fluid flow, ensuring

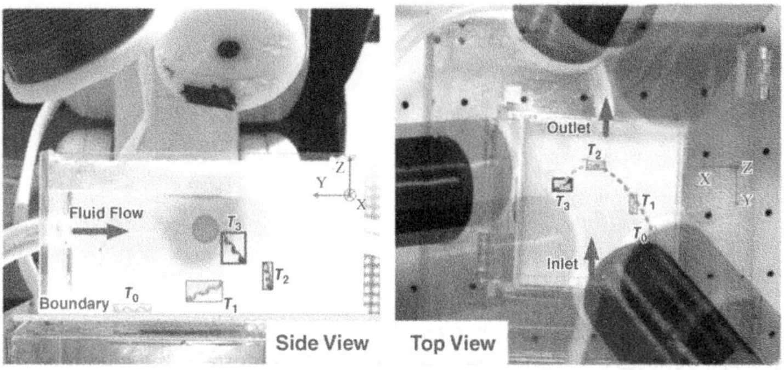

Figure 4.13 Top-view and side-view overlaid snapshots of the helical millirobot during the reference path tracking. Considering both the boundary effect and the influence of fluid flow.

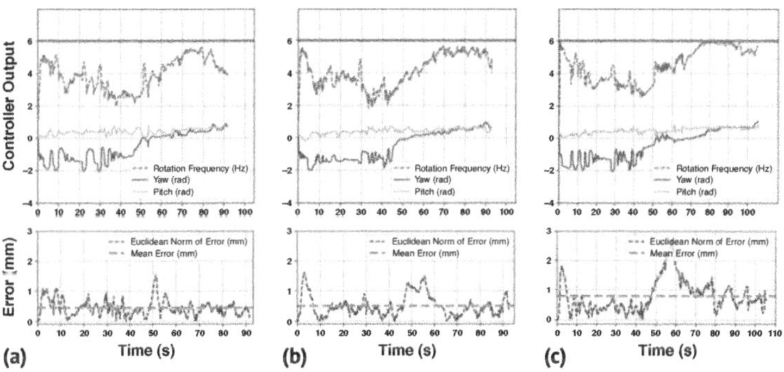

Figure 4.14 Results of the 3D path following under three different flow rates. (a) 75 mm/s. (b) 100 mm/s. (c) 125 mm/s. The first and second row represent the evolution of controller outputs and following errors under three flow rates, respectively.

tracking. Switching flow rates to 75, 100, and 125 mm/s, completion times were 91.8, 92.8, and 105.4 s, with mean errors at 0.445, 0.503, and 0.789 mm. Rotation frequency remained under 6 Hz. These experiments validate the controller's robustness and general applicability.

4.6.5 Scenario IV: Actuation of Helical Millirobot in 3D Tubular Environments

The Scenario IV, we explored the efficacy of our proposed method for targeted drug delivery using a helical millirobot in a simulated biofluid environment with fluid flow. The millirobot was tested in a silicon tube with a diameter of 4 mm, filled with 80 wt% glycerol to mimic the conditions of drug delivery in a dynamic fluid setting.

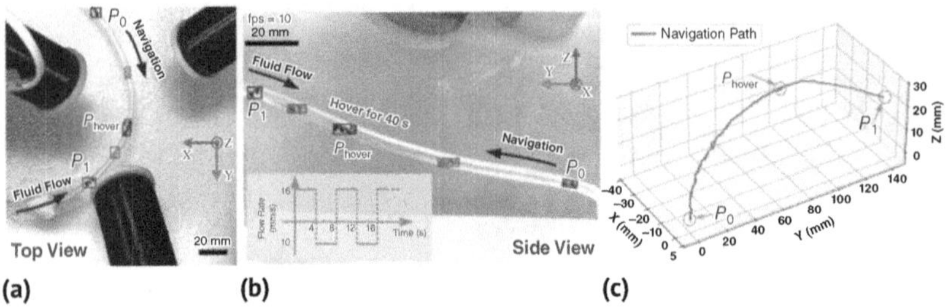

Figure 4.15 Top-view (a) and side-view (b) overlaid snapshots of the goal-reaching and hovering process. P_0 denotes the starting position and P_{hover} represents the targeted hovering position. Besides, P_1 shows the navigation endpoint that can be reached within the camera's field of view. (c) Trace of the helical millirobot's trajectory.

The glycerol flowed in a pulsatile pattern, featuring an 8-second cycle composed of 4 seconds of upstream flow at 16 mm/s followed by 4 seconds of downstream flow at 10 mm/s.

The experimental protocol, as illustrated in Figure 4.15a,b, consisted of three distinct phases. Initially, the helical millirobot traversed from the starting point P_0 to a hovering position P_{hover}. Subsequently, it simulated the drug release process while maintaining its position at P_{hover}. After completing the drug release over 40 s of hovering, the millirobot resumed forward motion and navigated the remaining distance to reach the endpoint P_1. Figure 4.15c provides a visualization of the entire navigation trajectory.

The experimental findings confirm the efficacy of our method in achieving precise hovering and goal-reaching capabilities in environments with pulsatile flow. This investigation offers meaningful insights into the potential of our approach for targeted drug delivery in the presence of biofluid flow influences.

4.7 Discussion and Conclusion

This study presents a robust 3D path following control framework for magnetic helical millirobots in complex fluids and a REMA for large workspace applications. The control framework integrates AFTSMC with a FTDOB to achieve precise trajectory tracking, addressing challenges such as fluid dynamics, input saturation, and disturbances. The REMA system uses three mobile electromagnetic coils mounted on robotic arms, optimized via MOO to generate flexible magnetic fields and gradients.

Experiments validated the control framework's ability to enable precise hovering and 3D path tracking of millirobots under varying flow rates and boundary effects. The REMA system demonstrated isotropic field generation and combined field-force control in scenarios involving helical swimmers, catheter navigation, and capsule propulsion, highlighting its potential for medical applications.

However, limitations remain. Future work for the control framework will focus on enhancing in vivo control accuracy through advanced imaging integration and addressing millirobot-tissue interactions. For REMA, improvements will target workspace coverage and coil configuration smoothness, with efforts to develop advanced control strategies for dynamic environments and explore clinical applications guided by fluoroscopy. Overall, these advancements mark a significant milestone in developing the autonomous magnetic microrobots for minimally invasive procedures. Future research will aim to refine these systems, overcoming limitations and expanding their biomedical applications.

Appendix Additional Algorithms

Algorithm 1: Pareto-optimized Coil Configuration Generation Using MOO Method

---

Require: Target position $\mathbf{p}_w$, desired field $\mathbf{b}_d(\mathbf{p}_w)$, desired force $\mathbf{f}_d(\mathbf{p}_w)$. Control mode: rotating field control or combined field and force control.
Ensure: The Pareto-optimized coil configuration $\mathbf{x}^*$
for optimization loop $k = 1$ to K **do**
    Sample a set of parameters $\mathbf{x}$ from $\mathcal{X}$ according to (4.5);
    Compute each coil's pose with respect to the $\mathbf{p}_w$;
    Obtain the field generated by each coil with unit current at $\mathbf{p}_w$ and construct the actuation matrix $\mathbf{A}_b = [\mathbf{b}_1(\mathbf{p}_w), \mathbf{b}_2(\mathbf{p}_w), \mathbf{b}_3(\mathbf{p}_w)]$;
    Switch control mode do
     case rotating field control do
      $f_1 \leftarrow \min(\mathbf{B}_{\text{iso}})$, $f_2 \leftarrow \min(\mathbf{B}_{\text{cosine}})$,
      $f_3 \leftarrow \sigma_3$, $f_4 \leftarrow 1/\kappa$;
     case combined field and force control do
      $f_1 \leftarrow \mathbf{e}_f$, $f_2 \leftarrow m_1$,
      $f_3 \leftarrow m_2$, $f_4 \leftarrow m_3$;
    End‘
    MOO procedure loop via the NSGA-II algorithm to optimize f_1, f_2, f_3, and f_4;
end for
Obtain the Pareto-fronts;
Obtain the Pareto-optimized coil configuration from Pareto-fronts.

References

1 Y. Wu, X. Dong, J.-K. Kim, C. Wang, and M. Sitti. Wireless soft millirobots for climbing three-dimensional surfaces in confined spaces. *Science Advances*, 8 (21): eabn3431, 2022.

2 L. Song, et al. Motion control of capsule robot based on adaptive magnetic levitation using electromagnetic coil. *IEEE Transactions on Automation Science and Engineering*, 20 (4): 2720–2731, 2023.

3 S. L. Charreyron, Q. Boehler, B. Kim, C. Weibel, C. Chautems, and B. J. Nelson. Modeling electromagnetic navigation systems. *IEEE Transactions on Robotics*, 37 (4): 1009–1021, 2021.

4 F. Ongaro, S. Pane, S. Scheggi, and S. Misra. Design of an electromagnetic setup for independent three-dimensional control of pairs of identical and nonidentical microrobots. *IEEE Transactions on Robotics*, 35 (1): 174–183, 2019.

5 J. J. Abbott, E. Diller, and A. J. Petruska. Magnetic methods in robotics. *Annual Review of Control, Robotics, and Autonomous Systems*, 3 (1): 57–90, 2020.

6 G. Pittiglio, M. Brockdorff, T. da Veiga, J. Davy, J. H. Chandler, and P. Valdastri. Collaborative magnetic manipulation via two robotically actuated permanent magnets. *IEEE Transactions on Robotics*, 39 (2): 1407–1418, 2023.

7 Z. Yang and L. Zhang. Magnetic actuation systems for miniature robots: a review. *Advanced Intelligent Systems*, 2 (9): 2000082, 2020.

8 F. Niu, J. Li, W. Ma, J. Yang, and D. Sun. Development of an enhanced electromagnetic actuation system with enlarged workspace. *IEEE/ASME Transactions on Mechatronics*, 22 (5): 2265–2276, 2017.

9 A. Oulmas, N. Andreff, and S. Régnier. 3D closed-loop swimming at low Reynolds numbers. *International Journal of Robotics Research*, 37 (11): 1359–1375, 2018.

10 S. Salmanipour, O. Youssefi, and E. D. Diller. Design of multi-degrees-of-freedom microrobots driven by homogeneous quasi-static magnetic fields. *IEEE Transactions on Robotics*, 37 (1): 246–256, 2021.

11 R. Chen, D. Folio, and A. Ferreira. Mathematical approach for the design configuration of magnetic system with multiple electromagnets. *Robotics and Autonomous Systems*, 135: 103674, 2021.

12 Y. Xu, K. Li, Z. Zhao, and M. Q.-H. Meng. Evaluation of different control strategies for trajectory following of a robotic capsule endoscope under rotating magnetic actuation. *IEEE Transactions on Automation Science and Engineering*, 20 (3): 2170–2181, 2023.

13 J. Sikorski, C. M. Heunis, F. Franco, and S. Misra. The ARMM system: an optimized mobile electromagnetic coil for non-linear actuation of flexible surgical instruments. *IEEE Transactions on Magnetics*, 55 (9): 1–9, 2019.

14 M. C. Hoang, et al. DEMA: Robotic dual-electromagnet actuation system integrated with localization for a magnetic capsule endoscope. *Sensors and Actuators A: Physical*, 361: 114596, 2023.

15 L. Çetin, A. Alasli, N. Akçura, A. Kahveci, F. C. Can, and Ö. Tamer. Dynamical electromagnetic actuation system for microscale manipulation. *Robotica*, 40 (10): 3586–3603, 2022.

16 D. Li, B. Chen, C. Huang, S. Xu, X. Wu, and T. Xu. TrinityMag: a novel magnetically actuated miniature robot manipulating system with a human-scale workspace. *Science China Technological Sciences*, 67: 3546–3558, 2024.

17 Y. Ko, H.-S. Lee, and C.-S. Kim. Model-based real-time simulator for robotic electromagnetic actuation. *IEEE Robotics and Automation Letters*, 9 (4): 3870–3877, 2024.

18 L. Yang, X. Du, E. Yu, D. Jin, and L. Zhang. DeltaMag: an electromagnetic manipulation system with parallel mobile coils. In *Proceedings of the IEEE International Conference on Robotics and Automation (ICRA)*, pages 9814–9820, 2019. IEEE.

19 X. Du, M. Zhang, J. Yu, L. Yang, P. W. Y. Chiu, and L. Zhang. Design and real-time optimization for a magnetic actuation system with enhanced flexibility. *IEEE/ASME Transactions on Mechatronics*, 26 (3): 1524–1535, 2021.

20 S. Sarabandi, J. M. Porta, and F. Thomas. Hand-eye calibration made easy through a closed-form two-stage method. *IEEE Robotics and Automation Letters*, 7 (2): 3679–3686, 2022.

21 X. Du, L. Yang, J. Yu, K. F. Chan, P. W. Y. Chiu, and L. Zhang. RoboMag: a magnetic actuation system based on mobile electromagnetic coils with tunable working space. In *Proceedings of the 5th International Conference on Advanced Robotics and Mechatronics (ICARM)*, pages 125–131, 2020. IEEE.

22 R. Yu, et al. Modeling electromagnetic navigation systems for medical applications using random forests and artificial neural networks. In *Proceedings of the IEEE International Conference on Robotics and Automation (ICRA)*, pages 9251–9256, 2020. IEEE.

23 Q. Boehler, S. Gervasoni, S. L. Charreyron, C. Chautems, and B. J. Nelson. On the workspace of electromagnetic navigation systems. *IEEE Transactions on Robotics*, 39 (1): 791–807, 2023.

24 A. J. Petruska and B. J. Nelson. Minimum bounds on the number of electromagnets required for remote magnetic manipulation. *IEEE Transactions on Robotics*, 31 (3): 714–722, 2015.

25 Z. Yang, H. Yang, Y. Cao, Y. Cui, and L. Zhang. Magnetically actuated continuum medical robots: a review. *Advanced Intelligent Systems*, 5 (6): 2200416, 2023.

26 N. Gunantara. A review of multi-objective optimization: methods and its applications. *Cogent Engineering*, 5 (1): 1502242, 2018.

27 K. Shang, H. Ishibuchi, L. He, and L. M. Pang. A survey on the hypervolume indicator in evolutionary multiobjective optimization. *IEEE Transactions on Evolutionary Computation*, 25 (1): 1–20, 2021.

28 J. O. A. L. Filho, E. L. F. Fortaleza, and M. C. M. M. de Campos. A derivative-free nonlinear Kalman filtering approach using flat inputs. *International Journal of Control*, 95 (11): 2900–2910, 2022.

29 G. Peng, C. Yang, W. He, and C. L. P. Chen. Force sensorless admittance control with neural learning for robots with actuator saturation. *IEEE Transactions on Industrial Electronics*, 67 (4): 3138–3148, 2020.

30 Y. Yu, C. Guo, and T. Li. Finite-time LOS path following of unmanned surface vessels with time-varying sideslip angles and input saturation. *IEEE/ASME Transactions on Mechatronics*, 27 (1): 463–474, 2022.

31 Z. Zheng and M. Feroskhan. Path following of a surface vessel with prescribed performance in the presence of input saturation and external disturbances. *IEEE/ASME Transactions on Mechatronics*, 22 (6): 2564–2575, 2017.

32 H. Castañeda, J. Rodriguez, and J. L. Gordillo. Continuous and smooth differentiator based on adaptive sliding mode control for a quad-rotor MAV. *Asian Journal of Control*, 23 (2): 661–672, 2021.

33 J. Yu, P. Shi, and L. Zhao. Finite-time command filtered backstepping control for a class of nonlinear systems. *Automatica*, 92: 173–180, 2018.

5

Performance-guided Optimization and Field-priority Force Control for Large Workspaces

This chapter proposes the performance-guided optimization method and magnetic field-priority force control method to further optimize the control algorithm for precise actuation in large workspaces based on the hardware conditions of the previous chapter, ensuring the isotropy of the rotating fields and efficient generation of the combined field and force, respectively.

5.1 Introduction

The rapid advancements in remotely controlled magnetic robots, which function within the human body, have significantly propelled the progress of magnetic actuation systems. This is attributed to their capacity to generate intricate magnetic fields and forces. Nevertheless, there remains a necessity to enhance the efficient production of these combined fields and forces, particularly in large workspaces. When a magnetic robot is propelled autonomously through the concurrent application of magnetic fields and forces, stringent demands are placed on the magnetic actuation system's ability to generate agile and precise field and force configurations. This is particularly evident when the robot is engaged in autonomous manipulations within large workspaces [1–6]. For achieving singularity-free control of both magnetic field and force at a point, the *OctoMag* system demonstrates the necessity of eight stationary electromagnetic coils, with control achieved solely through modulation of electric currents. Additionally, reconfigurable magnetic manipulation systems, as discussed in [7], can attain comparable control capabilities to static systems with reduced input requirements. Furthermore, research in [8] affirms that the performance of electromagnetic actuation (EMA) systems is influenced not only by the quantity of coils but also by their specific arrangements. Therefore, a more thorough investigation is warranted into the necessary number of mobile magnetic sources required to generate the desired field and force within large workspaces. Both the *DeltaMag* [9] and *RoboMag* [10] systems leverage three mobile coils to enable a relatively large workspace, wherein the motion of these coils is driven by a parallel mechanism and robotic arms, respectively. The intricacy of magnetic force control surpasses that of field control, largely attributed to the intricate mapping of the resultant three-component force onto a five-component gradient, as documented in [11] and [12]. While the *DeltaMag* and *RoboMag* primarily utilize three mobile coils for achieving desired field control, they have yet to demonstrate proficiency in the realm of combined field and force control. In fact, the design of a mobile multicoil system that meets clinical workspace requirements while ensuring sufficient flexibility in coil configurations and minimizing the coil count poses a significant challenge. Consequently, there is a compelling impetus for the development of a mobile EMA system with expansive workspaces, capable of adopting flexible coil configurations while optimizing the coil quantity. The problem of extending the design of the three-coil system to enable concurrent field and force control remains unresolved. By advancing the concept of a three-coil system that generates arbitrary 3D fields, we can delve into force control by exploring various coil configurations to identify suitable gradient distributions. This objective can be accomplished through the development of a novel EMA system that employs three coils mounted on three six-degree-of-freedom (6-DOF) robotic arms, allowing for decoupled movement in 3D space.

Besides, to precisely actuate magnetic robots in large workspaces, the efficient generation of isotropic rotating fields using EMA systems presents an enduring challenge. Recent scholarly endeavors have concentrated on the advancement of mobile magnetic actuation systems capable of operating within relatively expansive workspaces. Given the electromagnetic coil's (hereinafter referred to as the coil) capacity to swiftly vary fields through electric current control, EMA systems have been contemplated for generating rotating fields [13, 14]. Petruska and colleagues devised the *Omnimagnet*, an omnidirectional electromagnetic system comprising three orthogonally nested coils and a ferromagnetic core, to propel a mockup of a threaded capsule endoscope through a transparent lumen. This *Omnimagnet* could be affixed to a counterbalanced positioning arm for pose adjustments [15] or mounted onto a robotic arm to facilitate flexible magnetic manipulations across large workspaces. To actuate helical swimmers in three-dimensional (3D) large spaces by generating rotating fields, an EMA system employing three mobile coils in a coupled configuration was implemented using a parallel mechanism [9]. To enhance coil positioning flexibility, Du and team introduced the *RoboMag* system, which utilizes three mobile coils with decoupled movements to generate magnetic torque for actuating magnetic robots [10]. In the realm of rotating field generation using permanent magnets, Mahoney et al. presented a single rotating permanent magnet mounted on a 6-DOF robotic arm, enabling free movement of the actuator magnet while manipulating magnetic robots. However, when navigating magnetic robots through tortuous paths, such as the small intestines, the physical constraints of the robotic manipulator (including joint limits and singularities) impede the actuation of the magnetic robot and restrict the workspace. Furthermore, when a permanent magnet is used to create a rotating field, the magnitude and rotation rate of the instantaneous field at the robot's location vary elliptically, which also induces a magnetic gradient force on the magnetic robot [3]. To alleviate the undesirable oscillations in the helical swimmer's motion inside tubes caused by magnetic field gradients using a single rotating dipole field, a strategy employing two synchronized rotating dipole fields was developed to produce an approximately uniform field at the swimmer's position, thereby enhancing helical propulsion efficiency [16]. The aforementioned analysis of existing robotically actuated magnetic actuation systems primarily focused on field generation within large workspaces. However, the isotropy of the generated rotating fields has not been adequately addressed and may be constrained by potential collisions in practical applications. Moreover, existing mobile EMA systems equipped with three coils primarily concentrate on generating rotating fields at a single spatial position, with insufficient consideration given to field distributions within a targeted region. Additionally, this approach may result in gradients along its forward or lateral directions, potentially impeding motion. Furthermore, the rationale behind the selection of three coils has yet to be clearly elucidated.

In this chapter, we delineate two primary motivations behind selecting the three-coil configuration. Firstly, it facilitates the generation of a 3D field in any specified direction by enabling precise regulation of the electric currents flowing through each coil. Secondly, the mobile three-coil setup offers a diverse amalgamation of magnetic fields and gradients, alongside affording a substantial workspace. Following this, we introduce a field priority force control approach which enables the creation of a flexible coil arrangement to manipulate both the desired field and force. Additionally, this methodology incorporates Yoshikawa manipulability for each robotic arm, a crucial aspect for manipulating magnetic robots within expansive workspaces. Ultimately, we implemented a reconfigurable electromagnetic actuation (REMA) system, comprising three mobile coils mounted on three autonomous 6-DOF robotic arms, to substantiate the efficacy of the method. The EMA systems offer the capability to shut down promptly when necessary and produce rapid-response fields, making them suitable for applications requiring precise and timely control. In pursuit of generating the desired rotating magnetic fields for actuating magnetic robots, we have opted to employ mobile coils. To ensure safety and effectiveness, particularly as the robots navigate various depths within the human body, it is crucial to address two key challenges: collision avoidance between coils and the human body, and the generation of isotropic fields. To tackle these challenges, it is necessary to consider diverse coil configurations. This is essential not only to enhance the isotropy of the generated fields but also to mitigate the risk of collisions, both among the coils themselves and between the coils and the patient. Careful selection of coil configurations is vital for achieving the desired field characteristics while maintaining safety. In this chapter, we have established criteria for evaluating the isotropy of rotating fields and have developed optimized coil configurations that can produce such fields under practical operational constraints. This approach ensures that the EMA system not only functions effectively but also operates safely within the human body.

5.2 Performance-guided Rotating Field Control Method

Varying coil arrangements can yield distinct local field patterns, necessitating a discussion on techniques to ascertain the most effective coil configuration within a predefined search domain for achieving isotropic rotating field manipulation. To delve into this necessity, we present a schematic illustration of a REMA system, which employs mobile coils mounted on three independent robotic arms to facilitate diverse coil setups, as depicted in Figure 4.4. Furthermore, we introduce an optimization approach tailored for generating rotating fields exhibiting isotropic properties.

1. **Problems:** To assess the efficacy of various stationary systems, several evaluation metrics have been established, encompassing manipulability indices and the minimal singular values [4, 17]. However, no existing research has examined the field distributions of mobile EMA systems using corresponding metrics. An intriguing aspect lies in devising appropriate evaluation metrics to discern optimal coil configurations for mobile systems. Building upon four evaluation metrics, to achieve the objective of computing isotropy measurements, we uniformly sample 27 desired fields, denoted as $\mathbf{b}_d(\mathbf{p}_w)$, at a position $\mathbf{p}_w$, each with distinct directions within the 3D space. For computational simplicity while maintaining generality, we standardize the desired field as a unit field, denoted as $\hat{\mathbf{b}}(\mathbf{p}_w)$. It is imperative that all fields generated within the target region possess an amplitude and direction consistent with $\hat{\mathbf{b}}(\mathbf{p}_w)$. Guided by these principles, we have formulated two evaluation metrics to quantify the isotropy of the field distribution, which are specified as follows:

$$||\mathbf{b}||_{\text{iso}} = (1 - \frac{\sum_{i=0}^{i=N-1} ||||\mathbf{b}(\mathbf{p}_i)|| - ||\mathbf{b}_d(\mathbf{p}_w)||||}{N\,||\mathbf{b}_d(\mathbf{p}_w)||}) \times 100\% \tag{5.1}$$

$$\begin{cases} (\mathbf{b}_{\text{cosine}})_{\min} = \min(\dfrac{\mathbf{b}(\mathbf{p}_i)\cdot\mathbf{b}_d(\mathbf{p}_w)}{||\mathbf{b}(\mathbf{p}_i)||\,||\mathbf{b}_d(\mathbf{p}_w)||}), i = 0, \dots, N-1 \\ \measuredangle_{\text{err}}(\mathbf{b}, \mathbf{b}_d) = \arccos((\mathbf{b}\text{cosine})_{\min}) \end{cases} \tag{5.2}$$

The ability to perform a certain motion at any location and direction without singularities is of prime importance. This analysis can be performed by examining the field actuation matrix $A_b(\mathbf{p}_w)$. It provides information similar to a classic Jacobian matrix in robotics that can be utilized to characterize the control isotropy. For the singular values $\sigma_i (i = 1, 2, 3)$ of the matrix $A_b(\mathbf{p}_w)$, a larger smallest singular value σ_3 can reduce the maximal coil current required in the worst-case control scenarios near singularities. This is consistent with the temperature produced by the coil with a higher current value. Additionally, maximizing the inverse of the condition number k to be close to 1 can improve the control capabilities in different directions, which is defined as follows.

$$\frac{1}{k} = \frac{\sigma_{\min}}{\sigma_{\max}} = \frac{\sigma_3}{\sigma_1} \in [0, 1] \tag{5.3}$$

The fundamental approach involves assigning a score to each metric, serving as a benchmark for evaluating a specific coil configuration. It is important to note that each coil offers DOFs for control: 5-DOFs pertaining to its pose (excluding rotation along the principal axis) and 1-DOF related to electric current. The interplay between these multiple evaluation metrics and the DOFs of the three coils is intricate. The application of multi-objective optimization (MOO) techniques [18] could offer a viable strategy for navigating this complex landscape of multi-objective functions.

2. **Definition of Objective Functions:** To assess the system's performance in producing rotating fields across the workspace, it is crucial to establish comprehensive evaluation metrics for the optimization process. Notably, these metrics can vary in dimensions and units, which may influence the outcomes of the optimization. Normalization can expedite the quest for the optimal solution by scaling the values of the objective functions within a standardized range of [0, 1]. Based on the evaluation metrics described before, the multiple objective functions are defined as: $f_1 = \|\mathbf{b}_{iso}\|$, $f_2 = (\mathbf{b}_{cosine})_{min}$, $f_3 = \sigma_3$, $f_4 = 1/k$.
3. **Definition of Coil Configuration Space:** For the implementation of multi-objective optimization (MOO) aimed at generating a Pareto-optimized coil configuration, it is essential to establish a search space, specifically tailored to the coil configuration, incorporating design constraints on its orientation and position. To facilitate this, we define two coordinate frames: the world frame O_{w-xyz} and the coil frame O_{c-xyz}, as illustrated in Table 5.1. Primarily, each coil must maintain a safe distance from the patient's body to prevent any collisions. Furthermore, the overall arrangement of the three robotic arms must accommodate ample space for both the operating room bed and the imaging system. Consequently, we adopted a configuration where two robotic arms

Table 5.1 Position and orientation of each robotic arm's base relative to world frame.

	X(m)	Y(m)	Z(m)	Angle around Z(°)
Robotic arm♯1	−0.470	−0.310	−0.115	45
Robotic arm♯2	0.580	0	−0.115	180
Robotic arm♯3	−0.470	0.310	−0.115	−45

Table 5.2 Constraints of optimization parameters, (i = 1, 2, 3).

	L^i(m)	θ^i(°)	ϕ^1(°)	ϕ^2(°)	ϕ^3(°)	α^i(°)	β^i(°)
χ_0	0.15	45	225	0	130	0	0
$\chi_{\min}$	0.11	30	195	−45	90	−18	−18
$\chi_{\max}$	0.25	60	270	45	165	18	18

are positioned on one side of the bed, while the third robotic arm is situated on the opposite side, as outlined in Table 5.1.

As illustrated in Table 5.2, for the sake of generality, each coil frame's generalized coordinate system is characterized by a quintet of elements $x^i = (L^i, \theta^i, \alpha^i, \beta^i) \in R^5$, $i = 1, 2, 3$, referenced to the global frame. Additionally, specific constraints on these parameters are formulated to guarantee a collision-avoidant coil posture, as outlined in Table 5.2. An example of such constraints is the imposition of limits ($L^i \in (L^i_{\min}, L^i_{\max})$) on the L^i parameter, based on practical considerations, to maintain a safe spatial separation for each coil. We consider the set of pose parameters of the three coils as the coil configuration space (CCS). For the purposes of optimization, the CCS, denoted as χ, is formally defined as:

$$\chi := \mathbf{x} = L^1, \theta^1, \phi^1, \alpha^1, \beta^1, \dots, L^3, \theta^3, \phi^3, \alpha^3, \beta^3 \tag{5.4}$$

4. **The Performance-guided Optimization Method:** In the process of optimizing coil configurations, situating the coils in close proximity to the target area can potentially augment the efficiency of field magnitude relative to electric current, thereby enhancing the torque-generation capacity. However, this strategy might concurrently diminish the field isotropy performance or even result in coil collisions. This underscores the notion that advancements in one performance indicator can come at the expense of another. Consequently, a balanced approach tailored to the specific application requirements is essential.

The MOO methodology identifies optimal solutions for complex problems characterized by multiple, often conflicting objectives. By leveraging a predefined CCS and a suite of performance evaluation functions, MOO can derive the Pareto set, representing coil configurations that offer optimal compromises among various performance metrics. Among the prevalent algorithms within the MOO framework, the Non-dominated Sorting Genetic Algorithm II (NSGA-II) stands out due to its proficiency in exploring large search spaces, thereby facilitating efficient and effective resolution of MOO challenges.

Utilizing the four objective functions in conjunction with the defined CCS χ, the task of identifying the optimal coil configuration for rotating field control is formulated as a MOO problem, aiming to optimize the performance criteria to their fullest extent.

$$\begin{cases} x^* = \arg \max \{f_1, f_2, f_3, f_4\} \\ x \in \chi \\ \text{s.t.} \chi_{\min} \leq \chi \leq \chi_{\max} \end{cases} \tag{5.5}$$

The brief procedures of the performance-guided optimization method for coil configuration design are listed in Algorithms 1 and 2 in the Appendix. During the optimization, for a sampled coil configuration x in the CCS χ, we first uniformly collected 27 unit fields $\hat{\mathbf{b}}(\mathbf{p}_m)$ with different directions from 3D spaces. Next, each unit field was regarded as the desired field $\mathbf{b}_d$, and the corresponding actual fields under the given coil configuration x could be obtained according to:

$$\mathbf{i}_d = A_b^{-1}(\mathbf{p}_w)\mathbf{b}_d(\mathbf{p}_w) \tag{5.6}$$

To gather the precise magnetic field data, a cubic workspace denoted as Ω, situated within 3D space R^3, was established. This workspace, with a side length of 10 mm (aligning with the scale of the magnetic robot studied), was centered on the robot's position $\mathbf{p}_w$. The cube was subsequently partitioned into a grid of $N = 125$ points ($\mathbf{p}_i \in \Omega$), arranged in a $5 \times 5 \times 5$ 3D configuration, to facilitate the computation of the magnetic field b at each grid point $\mathbf{p}_i$. Once the 125 field values were obtained, the field evaluation metrics, namely $||\mathbf{b}||_{\text{iso}}$ and $(\mathbf{b}_{\text{cosine}})_{\min}$, were calculated and documented. This process was iteratively conducted for each of the 27 individual magnetic field components, ensuring that all were comprehensively analyzed. It is worth noting that the isotropy indexes σ_3 and $1/k$ of the actuation matrix A_b are solely determined by the specific arrangement of the coils in use.

After acquiring the values for four distinct objective functions, the NSGA-II algorithm was implemented in a parallel computing fashion to derive a collection of non-dominated solutions, commonly referred to as the Pareto front. It is noteworthy that the execution time of the NSGA-II algorithm on the REMA system for addressing our specific problem amounted to approximately 0.43 s, which was achieved in real-time. However, the task of narrowing down the Pareto front to a single, practical solution suitable for application—specifically, choosing an appropriate coil configuration from the obtained set for subsequent control—poses a significant challenge. In our research, we employed a pseudo-weight vector approach to select a solution from the Pareto front. The pseudo weight w_i for the ith objective function can be calculated as:

$$w_i = \frac{\frac{f_i^{\max} - f_i(x)}{f_i^{\max} - f_i^{\min}}}{\sum_{m=1}^{M} \frac{f_m^{\max} - f_m(x)}{f_m^{\max} - f_m^{\min}}} \tag{5.7}$$

where $f_i^{\max}$ and $f_i^{\min}$ denote the maximum and minimum values of the objective function in the obtained Pareto front. This equation calculates the normalized distance to the worst solution for each objective function f_i. The four objectives have different scales. Therefore, min-max normalization was applied to perform a linear transformation on the obtained Pareto front.

In the context of inverse design and parameter optimization for coil configurations, the NSGA-II method facilitates a swift and efficient exploration of design parameter values that align with the desired isotropic characteristics. For MOO scenarios, it is crucial to guarantee that each objective function evolves in a manner that augments the overall system performance throughout the optimization procedure. Figure 5.1 illustrates the optimization process for coil configurations, accompanied by the respective evaluation metrics. This figure underscores the progressive enhancement in the evaluation metrics as the iterations unfold, ultimately resulting in superior field isotropy performance. Due to the uncertainty regarding the actual Pareto front, we are unable to ascertain whether the algorithm has reached the true optimum. Hypervolume, a recognized performance metric in MOO, quantifies the volume enclosed between a predefined reference point and the obtained solution set. In our study, we aimed to maximize this hypervolume index using a specific reference point ([1, 1, 1, 1]). Figure 5.2 depicts the algorithm's convergence trajectory during the optimization process, suggesting that the optimization procedure has converged. A higher hypervolume value signifies a superior quality of the solution set, as outlined in [19].

In addition, according to the definitions of magnetic-feasible workspace (MFW) and $D - A$ distance provided by [5], we also evaluated these two evaluation metrics under the Pareto-optimized coil configuration.

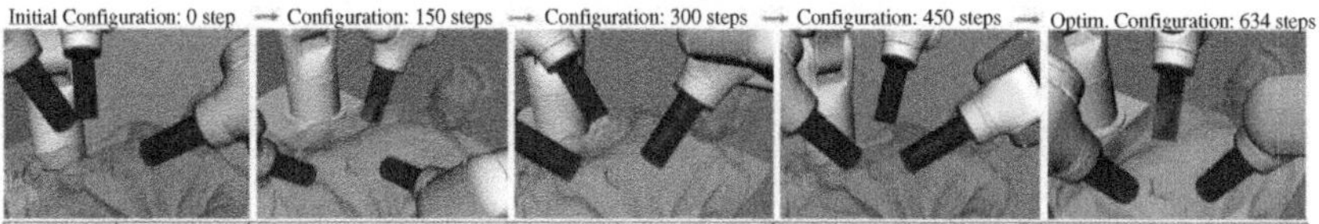

$\|\|\mathbf{b}\|\|_{iso}(\%)$	93.5	94.7	95.4	95.9	98.6
$\measuredangle_{err}(\mathbf{b}_d, \mathbf{b})(^\circ)$	32.4	24.5	15.7	10.8	6.2
σ_3	0.00085	0.0012	0.0016	0.0021	0.0027
σ_3/σ_1	0.13	0.46	0.57	0.74	0.85

Figure 5.1 The optimization process for coil configuration along with the corresponding evaluation metrics.

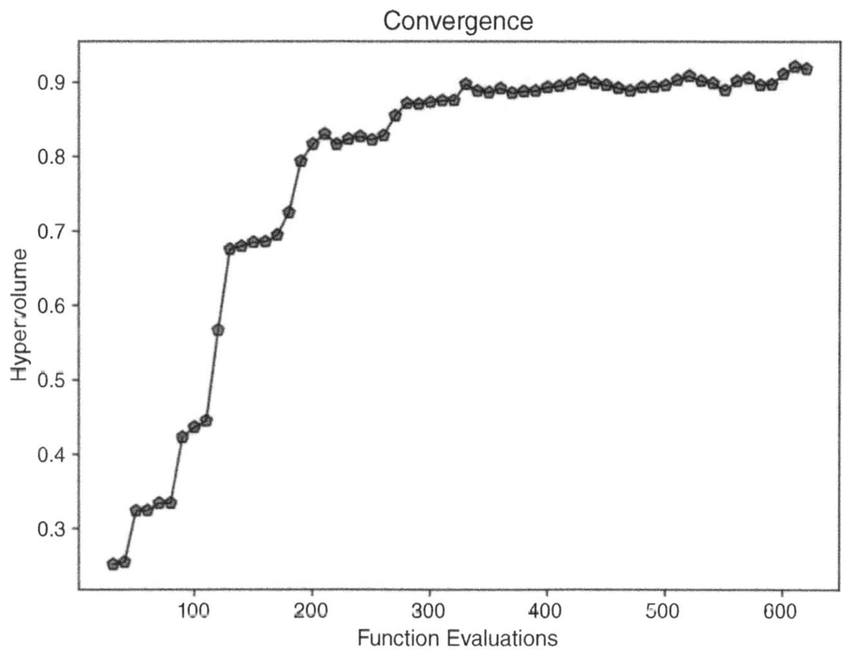

Figure 5.2 Analysis of the optimization process convergence by computing the hypervolume. The X-axis represents the number of optimization iterations.

We first illustrated the MFW obtained for a desired magnetic task set $D = \{\mathbf{b} \in R^3 \mid \|\mathbf{b}\|_2 \leq \bar{b}\}$ with $\bar{b} = 10\,\text{mT}$. Similarly, the target region is a cube of side 0.01 m with its center at the position (0, 0, 0). The results for the MFW presented in Figure 5.3a illustrated that a magnetic field of 10 mT in magnitude can be generated in any direction within this target region. We can also see the results of $D - A$ distance in Figure 5.3b.

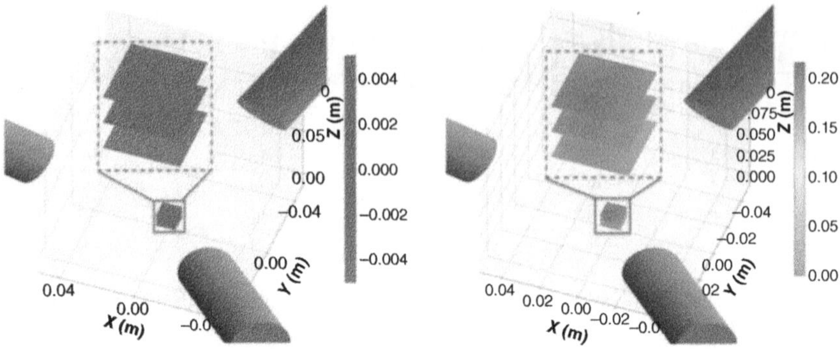

(a) (b)

Figure 5.3 Magnetic-feasible workspace (MFW) and D–A distance of the Pareto-optimized coil configuration. (a) MFW. The feasibility is tested within three horizontal slices (i.e., $z = -0.005, 0, 0.005$), and the scale represents the altitude z of the positions. (b) D–A distance distribution under the Pareto-optimized coil configuration.

5.3 Field-priority Force Control Method

1. **Problems:** It is important to note that each coil offers a comprehensive set of 6-DOFs for control purposes, comprising a 5-DOF pose (excluding the coil's rotation about its main axis) and an additional DOF related to electric current. However, managing fields and forces within the task space using the REMA system is non-trivial, owing to the intricate nature of establishing a clear relationship between the field, force, coil configuration, and electric currents. Additionally, to ensure effective field and force control across a broad workspace, it is essential to consider the manipulability of each robotic arm, thereby avoiding potential singularities. In light of various evaluation metrics that require enhancement, we delve into the utilization of MOO [18] techniques to address multifaceted functional objectives.
2. **Method:** It is noteworthy that the manipulation of any 3D field can be accomplished through the utilization of three coils, solely relying on variations in their electric currents. Hence, our platform's field control remains consistently viable, contingent upon the availability of a permissible range of electric currents ($\mathbf{i}_d \in [-10, 10]\,A$). The adjustment of various coil configurations serves as a means to explore force control. Furthermore, when considering both field and force control in expansive workspaces, it is imperative to ensure robust manipulability for each robotic arm to prevent singularities in specific joint configurations. Therefore, we incorporate the Yoshikawa manipulability measure for each robotic arm. As a result, for a specified field and force, our

proposed field-priority force control approach ensures field control and optimizes the actual force to a target value within acceptable deviations, while simultaneously enhancing the Yoshikawa manipulability measure for each robotic arm.

Leveraging the ideas of MOO technology, we have developed a field-priority force control approach, which aims to concurrently reduce force errors and elevate the Yoshikawa manipulability measure for each robotic arm. Consequently, the MOO problem can be articulated as follows:

$$\begin{cases} x^* = \arg\max \{e_{\mathbf{f}}, m_1, m_2, m_3\} \\ x \in \chi \\ \text{s.t.} \chi_{\min} \leq \chi \leq \chi_{\max} \end{cases} \tag{5.8}$$

where $e_{\mathbf{f}} = - \parallel \mathbf{f}_d - \mathbf{f}_a \parallel$, $\mathbf{f}_d$ denotes the desired force and $\mathbf{f}_a$ represents the actual force for a specific coil configuration x. The Yoshikawa manipulability $m_i = \sqrt{\det(J_i(\mathbf{q})J_i^T(\mathbf{q}))}$, $i = 1, 2, 3$, which describes how well-conditioned the robotic arm is to achieve an arbitrary velocity in the Cartesian space. $J(\mathbf{q})$ represents the Jacobian matrix of the robotic arm, which is a function of the joint position vector $\mathbf{q}$.

Figure 5.4 illustrates the spatial arrangement of three coils alongside the Yoshikawa manipulability indices for each robotic arm, considering their specified base positions outlined in Table 5.1 and the constraints imposed by the three-coil setup detailed in Table 5.2. The asymmetry observed in the manipulability profiles of the robotic arms can be attributed to the irregular workspace, installation layout, and sampling techniques employed. Furthermore, our optimization endeavor aims to position the coils in areas that maximize manipulability to the fullest extent possible. Concurrently, we identified sufficient spatial capacity for imaging magnetic robots, ensuring no occlusion occurs. For addressing the MOO problem, we selected the NSGA-II, given its proficiency in exploring vast search domains, as delineated in Algorithm 3 in the Appendix.

After acquiring the values for the four objective functions, namely $e_{\mathbf{f}}$, m_1, m_2, and m_3, we utilized the NSGA-II algorithm to derive a collection of non-dominated solutions, commonly referred to as the Pareto front. It's worth noting that the NSGA-II algorithm, when applied to our problem on the REMA system, executes swiftly within approximately 0.5 s, facilitating real-time online operations. However, the task of refining the Pareto front to a solitary, practical solution, which entails choosing an apt coil configuration from the resultant set for subsequent control measures, poses a significant challenge. In our research, we employed a pseudoweight vector approach to select

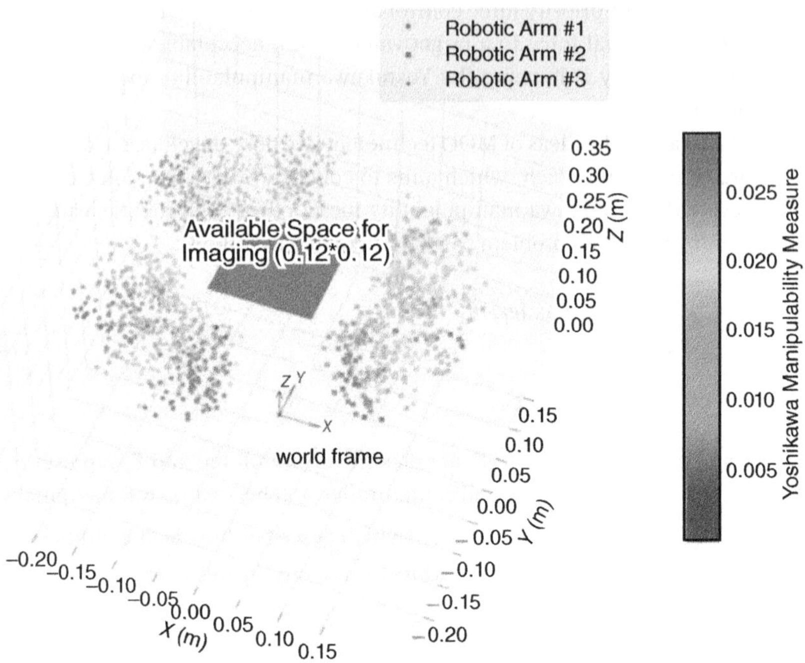

Figure 5.4 Distributions of three coils in the workspace with three robotic arms' Yoshikawa manipulability measure.

a solution from the Pareto front. Specifically, the pseudoweight w_i for the ith objective function was computed according to the prescribed formula (5.7).

3. **Feasibility Evaluation:** Given the intricate nature of the high-dimensional field-priority force control approach, assessing its practical applicability—specifically, determining whether a magnetic robot can be maneuvered with a prescribed set of fields and forces while maintaining any arbitrary pose during autonomous magnetic manipulation—is highly challenging. To facilitate a comprehensive yet straightforward quantitative analysis of our proposed method, we introduced a direction vector space that encompasses the desired fields and forces. Subsequently, we evaluated the feasibility of our method by establishing a direction vector space S, comprising 50 direction vectors uniformly distributed across the 3D space. Each of these direction vectors is represented by $\hat{s}$, an example of which is $\hat{s} = [1, 0, 0]T$. Therefore, the S is intuitively defined as follows:

$$S = \hat{s} \in R^3 \tag{5.9}$$

Based on the design S, we systematically selected two directional vectors, labeled as $\hat{s}_b$ and $\hat{s}_f$, to function as the orientation for the applied magnetic field and the force, respectively. It is important to recognize that the magnitude of the magnetic dipole moment $\| \mathbf{m}_r \|$ of the magnetic robot solely modulates the intensity of the produced force. Consequently, in our analysis, we normalized our considerations to the unit dipole moment $\hat{\mathbf{m}}_r$. Once these two directional vectors were chosen, along with the predefined desired amplitudes for both the field and the force, we employed the approach outlined in Algorithm 2 to ascertain the configuration of the coil and the necessary electric current. Given that field control is a priority and is ensured throughout the optimization process, our emphasis was on assessing the viability of force generation. To quantify the discrepancy between the intended and realized force, we formulated an evaluation criterion, the force percentage error ε, as detailed below:

$$\varepsilon = \frac{\| \mathbf{f}_d - \mathbf{f}_a \|}{\| \mathbf{f}_d \|} \times 100\% \tag{5.10}$$

where $\mathbf{f}_d$ and $\mathbf{f}_a$ are the desired and actual forces, respectively.

The comprehensive outcomes are illustrated in Figure 5.5. Utilizing the unit dipole moment as the basis, we examined the actual forces achieved under 50 randomly selected directional vectors, in accordance with the specified

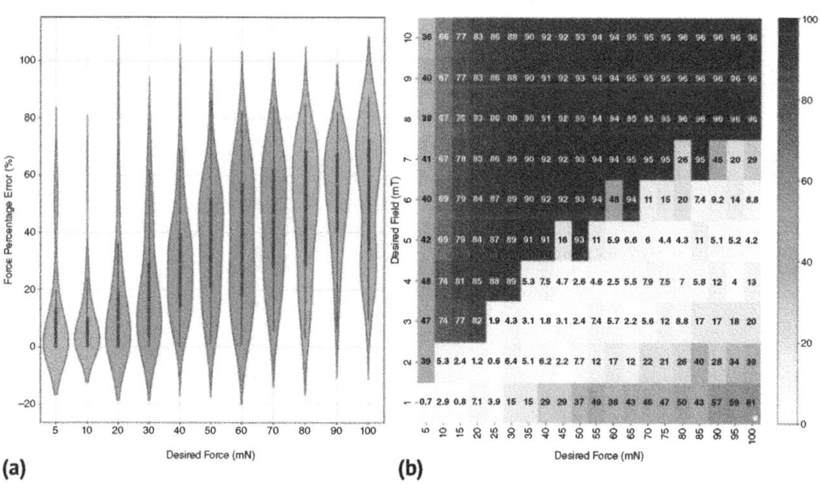

Figure 5.5 Performance of the field-priority force control method. (a) Violin plots represent the force percentage error distribution under the desired field 1 mT with desired force range [5–100 mN]. Width of each density curve corresponds with the approximate frequency of data points in each region. Small box plot in the middle of each density curve shows the ends of the first and third quartiles and the central dot of the median. (b) Median distribution of force percentage error under desired field range [1–10 mT] with desired force range [5–100 mN].

desired fields and forces. Specifically, Figure 5.5a depicts the distribution of force percentage errors across these 50 samples, when the target field strength was set to 1 mT, within a predefined force range of [5–100 mN] in the coordinate system CCS χ. The findings revealed a close alignment between the actual force fa and the desired values, particularly when the desired force was within 20 mN; under these conditions, the force percentage error remained below 10%. Notably, the most favorable distribution of force percentage errors was observed when the desired force ranged between [5–10 mN], achieving an error rate of less than 3%.

Furthermore, we analyzed the distribution of force percentage errors for a different desired force range of [5–10 mN], as shown in Figure 5.5b. To evaluate the performance of our method, we computed the median force percentage error across all 50 directional vectors for each combination of desired field and force. Our observations indicated that as the desired field intensity increases, a correspondingly larger desired force is necessary to maintain a lower force percentage error. This can be attributed to the physical constraint that, given a specific applied field and dipole moment for a magnetic robot, the actual force is confined to a certain range within the defined CCS χ. Additionally, during the optimization process, we incorporated the Yoshikawa manipulability metric for each robotic arm, which excluded certain unfavorable joint configurations, thereby potentially compromising the force percentage error. Figure 5.6 compares the median force percentage error distribution

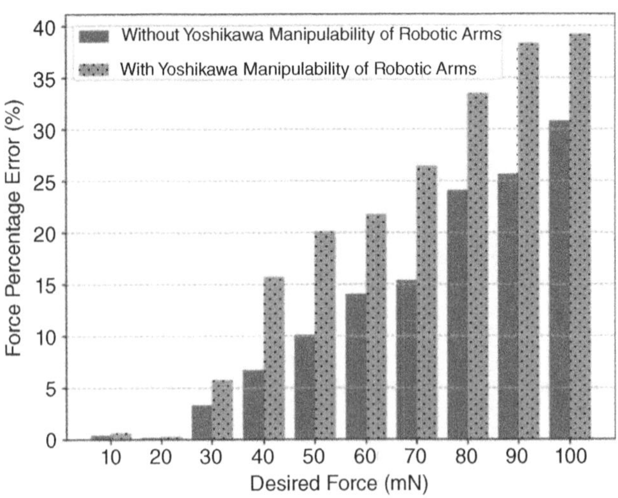

Figure 5.6 Comparison of the median of field-force percentage error distribution for considering and not considering the robotic arms' Yoshikawa manipulability when the desired field is 1 mT.

with and without considering the Yoshikawa manipulability during optimization, with a fixed desired field amplitude of 1 mT. This also implies that the Yoshikawa manipulability metric is explored and enhanced at the expense of the force control error.

4. **Analysis:** The examination of field-priority force control offers valuable understanding on manipulating both field and force utilizing the three-coil REMA system. Two primary obstacles hinder the reduction of field percentage error. Firstly, there are constraints imposed by the CCS to ensure safe, non-colliding manipulations. Secondly, the Yoshikawa manipulability of each robotic arm plays a role. As an optimization objective, incorporating Yoshikawa manipulability may slightly compromise the precision of force control but enhances the workspace range for manipulation. By consulting the field-force heat map, an appropriate field can be selected to minimize the force percentage error while achieving the desired force. For instance, to apply a force of 50 mN to a magnetic robot with a unit dipole moment, a field with an amplitude of 3 mT would yield a lower force percentage error.
5. **Simulation:** The proposed algorithm's efficacy was initially confirmed in a simulation environment built using ROS software. The virtual capsule robot, illustrated in Figure 5.7a, was tasked with following a reference path across a broad workspace. To replicate real-world motion, we factored in the robot's gravity and dynamics, necessitating the continuous application of a vertical magnetic force to counteract gravity. Real-time adjustments to the coil

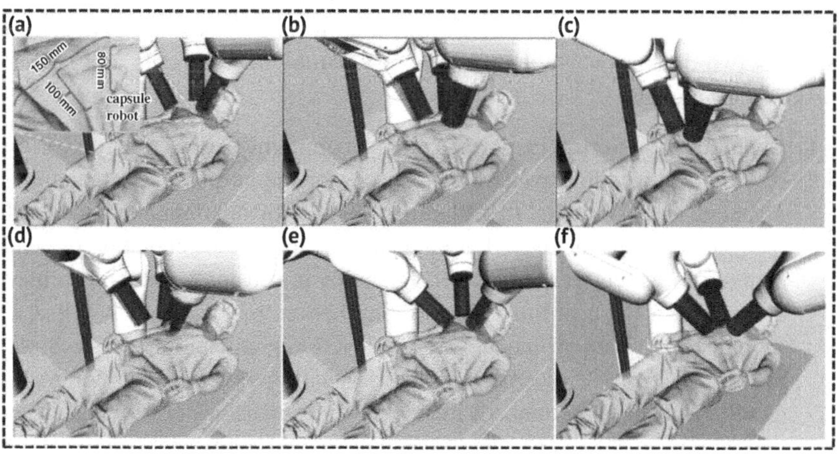

Figure 5.7 Virtual capsule robot in the simulation environment is magnetically propelled to track a 3D helix path under the guidance of the field-priority force control method and pure pursuit path tracking algorithm. (a) $t = 0$ s. (b) $t = 20$ s. (c) $t = 40$ s. (d) $t = 60$ s. (e) $t = 80$ s. (f) $t = 100$ s.

configuration for path tracking could potentially disrupt the control process's smoothness. Hence, we devised a compromise approach by dividing the travel path into segments based on its curvature. For each segment, Algorithm 2 was employed to generate the coil configuration needed to produce the desired field and force. It's worth noting that during transitions between different coil configurations, the fields were deactivated to prevent unintended capsule movement, as coordinating the three robotic arms and adjusting the three-coil currents to maintain the desired field and force was challenging. As a result, the robot would temporarily descend vertically under gravity's influence. As demonstrated in Figure 5.7a–f, the capsule robot effectively traced the reference path using the field-priority force control method and the pure pursuit path tracking algorithm.

5.4 Experimental Validation and Analysis

To assess the efficacy of the novel rotating field control approach utilizing three mobile coils, we initially devised the REMA system, which comprises three robotic arms, each fitted with a coil. This rotating field has the capability to elicit a corkscrew-like movement in a magnetic robot. By exerting torques, the field aims to orient the dipole moment of the magnetic robot in accordance with the field's direction. As an illustrative example, a helical swimmer is an apt choice for navigating through high-viscosity media, given its ability to execute drilling maneuvers under the influence of a rotating field. Subsequently, we undertook a series of experimental trials, encompassing the operation of helical swimmers in diverse settings: two-dimensional tube environments, 3D tube environments, and 3D free-space environments.

1. **Experimental Setup:** Figure 5.8a offers a comprehensive view of our REMA system, deployed in expansive workspaces for the purpose of image-guided magnetic manipulation. This platform consists of three mobile coils, each mounted on a distinct 6-DOF robotic arm, enabling the generation of fields across these extensive workspaces. Flexibility in coil configuration is achievable through the coordinated operation of these robotic arms. A full-scale human body phantom model was employed to simulate human-sized scenarios. Figures 5.8b,c depict two helical swimmers, each incorporating a Grade-N52 NdFeB permanent magnet (with dimensions of 2 mm diameter and 2.5 mm height for one, and 1 mm diameter and 2 mm height for the other) positioned at their head, where the dipole moment is oriented perpendicular to the swimmer's principal axis. To more accurately emulate the tubular tissues found in the human body, PVC tubes (as shown in Figure 5.8d) were filled with 80% glycerol, mimicking an environment characterized by a low Reynolds number. Furthermore, in Figure 5.8e, a cubic container phantom

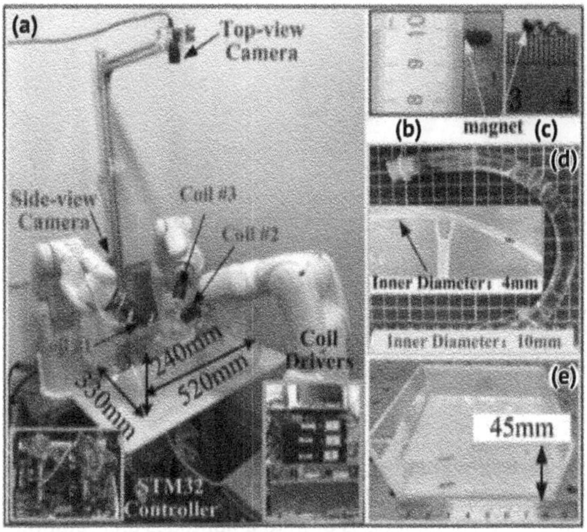

Figure 5.8 Overview of the setup for the real-world experiments and the developed REMA prototype. (a) The REMA system comprises three mobile coils actuated by three independent 6-DOF robotic arms. A low-level STM32 controller and three coil drivers are used to manipulate the electric currents flowing into the coils. (b) Helical swimmer with a 2 mm-diameter and 2.5 mm-high permanent magnet. (c) Helical swimmer with a 1 mm-diameter and 2 mm-high permanent magnet. (d) PVC tube phantom. (e) Cubic container phantom.

filled with 80% glycerol was utilized for the actuation of the helical swimmer. While these phantoms may not perfectly replicate actual medical conditions, they suffice for our proof-of-concept experiments. Given the uniformity of the field in the local area and the slower movement speed of the magnetic robot compared to the robotic arms, the motion resolution for each robotic arm was set at 3 mm; this means that whenever the position of the magnetic robot shifts by more than 3 mm, the robotic arms adjust to track its updated position. This approach minimizes the computational resources required for real-time trajectory planning of the robotic arms.

2. **Actuation of a Helical Swimmer via Different Pareto-optimized Coil Configurations Under Practical Considerations**: To assess the efficacy of a generated rotating field, a direct and effective approach involves utilizing a helical swimmer as an actuator. We conducted two distinct scenarios, corresponding to the helical robot operating at varying depths within the human body. These implementations served to guide the helical swimmer and thereby evaluate the performance of the methodology.

   In the first scenario depicted in Figure 5.9a1, the helical swimmer was submerged within a curved tube securely mounted on the interior wall of the

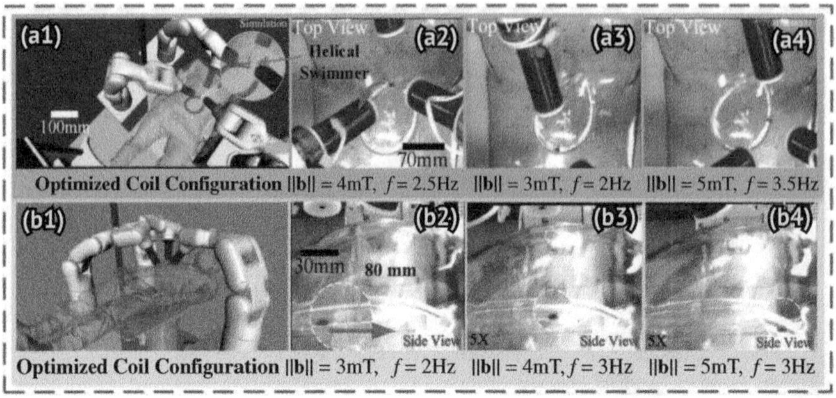

Figure 5.9 Results of magnetically actuating helical swimmers under the guidance of the proposed rotating field control method. Case I (a1–a4): the actuation of a helical swimmer in a curved tube under different combinations of field amplitudes and rotating frequencies generated by the Pareto-optimized coil configuration. Case II (b1–b4): the actuation of a helical swimmer in a tube with a distance of 80 mm to the upper surface of the phantom. Specifically, (b1) delivers the newly Pareto-optimized coil configuration, which differs from the one in Case I.

human body model phantom, facilitating stable fixation and image acquisition, as illustrated in Figure 5.9a2. For the second scenario, a tube was positioned inside the human body model phantom, situated 80 mm beneath its upper surface (refer to Figure 5.9b2), with the aim of verifying the swimmer's continued optimal swimming performance. Prior to this experiment, the centerline of the tube was reconstructed to serve as the intended path for the helical swimmer. Subsequently, the position of the propulsion wire ($\mathbf{p}_w$) was determined using images captured by a top-view camera. In both scenarios, the coil configuration was adjusted in accordance with Algorithm 1. At each time point, the necessary electric currents were computed to guarantee that the rotating field's direction remained consistently aligned with the tube's centerline.

Figure 5.9a1 exhibits the Pareto-optimized coil setup for the initial scenario, while Figures 5.9a2–a4 present a sequence of images documenting the helical swimmer's progression to its target position within the tube, under varying field amplitudes and rotational frequencies. The findings reveal that the Pareto-optimized coil configuration is capable of producing a rotating field with robust performance across a spectrum of amplitudes and frequencies. To showcase the method's proficiency in achieving coil-human body collision avoidance and generating isotropic fields, a tailored coil configuration was optimized for the second scenario, as depicted in Figure 5.9b1.

Notably, the new coil configuration diverged from the first due to the altered distance between the coils and the helical swimmer, stemming from the inclusion of potential coil-phantom collisions in the optimization process. Analogous to the first scenario, the experimental outcomes are illustrated in Figures 5.9b2–b4. These results underscore the helical swimmer's impressive swimming performance. In summary, both scenarios confirm that the method adeptly determines the appropriate Pareto-optimized coil configurations, taking into account the collision constraints specific to each scenario.

3. **The Impact of Gradient Force on the Motion of a Helical Swimmer:** It is important to recognize that a coil configuration that has not been optimized might induce a gradient force along the reverse or side directions of the helical swimmer, which could impede its typical motion. To verify that this hindrance is specifically due to the gradient force, rather than other variables like friction, we performed experimental verification in a 3D, unobstructed environment. Initially, the helical swimmer was submerged in an 80% glycerol solution within a cubic phantom container, as illustrated in Figure 5.10. We analyzed two types of coil configurations: a non-optimized one and a Pareto-optimized one. Both configurations were configured with identical desired field parameters to propel the helical swimmer in the 3D, unobstructed space. By examining the field distribution generated by these two configurations, we could compute the gradient force acting on the helical swimmer. As shown in Figures 5.10a–c (first row), because of the gradient force

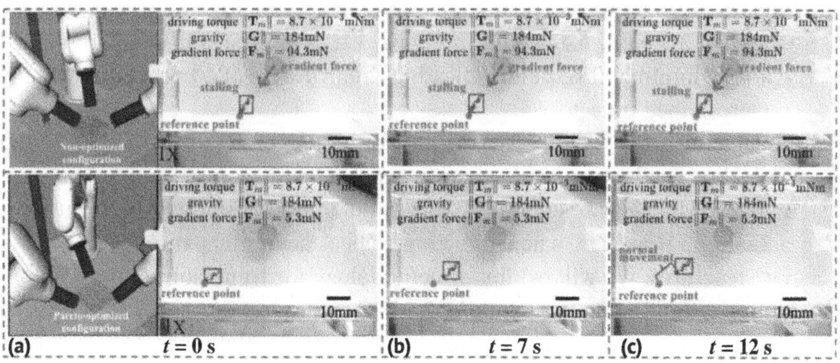

Figure 5.10 Comparison of actuating a helical robot in 3D free space using a non-optimized coil configuration and a Pareto-optimized coil configuration, respectively. The same desired fields are used as the inputs for both coil configurations. (a–c) (the first row) Demonstration of actuating the helical swimmer under the influence of unexpected gradient forces. (a–c) (the second row) Validation of the Pareto-optimized coil configuration producing near-isotropic fields and avoiding the influence of gradient forces on the robot's motion.

pointing in the opposite direction of the robot's motion, the resultant driving torque (8.7×10^{-3} mN m) was insufficient to counteract the robot's gravity (184 mN) and the gradient force (94.3 mN), leading to ineffective movement from 0 to 12 s in the 3D space. To tackle this issue, we implemented a Pareto-optimized coil configuration to generate the same desired field. The outcomes for the Pareto-optimized configuration are displayed in Figures 5.10a–c (second row). Notably, due to the presence of a comparatively minor gradient force (5.3 mN), the helical swimmer was able to overcome gravity and exhibit normal movement. This finding underscores that the primary factor impeding the helical swimmer's motion is the gradient force produced by the non-optimized coil configuration.

4. **3D Path Tracking of a Helical Swimmer in Free Spaces:** We additionally exhibited the proficiency of the introduced Pareto-optimized coil configuration in executing 3D path tracking for the helical swimmer in unimpeded areas. The effectiveness of controlling the helical swimmer in 3D space is influenced by its weight. Consequently, we employed a pure-pursuit approach to regulate the pitch and yaw movements of the helical swimmer, enabling it to accurately follow a predefined 3D path.

   For our experimental setup, we utilized a cubic phantom as the test environment. To visualize the positional changes of the helical swimmer during the path tracking process, we compiled overlaid images captured from both the top view (depicted in Figure 5.11a) and the side view (shown in Figure 5.11b). The path tracking performance of the helical swimmer is illustrated in Figure 5.11c, demonstrating its ability to closely adhere to the intended 3D path. As evident in Figure 5.11d, the tracking errors remained within a minimal threshold, indicating a high degree of accuracy in following the prescribed trajectory.
5. **Actuation of a Helical Swimmer Subjected to Gradient Force in 3D Tubular Environments:** In prior experimental endeavors, the coil configuration of helical swimmers was maintained constant throughout their actuation. However, when directing helical swimmers through 3D tubular structures, we will exhibit a methodology for transitioning from a non-optimized coil setup to a Pareto-optimized configuration, aimed at mitigating the effects of gradient forces. Figure 5.12 portrays the navigation of a helical swimmer along a curved 3D tube, where gradient forces are in play. To evaluate performance improvements, two experimental conditions (Cases I and II) were established, highlighting how a Pareto-optimized coil configuration can effectively counteract these gradient forces.

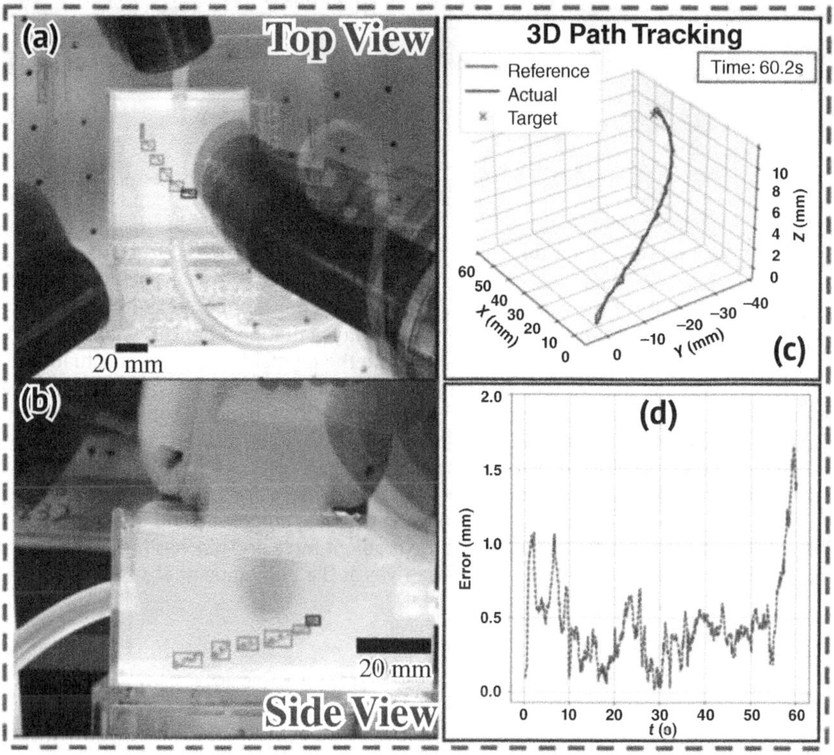

Figure 5.11 Results of 3D path tracking of a helical swimmer in a cubic container under the rotating field control. (a–b) The top-view and side-view overlaid snapshots of actuating the helical swimmer to track the reference path. (c) Path tracking control results of the actual path. (d) Evolution of tracking error over time.

During the operation of the helical swimmer within 3D tubular environments, the instantaneous position of the swimmer was computed using data from both overhead and side-view cameras, enabling 3D visual servo feedback control. As shown in Figure 5.12a, identical coil configurations were employed in both Case I and Case II to produce consistent rotating fields for actuating the helical swimmer. Figure 5.12b illustrates that in both scenarios, the helical swimmers encountered gradient forces, impeding their smooth progression. Notably, in Case II, upon switching to the new Pareto-optimized coil configuration and maintaining the same desired field input at $t = 26$ s, a substantial decrease in the gradient force was observed, from 61.4 to 4.2 mN. Specifically, Figures 5.13a,b depict the temporal evolution of driving torque and gradient force, respectively, during the actuation of the helical

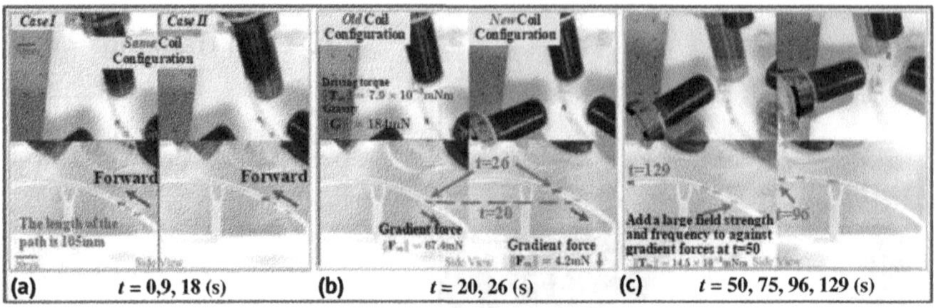

Figure 5.12 Results of the comparative analysis of the non-optimized and the Pareto-optimized coil configurations for magnetically actuating a helical swimmer in 3D tube environments. (a) Case I and Case II illustrate the overlaid snapshots of actuating the helical swimmer in a 3D tube environment under the rotating field generated via the same coil configuration. (b) At $t = 20$ s, (Case I) the helical swimmer is subject to gradient forces (67.4 mN) under the old coil configuration and struggles to overcome their influence. In contrast (Case II), with the newly generated coil configuration, the helical swimmer readily overcomes the influence of gradient forces and continues to move forward. (c) At $t = 50$ s, in Case I, upon increasing the field amplitude and frequency, the helical swimmer successfully counteracts the gradient force and reaches the ending point at $t = 129$ s. In Case II, the helical swimmer arrives at the ending point at $t = 96$ s.

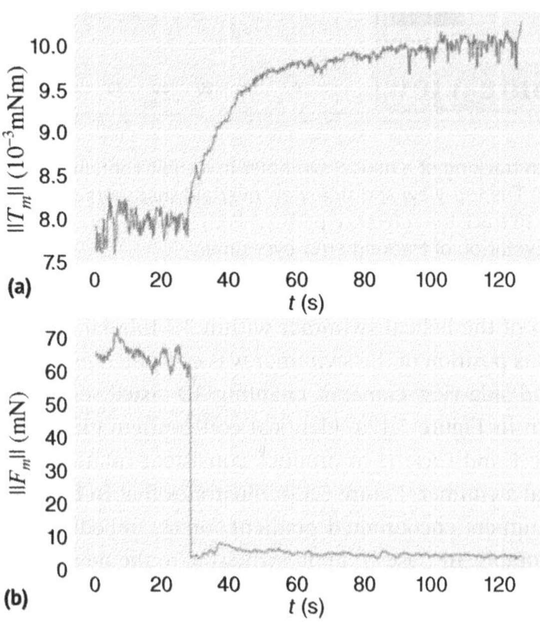

Figure 5.13 Time evolution of the amplitude of magnetic driving torque T_m (a) and gradient force F_m (b) during the actuation of the helical swimmer in Case II.

swimmer in Case II. This reduction was attributed to the Pareto-optimized coil configuration's effectiveness in minimizing the impact of unforeseen gradient forces. Furthermore, the transition from the original coil configuration to the Pareto-optimized one facilitated the helical swimmer's transition from a stalled state to normal motion. As evidenced in Figure 5.12c, by enhancing the field amplitude and frequency at $t = 50$ s, the helical swimmer successfully overcame the gradient force and continued to advance until reaching its intended final position at $t = 129$ s. These experimental findings underscore the proposed method's efficacy in ensuring the isotropy of rotating fields, thereby preventing abnormal robotic movements induced by gradient forces.

To assess the efficacy of the proposed field-priority force control approach for autonomous magnetic manipulations utilizing the REMA system, we initially constructed the REMA prototype, which incorporated three robotic arms (specifically, the SJ-603-A model from Anno Robot). Following this, a series of experiments were carried out, encompassing tasks such as catheter navigation and the propulsion of a mock capsule. Additionally, a simulation setup tailored for REMA was developed to facilitate the debugging and fine-tuning of pertinent parameters through the conduct of these experiments.

1. **Experimental Setup:** The detailed experimental setup is consistent with the REMA system in the previous experiment, as shown in Figure 5.8.
2. **Evaluation of Field-priority Force Control Method in 2D Tubular Environments:** To evaluate the efficacy of the introduced field-priority force control approach, a test was conducted involving the propulsion of a mockup capsule within a C-shaped PVC tube. This tube was securely attached to the internal surface of a human-body model phantom, as illustrated in Figure 5.14a. The field-priority force control method ensured that the capsule followed the tube's centerline by employing targeted magnetic fields, while simultaneously generating propulsive forces to advance the capsule through the tube. Notably, both the intended magnetic field and the propulsive force were oriented in the same direction, aligning with the tube's centerline.

   The experimental outcomes are presented in Figure 5.14a–f. Throughout the actuation process, the intended direction of the capsule was determined based on its current position $\mathbf{p}_w$ for the purpose of directional control—specifically, to maintain alignment with the tube's centerline. Failure to do so could result in the capsule becoming lodged due to the tube's internal wall constraints. It is important to note that switching from one coil configuration to another necessitates a period of time (occasionally exceeding 2 s), which is contingent upon the robotic arms' trajectory execution duration. As an illustration, the transition from the coil configuration depicted in Figure 5.14a to that in Figure 5.14b (from a real-world perspective) requires

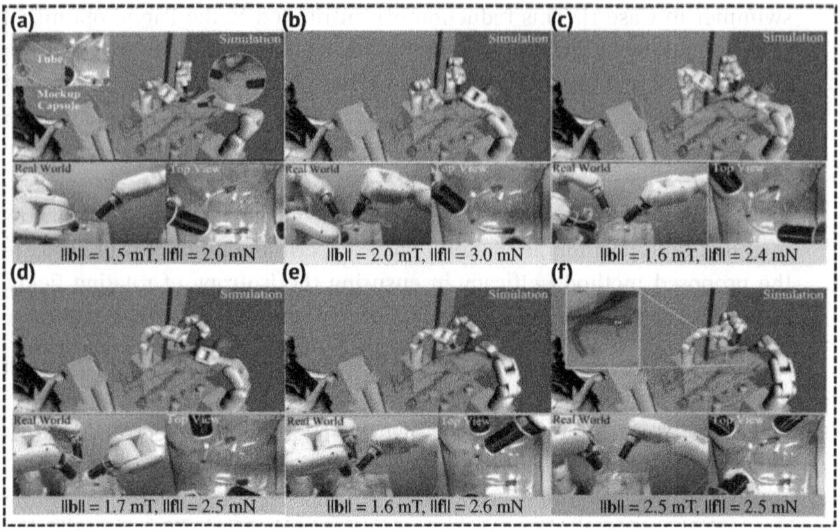

Figure 5.14 Capsule is magnetically propelled through a C-shaped tube under the guidance of the field-priority force control method. Six combinations of field and force are applied to propel the capsule through the tube. Simulation and real-world views show the real-time capsule poses and the corresponding coil configurations. The arrow in the top view indicates the capsule heading direction. (a) $t = 0$ s. (b) $t = 18$ s. (c) $t = 30$ s. (d) $t = 38$ s. (e) $t = 46$ s. (f) $t = 70$ s.

approximately 3 s for the robotic arms to adjust their positions. While it is feasible to increase the motion speed of the robotic arms, this may introduce instability and safety hazards. The findings demonstrate that six distinct combinations of magnetic field and force were effectively utilized to propel the capsule through the tube. Figure 5.14f illustrates the simulated path traversed by the capsule.

3. **Evaluation of Field-priority Force Control Method in 3D Spaces:** To further validate the efficacy of our proposed field-priority force control approach, we designed an experiment where a capsule was tasked with following a predefined trajectory in 3D space, navigating through various heading orientations. The capsule was immersed and suspended within a cubic vessel filled to 80% capacity with glycerol. Utilizing both overhead and lateral cameras, we were able to monitor and record the capsule's precise 3D position throughout the experiment.

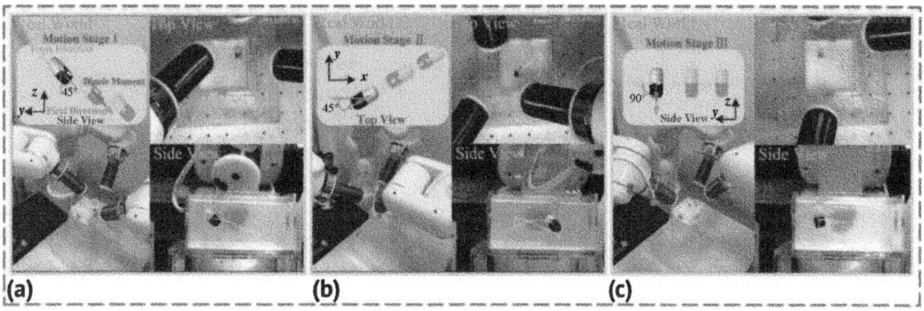

Figure 5.15 Results of magnetically steering a mockup capsule under the guidance of the field-priority force control method to track an N-shaped path in 3D spaces. The overall process is divided into three motion stages. (a) Motion Stage I with the capsule heading maintaining 45° with respect to the gravity direction and the force along the y–z plane ($t = 30$ s). (b) Motion Stage II, wherein the angle between the capsule's heading and the force direction is the same ($t = 60$ s). (c) Motion Stage III, wherein the capsule heading maintains vertical (along the negative z-axis) and the force along the y-axis ($t = 80$ s).

Figure 5.15 displays the experimental outcomes where a capsule navigated an N-shaped trajectory within a cubic container, adhering to specified headings. The entire experiment was divided into three distinct motion phases. Notably, during motion phases I and III, the intended force direction vector deviated from the magnetic field direction vector, indicating a mismatch between the heading and motion directions. A tailored coil configuration was employed at each motion phase to generate the required field and force according to the planned path. The capsule's velocity was primarily governed by the magnitude of the applied force, which necessitated reduction prior to transitioning to the subsequent motion phase. Specifically, when both the field and force direction vectors remained constant, adjustments in their magnitudes with a uniform scaling factor could be achieved solely by manipulating the electric currents under the same coil setup. Implemented the proportional-integral-derivative (PID) algorithm to determine this scaling factor based on the capsule's real-time position, thereby concurrently regulating the amplitudes of both the field and force. Guided by this control approach, the capsule successfully traced the planned path across the three motion phases, as illustrated in Figure 5.16. Furthermore, the introduced field-priority force control method demonstrated its efficacy in balancing field-force production and the maneuverability of the three robotic arms. Considering the REMA system and control strategy, the experiments conducted herein underscored the feasibility of steering a capsule in 3D spaces.

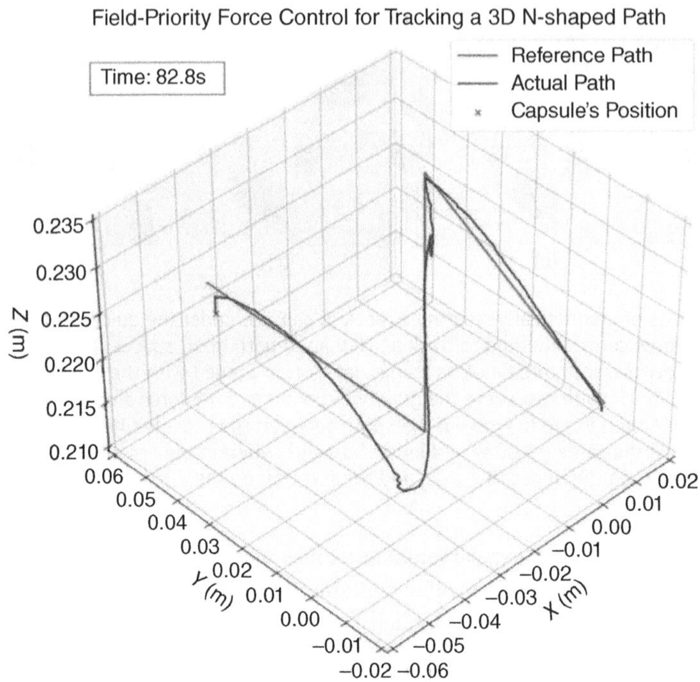

Figure 5.16 Results of 3D N-shaped path tracking.

5.5 Discussion and Conclusion

They have devised an innovative strategy to guarantee the uniformity of rotating magnetic fields produced through Pareto-optimized coil arrangements. Initially, we investigated a portable three-coil setup capable of swiftly generating rotating fields across expansive workspaces, while preserving field uniformity and minimizing gradient effects. Additionally, to enhance the precision of field generation, we employed a performance-driven optimization technique to adjust the coil configuration, ensuring the production of rotating fields with isotropic properties. To substantiate our approach, we developed a mobile EMA system comprising three robotic arms with 6-DOFs, enabling autonomous control of magnetic robots in extensive workspaces. Experimental outcomes demonstrated that our system and optimization technique exhibited commendable performance in magnetic manipulation tasks and demonstrated adaptability to scalable, diverse environments.

While the research has marked significant advancements, the existing methodologies encounter specific constraints that require deeper exploration and resolution. Notably, the pulsatile liquid flow prevalent in physiological settings remains a formidable hurdle for current approaches. A promising avenue to address this limitation could be the exploration of learning-based methodologies aimed at mitigating the impact of unknown flow rates, thereby enabling more accurate motion control, as suggested in references [20–22].

Additionally, we presented a novel technique for controlling magnetic fields and forces employing a mobile EMA platform. This chapter delved into the determination of the necessary number of mobile coils for the combined generation of fields and forces. To ensure versatility in controlling both fields and forces, we adopted a three-coil setup, which facilitated a diverse range of field and force combinations while providing an expansive workspace through a REMA system comprising three robotic arms. Following this, we introduced a field-priority force control approach to derive a Pareto-optimized coil configuration, thereby ensuring robust field control and optimizing the actual force to a specified value within acceptable margins of error. Besides, we enhanced the Yoshikawa manipulability index for each robotic arm. Developed a REMA prototype utilizing three 6-DOF robotic arms for autonomous magnetic robot control in extensive workspaces. Experimental validation underscored the proficiency and efficacy of the magnetic manipulation method.

In our future work, it's necessary to emphasize system design improvements and demonstrate the clinical feasibility of the proposed system. Furthermore, based on this system, aiming to develop advanced control strategies that will enhance the efficiency and safety of autonomous robot navigation using our system in dynamic environments.

Appendix Additional Algorithms

Algorithm 1: The Performance-guided Optimization Method for Rotating Field Control

Data: Given a target position $\mathbf{p}_w$ and generate a 3D grid with 125 points, centered at $\mathbf{p}_w$. Sample 27 desired unit field vector $\hat{\mathbf{b}}$ from space to form set $\hat{\mathbf{B}}$. Initialize actual field buffer $\mathbf{B}$, field amplitude isotropy buffer $\mathbf{B}_{\text{iso}}$, and field direction similarity buffer $\mathbf{B}_{\text{cosine}}$

Output: optimized coil configuration parameters x^*.

Compute the field amplitude isotropy $||\mathbf{b}||_{\text{iso}}$ and the field direction similarity $(\mathbf{b}_{\text{cosine}})_{\min}$ based on the field buffer $\mathbf{B}$;

Put the $||\mathbf{b}||_{\text{iso}}$ into the buffer $\mathbf{B}_{\text{iso}}$;

Put the smallest field direction similarity $(\mathbf{b}_{\text{cosine}})_{\min}$ into the buffer $\mathbf{B}_{\text{cosine}}$;

end

$f_1 \leftarrow \min(\mathbf{B}_{\text{iso}})$, $f_2 \leftarrow \min(\mathbf{B}_{\text{cosine}})$,

$f_3 \leftarrow \sigma_3$, $f_4 \leftarrow 1/k$; **MOO** procedure loop via the NSGA-II algorithm;

end

Obtain the Pareto-fronts;

Normalize f_1, f_2, f_3, f_4;

Compute the pseudo-weight according to Eq. (5.7);

Obtain the optimized results from Pareto-fronts based on the result of pseudo-weight

---

Algorithm 2: Field-priority Force Control Method

Input: Target position $\mathbf{p}_w$, desired field $\mathbf{b}_d(\mathbf{p}_w)$, desired force $\mathbf{b}_d(\mathbf{p}_w)$.

Output: The Pareto-optimized coil configuration x^* and electric currents $\mathbf{i}_d$

1: **for** optimization loop $k = 1$ to K **do**

2: Sample a set of parameters $\mathbf{x}$ from χ according to Eq. (5.4);

3: Compute each coil's pose with respect to the $\mathbf{p}_w$ using the sampled parameter $\mathbf{x}$;

4: Obtain the field generated by each coil with $1A$ current at pw and construct the actuation matrix $A_b = [\mathbf{b}_1(\mathbf{p}_w), \mathbf{b}_2(\mathbf{p}_w), \mathbf{b}_3(\mathbf{p}_w)]$;

5: Compute the desired electric currents id and apply them to the coils $\mathbf{i}_d = A_b^{-1}\mathbf{b}_d(\mathbf{p}_w)$;

6: Compute resultant magnetic gradients $\nabla\mathbf{b}(\mathbf{p}_w)$;

7: Compute actual magnetic force $\mathbf{f}_a = (\mathbf{m} \cdot \nabla)\mathbf{b}(\mathbf{p}_w)$;

8: Compute the manipulability for each robotic arm $m_i(i = 1, 2, 3)$;

9: MOO procedure loop via the NSGA-II algorithm to optimize objective function according to Eq. (5.8);

10: **end**

11: Obtain the Pareto-fronts;

12: Normalize $\boldsymbol{ef}$, m_1, m_2, m_3;

13: Compute the pseudo-weight according to Eq. (5.7);

14: Obtain the Pareto-optimized coil configuration from Pareto-fronts based on the pseudo-weight.

Algorithm 3: The Performance-guided Optimization Method for Rotating Field Control

Data: Given a target position $\mathbf{p}_w$ and generate a 3D grid with 125 points, centered at $\mathbf{p}_w$. Sample 27 desired unit field vector $\hat{\mathbf{b}}$ from space to form set $\hat{\mathbf{B}}$. Initialize actual field buffer $\mathbf{B}$, field amplitude isotropy buffer $\mathbf{B}_{iso}$, and field direction similarity buffer $\mathbf{B}_{cosine}$

Output: optimized coil configuration parameters x^*.

1: **for** optimization loop $m = 1$ to M **do**

2: Sample a coil configuration parameters $\mathbf{x}$ from χ according to Eq. (5.4);

3: Compute each coil's pose with respect to the target point $\mathbf{p}_r$ using parameter $\mathbf{x}$;

4: Compute the actuation matrix $A_b(\mathbf{p}_w)$ and obtain its singular values $\sigma_1, \sigma_2, \sigma_3$;

5: **for** $\hat{\mathbf{b}}_k \in \hat{\mathbf{B}}$ $(k = 1, 2, \dots, 27)$

6: **do**

7: $\mathbf{b}_d \leftarrow \hat{\mathbf{b}}_k$;

8: Compute the desired electric currents $\mathbf{i}_d$ according to Eq. (5.6);

9: **for** $\mathbf{p}_i \in \mathbf{P}$ $(i = 1, 2, \dots, 125)$

10: **do**

11: Compute resultant magnetic field $\mathbf{b}(\mathbf{p}_i)$;

12: Put the $\mathbf{b}(\mathbf{p}_i)$ into the field buffer $\mathbf{B}$;

13: **end**

References

1 S. Salmanipour and E. Diller. Eight-degrees-of-freedom remote actuation of small magnetic mechanisms. In *2018 IEEE International Conference on Robotics and Automation (ICRA)*, pages 3608–3613, 2018. IEEE.

2 J. Sikorski, C. M. Heunis, F. Franco, and S. Misra. The ARMM system: an optimized mobile electromagnetic coil for non-linear actuation of flexible surgical instruments. *IEEE Transactions on Magnetics*, 55 (9): 1–9, 2019.

3 A. J. Petruska and J. J. Abbott. Omnimagnet: an omnidirectional electromagnet for controlled dipole-field generation. *IEEE Transactions on Magnetics*, 50 (7): 1–10, 2014.

4 S. L. Charreyron, Q. Boehler, B. Kim, C. Weibel, C. Chautems, and B. J. Nelson. Modeling electromagnetic navigation systems. *IEEE Transactions on Robotics*, 37 (4): 1009–1021, 2021.

5 Q. Boehler, S. Gervasoni, S. L. Charreyron, C. Chautems, and B. J. Nelson. On the workspace of electromagnetic navigation systems. *IEEE Transactions on Robotics*, 39 (1): 791–807, 2022.

6 Z. Yang, L. Yang, M. Zhang, N. Xia, and L. Zhang. Ultrasound-guided wired magnetic microrobot with active steering and ejectable tip. *IEEE Transactions on Industrial Electronics*, 70 (1): 614–623, 2022.

7 A. J. Petruska and B. J. Nelson. Minimum bounds on the number of electromagnets required for remote magnetic manipulation. *IEEE Transactions on Robotics*, 31 (3): 714–722, 2015.

8 R. Chen, D. Folio, and A. Ferreira. Mathematical approach for the design configuration of magnetic system with multiple electromagnets. *Robotics and Autonomous Systems*, 135: 103674, 2021.

9 L. Yang, X. Du, E. Yu, D. Jin, and L. Zhang. Deltamag: an electromagnetic manipulation system with parallel mobile coils. In *2019 International Conference on Robotics and Automation (ICRA)*, pages 9814–9820, 2019. IEEE.

10 X. Du, L. Yang, J. Yu, K. F. Chan, P. W. Y. Chiu, and L. Zhang. Robomag: a magnetic actuation system based on mobile electromagnetic coils with tunable working space. In *2020 5th International Conference on Advanced Robotics and Mechatronics (ICARM)*, pages 125–131, 2020. IEEE.

11 S. Hwang, A. Ramos-Sebastian, and S. H. Kim. Feedbackless automatic control of magnetic milli/microrobots: generation and control of a trapping point using a single coil electromagnetic system. *IEEE/ASME Transactions on Mechatronics*, 27 (5): 2997–3007, 2021.

12 L. Yang, M. Zhang, Z. Yang, and L. Zhang. Multimode control of a parallel-mobile-coil system for adaptable large-workspace microrobotic actuation. *IEEE/ASME Transactions on Mechatronics*, 28 (3): 1662–1673, 2022.

13 F. Ongaro, S. Pane, S. Scheggi, and S. Misra. Design of an electromagnetic setup for independent three-dimensional control of pairs of identical and nonidentical microrobots. *IEEE Transactions on Robotics*, 35 (1): 174–183, 2018.

14 B. E. Kratochvil, M. P. Kummer, S. Erni, R. Borer, D. R. Frutiger, S. Schürle, and B. J. Nelson. Minimag: a hemispherical electromagnetic system for 5-DOF wireless micromanipulation. In *Experimental Robotics: The 12th International Symposium on Experimental Robotics*, pages 317–329, 2014. Springer.

15 T. L. Bruns, et al. Magnetically steered robotic insertion of cochlear-implant electrode arrays: system integration and first-in-cadaver results. *IEEE Robotics and Automation Letters*, 5 (2): 2240–2247, 2020.

16 A. Hosney, A. Klingner, S. Misra, and I. S. Khalil. Propulsion and steering of helical magnetic microrobots using two synchronized rotating dipole fields in three-dimensional space. In *2015 IEEE/RSJ International Conference on Intelligent Robots and Systems (IROS)*, pages 1988–1993, 2015. IEEE.

17 A. Pourkand and J. J. Abbott. A critical analysis of eight-electromagnet manipulation systems: the role of electromagnet configuration on strength, isotropy, and access. *IEEE Robotics and Automation Letters*, 3 (4): 2957–2962, 2018.

18 N. Gunantara. A review of multi-objective optimization: methods and its applications. *Cogent Engineering*, 5 (1): 1502242, 2018.

19 K. Shang, H. Ishibuchi, L. He, and L. M. Pang. A survey on the hypervolume indicator in evolutionary multiobjective optimization. *IEEE Transactions on Evolutionary Computation*, 25 (1): 1–20, 2020.

20 M. Cai, Q. Wang, Z. Qi, D. Jin, X. Wu, T. Xu, and L. Zhang. Deep reinforcement learning framework-based flow rate rejection control of soft magnetic miniature robots. *IEEE Transactions on Cybernetics*, 53 (12): 7699–7711, 2022.

21 S. Xu, J. Liu, C. Yang, X. Wu, and T. Xu. A learning-based stable servo control strategy using broad learning system applied for microrobotic control. *IEEE Transactions on Cybernetics*, 52 (12): 13727–13737, 2021.

22 M. E. Tiryaki, S. O. Demir, and M. Sitti. Deep learning-based 3D magnetic microrobot tracking using 2D MR images. *IEEE Robotics and Automation Letters*, 7 (3): 6982–6989, 2022.

6

Multimode Control and Optimal Design of a Parallel-mobile-coil System

This chapter introduces the parallel-mobile-coil system (PMCS) and presents three distinct control modes. By leveraging the coils' mobility, the PMCS can adapt to various actuation conditions and different microrobot types. In addition, a computational design approach is developed to optimally build a prototype PMCS for microrobotic navigation.

6.1 Introduction

Magnetic microrobots, ranging in size from the microscale to the mesoscale [1], have attracted wide attention in biomedical applications [2]. The recognition of their significant potential in minimally invasive medicine [3] is largely attributed to the distinctive benefits of magnetic fields, including strong penetration through biological tissues and high biocompatibility. In particular, microscale magnetic microrobots have demonstrated notable progress in various tasks such as targeted delivery [4, 5], intracellular measurements [6], accurate in vivo drug administration [7], and in vivo cell transport [8].

In millimeter-scale robotics, actively controlled magnetic medical devices—such as endoscopes [9] and catheters [10], are widely regarded as emerging technologies that are transforming conventional diagnostic and therapeutic practices. These magnetic capsule endoscopes significantly decrease both the duration of examinations and the rate at which capsules are retained within the gastrointestinal (GI) tract. Through active manipulation, these endoscopic capsules enable a thorough inspection of the entire GI tract [11]—a capability that traditional endoscopy methods cannot match. Moreover, in the treatment of vascular diseases, magnetic catheter-assisted endovascular interventions offer significant benefits compared with traditional manual approaches, especially by improving operational precision and enhancing the efficiency of both surgical and diagnostic procedures [1].

The advancement of various magnetic actuation systems [12] has played a crucial role in enabling the effective operation of magnetic microrobotic devices. These systems are complemented by control schemes designed to achieve precise motion control of magnetic microrobots [13]. By employing multiple fixed electromagnetic coils, a range of system configurations has been developed to enable magnetic actuation. For instance, Helmholtz coils are commonly used to control a uniform magnetic field for single [14], multiple [15], and swarm [16, 17] microrobots. Custom stationary coil-based systems have been tailored for specific applications, with the OctoMag being a notable example designed for minimally invasive retinal therapy. Building on the foundational design and control principles of OctoMag, numerous other systems have been developed [10, 18]. Among these systems, several have effectively incorporated multiple fixed electromagnetic coils for clinical purposes. In particular, the CardioMag system [18] has been developed for in vivo large-animal experiments, whereas the Aeon Phocus system [10] is intended for human clinical applications.

The practical success of these systems has demonstrated that, because magnetic fields diminish rapidly with distance, stationary electromagnetic coils need to be relatively large in order to produce sufficiently strong magnetic fields across wide workspaces, such as the $200 \times 300 \times 500\ \text{mm}^3$ required for CardioMag. This scalability challenge results in systems that are heavy, which can further increase design complexity and overall cost. Moreover, the large inductance associated with oversized coils limits the achievable control bandwidth and leads to higher power consumption. To address these limitations, alternative approaches using permanent magnets to produce strong magnetic fields over large workspaces have also been investigated [19, 20]. These approaches enable magnetic-field regulation by changing the magnets' position and orientation through a serial robotic manipulator. However, a key limitation is that the relatively low mechanical motion speed makes it difficult to generate high-frequency magnetic fields (e.g., greater than 10 Hz). This limitation underscores the need for innovative solutions to enhance the flexibility and efficiency of magnetic actuation systems.

To address the scalability limitations of stationary coil-based systems, a novel approach has been proposed that employs multiple mobile coils to achieve magnetic actuation. This approach has led to the realization of 1D rotation [21], planar motion [22], and 3D motion [23] of coils. More recently, a new type of mobile-coil system, the PMCS [24], has been proposed, as depicted in Figure 6.1. Yet, it is still unclear if the prototype's system parameters in [24] result in the best actuation performance. Furthermore, there are unresolved questions regarding how to redesign such a system for different workspace requirements and how to develop controllers that can achieve precise navigation of microrobots.

Investigations into optimizing the parameters of magnetic actuation systems for stationary coil-based configurations have been conducted to achieve superior performance. In systems such as OctoMag, the mutual intersection angles between

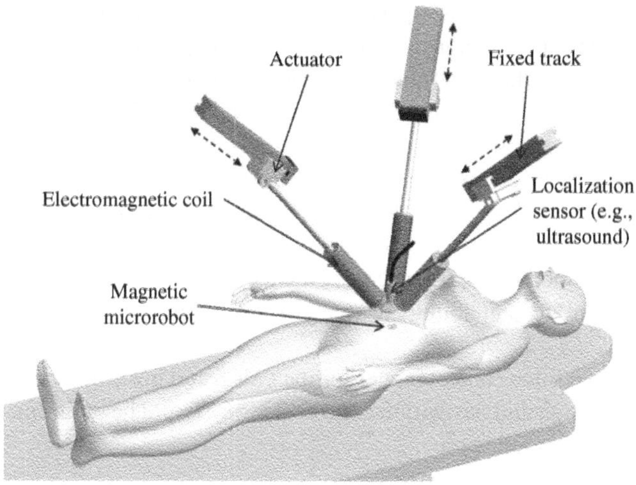

Figure 6.1 Conceptual illustration of magnetic actuation using the PMCS. The three coils are driven by a parallel mechanism, enabling their coordinated motion to achieve a large 3D workspace.

coils are carefully optimized to increase the minimum singular value of the magnetic actuation matrix. Moreover, Pourkand and Abbott [25] utilized the actuation condition number to determine an optimal configuration for stationary eight-coil systems, with the goal of improving actuation isotropy. In addition, many studies have further adjusted key structural parameters to enhance the uniformity of actuation performance throughout the workspace [26]. In a comprehensive study, Chen et al. [27] introduced various metrics to assess magnetic actuation performance, including manipulability, which serves as a valuable tool for parameter optimization. However, it is important to note that the optimization of parameters for magnetic actuation systems featuring multiple mobile coils has not been thoroughly explored. These systems necessitate additional considerations regarding the coils' pose variations. In contrast to stationary coil setups, mobile-coil systems operating over a large workspace require the optimization process to jointly consider both mechanical performance and magnetic actuation criteria.

For mobile coil-based systems, a major challenge lies in re-optimizing structural parameters to meet varying workspace demands. Therefore, it is important to provide end users with a convenient way to adjust and reconfigure these parameters when the required workspace dimensions change. This highlights the importance of developing user-friendly design strategies that can accommodate varying operational environments.

The mobile coil-based system enables microrobot navigation and control over a large workspace, as demonstrated by automated microcatheter steering using quasi-static magnetic torque [24, 28–30]. Dynamic-torque-based navigation of magnetic microrobots has been demonstrated under open-loop control schemes [23] and PID control [31], yet these methods do not achieve high trajectory tracking accuracy. To overcome this limitation, an observer-based control strategy can be adopted to estimate and compensate for system uncertainties, which may originate from the magnetic field generation of the mobile coils as well as the microrobot's operating environment [5, 13, 32–34]. However, this approach necessitates meticulous design and tuning of both the observer and the controller. Therefore, designing a direct control method that can achieve highly accurate navigation and motion regulation is essential for enabling the practical deployment of mobile coil-based systems.

In this study, we present three control modes for the PMCS and investigate both the optimal parameter design and navigation control of the system. The main contributions are summarized as follows:

1. A magnetic force model for multiple mobile coils within the PMCS is established.
2. A unified multimode control framework has been developed with a focus on optimizing the motion of coils. This framework, featuring three distinct control modes, enhances the adaptability of the PMCS across various actuation scenarios and microrobot types.
3. We propose a structural parameter design approach that allows a PMCS to be constructed based on a prescribed cylindrical workspace. By analyzing the workspace, performance indices related to both mechanical motion and magnetic actuation are jointly optimized.
4. For microrobotic navigation with the PMCS, we develop control strategies for both dynamic torque-driven and force-based actuation. Experimental validations demonstrate that the proposed controllers enable accurate trajectory tracking.

6.2 System Design and Modeling

In the PMCS, three coils are driven by a parallel robotic mechanism to move close to the magnetic microrobots as required. With eye-in-hand localization methods—such as vision or ultrasound—the system can generate the desired dynamic magnetic fields directly at the microrobot's position for magnetic actuation. This design benefits from keeping the coils in close proximity to the controlled device, thereby reducing coil requirements, including the driving current. Notably, the three mobile coils in the PMCS are inherently constrained to nonoverlapping physical motions, ensuring safe operation. In addition, the parallel actuation mechanism [35] provides high payload capability and stiffness, while also lowering the peak power demand of the actuators.

6.2.1 Design of the PMCS System

The parallel-mobile-coil mechanism consists of three coil branches, each driven by an actuator moving along a fixed linear guide. The other ends of the branches are symmetrically connected to the so-called coil-end plate through spherical joints. The geometric representation of the kth (k = 1, 2, 3) coil branch is illustrated in Figure 6.2. In this figure, vectors (e.g., $\hat{\mathbf{n}}_{\mathrm{P}}^{\mathcal{G}}$) are written in lowercase bold font, where the superscript and subscript indicate the coordinate frame and the corresponding identifier, respectively. The hat symbol denotes a unit vector. Spatial points are denoted using uppercase regular font; for example, points P and W represent the center of the coil-end plate and the working point where the microrobot is located, respectively.

In the system, the three coil branches are symmetrically arranged around the coil-end plate's normal vector $\hat{\mathbf{n}}_{\mathrm{P}}^{\mathcal{G}}$. To ensure unique deterministic kinematics, three constraints are imposed: (1) $\hat{\mathbf{n}}_{\mathrm{P}}^{\mathcal{G}}$ must always be parallel to $\hat{\mathbf{z}}^{\mathcal{G}}$, where $\{\mathcal{G} : \mathrm{O}^{\mathcal{G}} - \hat{\mathbf{x}}^{\mathcal{G}}\hat{\mathbf{y}}^{\mathcal{G}}\hat{\mathbf{z}}^{\mathcal{G}}\}$; (2) the normal vector $\hat{\mathbf{n}}_{\mathrm{t}}^{\mathcal{G}}$ of the linear actuator's sliding surface must be perpendicular to $\hat{\mathbf{z}}^{\mathcal{G}}$; (3) the coil-end plate is constrained from rotating about $\hat{\mathbf{n}}_{\mathrm{P}}^{\mathcal{G}}$. With these constraints, the motion of the coil-end plate is

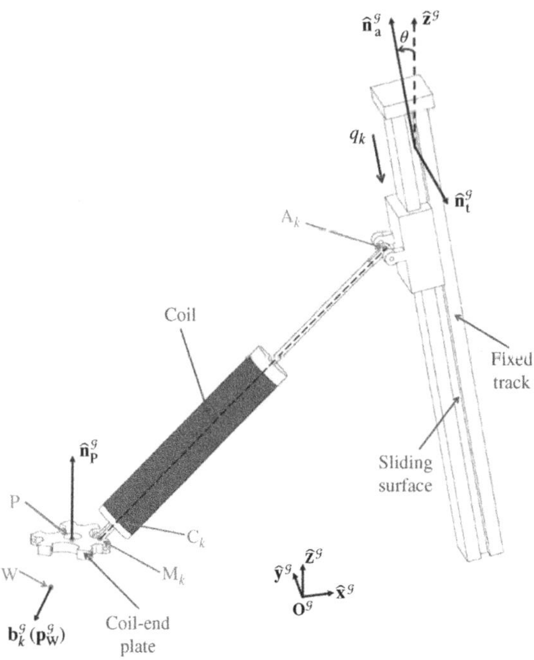

Figure 6.2 Geometric representation of the kth branch in the parallel-mobile-coil mechanism. Key points and vectors used for kinematic modeling and magnetic field computation are highlighted in the figure.

limited to three degrees of freedom (DOF), corresponding to translations along *xyz* axes. The mechanism design satisfying these constraints is depicted in Figure 6.3. The second constraint is satisfied by aligning the fixed track to be coplanar with $\hat{\mathbf{n}}_{\mathrm{P}}^{\mathcal{G}}$. To meet the first and third constraints, a parallelogram mechanism—similar to that used in the Delta robot [36]—is incorporated into each coil branch. Specifically, this mechanism is formed by the slider, the coil-end plate, and two linkages (Linkage I), which are interconnected via four spherical joints. This configuration imposes two point constraints on the coil-end plate for each coil branch. As a result, three branches (i.e., a total of six point constraints) are sufficient to completely constrain the motion of the coil-end plate. Moreover, the three parallelogram mechanisms collectively enforce the first and third constraints. To avoid mechanical interference between branches while maintaining reliable attachment to the coil-end plate, the distance between points P and M_k should be chosen large enough. For coil positioning, each coil is incorporated into the parallelogram mechanism through two spherical joints, an additional linkage (Linkage II), and a connector, as illustrated in Figure 6.3. This design embeds two identical parallelogram mechanisms within each coil branch, thereby guaranteeing deterministic coil motion. A spring is installed between the connector and Linkage II to

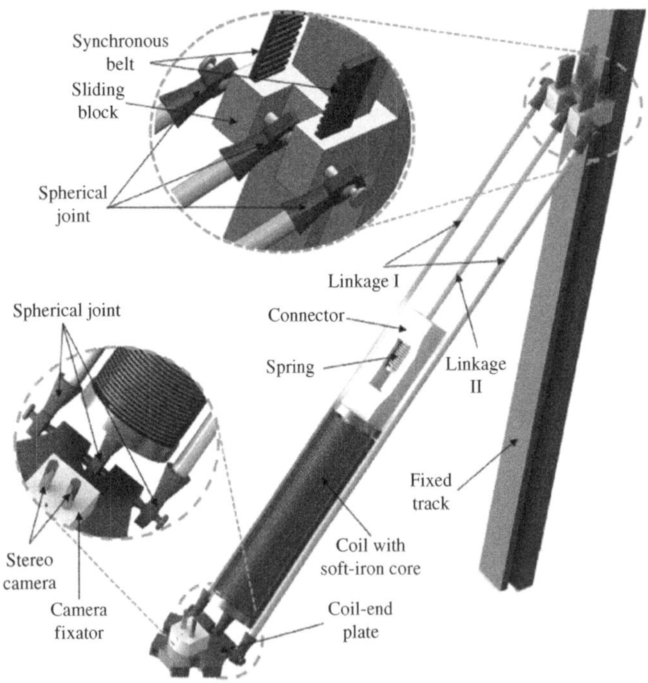

Figure 6.3 Mechanical design of the parallel-mobile-coil mechanism.

improve tolerance to fabrication inaccuracies. The sliders are actuated by linear drives—such as synchronous belt mechanisms—and travel along fixed tracks via a roller-rail system.

6.2.2 Magnetic Field and Force Modeling of PMCS

Considering that magnetic microrobots typically operate at low speeds, the dynamics of the parallel-mobile-coil mechanism are neglected under the reasonable assumption that the linear actuators can accurately drive the mechanism to execute the required motions throughout the workspace. With this design, the parallel structure can be classified as a 3-PP(SS)2 mechanism, and its kinematic properties are derived as follows:

The position of the coil-end plate center, denoted by $-\mathbf{p}_{\mathrm{P}}^{\mathcal{G}} = [x_{\mathrm{P}}, y_{\mathrm{P}}, z_{\mathrm{P}}]^{\mathrm{T}} \in \mathbb{R}^3$ has a unique and deterministic mapping to the actuator positions$-\mathbf{q} = [q_1, q_2, q_3]^{\mathrm{T}} \in \mathbb{R}^3$:

$$\mathbf{p}_{\mathrm{P}}^{\mathcal{G}} = \mathcal{F}(\mathbf{q}) \tag{6.1}$$

In the following, the inverse kinematics is derived, i.e., $\mathbf{q} = \mathcal{F}^{-1}(\mathbf{p}_{\mathrm{P}}^{\mathcal{G}})$, which is required for the closed-loop control of the parallel-mobile-coil mechanism.

As illustrated in Figure 6.2, the coordinates of M_k can be computed as:

$$\mathbf{p}_{\mathrm{M}_k}^{\mathcal{G}} = \begin{bmatrix} x_{\mathrm{M}_k} \\ y_{\mathrm{M}_k} \\ z_{\mathrm{M}_k} \end{bmatrix} = \begin{bmatrix} x_{\mathrm{P}} + L_{\mathrm{MP}} \cos \alpha_k \\ y_{\mathrm{P}} + L_{\mathrm{MP}} \sin \alpha_k \\ z_{\mathrm{P}} \end{bmatrix} \tag{6.2}$$

where L_{MP} denotes the constant distance between points P and point M_k, and α_k is defined as:

$$\alpha_k = (k-1) \cdot 120^\circ \tag{6.3}$$

Likewise, the coordinates of A_k can be determined as:

$$\mathbf{p}_{\mathrm{A}_k}^{\mathcal{G}} = \begin{bmatrix} x_{\mathrm{A}_k} \\ y_{\mathrm{A}_k} \\ z_{\mathrm{A}_k} \end{bmatrix} = \begin{bmatrix} (L_{\mathrm{h}} + q_k \sin \theta) \cos a_k \\ (L_{\mathrm{h}} + q_k \sin \theta) \sin a_k \\ L_{\mathrm{v}} - q_k \cos \theta \end{bmatrix} \tag{6.4}$$

where L_{h} and L_{v} denote the constant distance between $\mathrm{O}^{\mathcal{G}}$ and A_k, respectively when $q_k = 0$. According to the mechanical constraint, the distance $\|\mathbf{p}_{\mathrm{A}_k}^{\mathcal{G}} - \mathbf{p}_{\mathrm{M}_k}^{\mathcal{G}}\|$ remains constant, i.e.,

$$\|\mathbf{p}_{\mathrm{A}_k}^{\mathcal{G}} - \mathbf{p}_{\mathrm{M}_k}^{\mathcal{G}}\| - L_{\mathrm{AM}} = 0 \tag{6.5}$$

where L_{AM} is the constant length between A_k and M_k. By Eqs. (6.2)–(6.5), the inverse kinematics of the parallel-mobile-coil mechanism can be derived as:

$$q_k = \mathcal{F}^{-1}(\mathbf{p}_{\mathrm{P}}^{\mathcal{G}}) = \begin{cases} c_1 - \sqrt{c_1^2 - c_2}, & 0° \le \theta \le 45° \\ c_1 + \sqrt{c_1^2 - c_2}, & 45° < \theta \le 90° \end{cases} \tag{6.6}$$

where

$$\begin{cases} c_1 = (x_{\mathrm{P}} \cos a_k + y_{\mathrm{P}} \sin a_k - L_{\mathrm{h}} + L_{\mathrm{MP}}) \sin\theta \\ \qquad + (L_{\mathrm{v}} - z_{\mathrm{P}}) \cos\theta \\ c_2 = x_{\mathrm{P}}^2 + y_{\mathrm{P}}^2 + (z_{\mathrm{P}} - L_{\mathrm{v}})^2 + (L_{\mathrm{h}} - L_{\mathrm{MP}})^2 - L_{\mathrm{AM}}^2 \\ \qquad - 2(L_{\mathrm{h}} - L_{\mathrm{MP}})(x_{\mathrm{P}} \cos a_k + y_{\mathrm{P}} \sin a_k) \end{cases} \tag{6.7}$$

With the inverse kinematics established, the actuator motions can be regulated to position the coil-end plate at a desired location, enabling microrobot feedback and magnetic actuation.

To compute the magnetic field/gradient produced by the PMCS at the working point $\mathbf{p}_{\mathrm{W}}^{\mathcal{G}}$, i.e., the microrobot position, three main steps are required.

At the first step, $\mathbf{p}_{\mathrm{W}}^{\mathcal{G}}$ is converted into the local coordinate frames of the coils so that the magnetic field/gradient contributed by each individual coil can be evaluated. This coordinate transformation is given by:

$$\mathbf{p}_{\mathrm{W}}^{\mathcal{C}_k} = {}^{\mathcal{C}_k}\mathcal{T}_{\mathcal{G}}(\mathbf{p}_{\mathrm{W}}^{\mathcal{G}}, \mathbf{p}_{\mathrm{P}}^{\mathcal{G}}),\ k = 1, 2, 3 \tag{6.8}$$

where $\{\mathcal{C}_k : \mathrm{O}^{\mathcal{C}_k} - \hat{\mathbf{x}}^{\mathcal{C}_k}\hat{\mathbf{y}}^{\mathcal{C}_k}\hat{\mathbf{z}}^{\mathcal{C}_k}\}$ represents the local coordinate frame of the kth coil. Owing to the nonlinearity of $\mathcal{F}(\mathbf{q})$ and the changing position and orientation of the coil with respect to the coil-end plate, the transformation ${}^{\mathcal{C}_k}\mathcal{T}_{\mathcal{G}}(\mathbf{p}_{\mathrm{W}}^{\mathcal{G}}, \mathbf{p}_{\mathrm{P}}^{\mathcal{G}})$ is composed of a sequence of nonlinear functions:

$$\begin{cases} z_{\mathrm{W}}^{\mathcal{C}_k} = \dfrac{\left(\mathbf{p}_{\mathrm{AC}_k}^{\mathcal{G}}\right)^{\mathrm{T}} \cdot \mathbf{p}_{\mathrm{CW}_k}^{\mathcal{G}}}{\|\mathbf{p}_{\mathrm{AC}_k}^{\mathcal{G}}\|} \\ y_{\mathrm{W}}^{\mathcal{C}_k} = 0 \\ x_{\mathrm{W}}^{\mathcal{C}_k} = \sqrt{\|\mathbf{p}_{\mathrm{CW}_k}^{\mathcal{G}}\|^2 - \left(z_{\mathrm{W}}^{\mathcal{C}_k}\right)^2} \end{cases} \tag{6.9}$$

In the second step, the magnetic field and gradient produced by each coil at $\mathbf{p}_{\mathrm{W}}^{\mathcal{C}_k}$ are determined. Assuming that the kth coil, equipped with a soft iron core, operates within its linear magnetization regime, the resulting magnetic flux density $-\mathbf{b}_k^{\mathcal{C}_k} \in \mathbb{R}^3$ (unit: mT) and the gradient matrix $-\mathbf{G}_k^{\mathcal{C}_k}$ at $\mathbf{p}_{\mathrm{W}}^{\mathcal{C}_k}$ vary linearly with the coil current i_k. They can be computed as:

$$\begin{cases} \mathbf{b}_k^{\mathcal{C}_k}\left(\mathbf{p}_{\mathrm{W}}^{\mathcal{C}_k}\right) = i_k \cdot \bar{\mathbf{b}}_k^{\mathcal{C}_k}\left(\mathbf{p}_{\mathrm{W}}^{\mathcal{C}_k}\right) = i_k \cdot \bar{\mathcal{M}}_k^{\mathcal{C}_k}\left\{\mathbf{p}_{\mathrm{W}}^{\mathcal{C}_k}\right\} \\ \mathbf{G}_k^{\mathcal{C}_k}\left(\mathbf{p}_{\mathrm{W}}^{\mathcal{C}_k}\right) = i_k \cdot \bar{\mathbf{G}}_k^{\mathcal{C}_k}\left(\mathbf{p}_{\mathrm{W}}^{\mathcal{C}_k}\right) \\ \qquad = i_k \left[\dfrac{\partial \bar{\mathcal{M}}_k^{\mathcal{C}_k}\left\{\mathbf{p}_{\mathrm{W}}^{\mathcal{C}_k}\right\}}{\partial x^{\mathcal{C}_k}} \; \dfrac{\partial \bar{\mathcal{M}}_k^{\mathcal{C}_k}\left\{\mathbf{p}_{\mathrm{W}}^{\mathcal{C}_k}\right\}}{\partial y^{\mathcal{C}_k}} \; \dfrac{\partial \bar{\mathcal{M}}_k^{\mathcal{C}_k}\left\{\mathbf{p}_{\mathrm{W}}^{\mathcal{C}_k}\right\}}{\partial z^{\mathcal{C}_k}} \right]^{\mathrm{T}} \end{cases} \tag{6.10}$$

where $\bar{\mathcal{M}}_k^{\mathcal{C}_k}(\cdot)$ denotes the unit-field map of the kth coil, i.e., the magnetic field/gradient distribution generated when the coil is driven by a 1 A current. The overbar symbol indicates unit-field quantities. $\bar{\mathcal{M}}_k^{\mathcal{C}_k}(\cdot)$ can be obtained by fitting a dipole model to finite element method (FEM) simulation data; however, this approach provides adequate accuracy mainly within a small workspace (20 × 20 mm) and at relatively large distances [19]. Alternatively, multipole expansion models can be adopted after calibrating the coefficients using experimental measurements [37, 38]. A drawback of this method is its relatively high computational complexity, particularly for non-axisymmetric coils, which require more general multipole expansions [37].

We propose a calibration method that employs an artificial neural network (ANN) to derive the coil's unit-field map based on FEM simulations in COMSOL Multiphysics 5.3. This approach is simple to implement and serves as an alternative field-modeling strategy for electromagnetic coils. Since the coil is axisymmetric, simulating only the axisymmetric plane is sufficient to obtain the magnetic field distribution of the cylindrical coil. To reduce the mismatch between FEM predictions and experimental measurements, an ANN is trained to capture the modeling error across the field-map domain. Magnetic field data are collected using a 3D magnetic field sensor (Model: TLE493D W2B6, Infineon Inc.), as shown in Figure 6.4 of [24]. This method enhances error modeling, reducing mean error and standard deviation from 6.33% and 5.95% to 2.26% and 4.31%, respectively, compared to the function-based modeling in [24]. Alternatively, an ANN can be trained purely from measurement data to achieve high accuracy as well [39]. However, this requires a much larger dataset and longer training time. With the trained ANN, the dense FEM field map can be calibrated, which further supports accurate computation of the field gradient in subsequent steps.

In the third step, coordinate transformations are performed to express $\mathbf{b}_k^{\mathcal{C}_k}\left(\mathbf{p}_{\mathrm{W}}^{\mathcal{C}_k}\right)$ and $\mathbf{G}_k^{\mathcal{C}_k}\left(\mathbf{p}_{\mathrm{W}}^{\mathcal{C}_k}\right)$ in the global coordinate frame, which is required to compute the combined magnetic field generated by all three coils. Specifically, the magnetic field contributions of the three coils are mapped into the global frame through rotational coordinate transformations.

$$\mathbf{b}_k^{\mathcal{G}} = {}^{\mathcal{G}}\mathbf{R}_{\mathcal{C}_k}\left(\mathbf{p}_{\mathrm{W}}^{\mathcal{C}_k}, \mathbf{p}_{\mathrm{P}}^{\mathcal{G}}\right) \cdot \mathbf{b}_k^{\mathcal{C}_k}\left(\mathbf{p}_{\mathrm{W}}^{\mathcal{C}_k}\right) \tag{6.11}$$

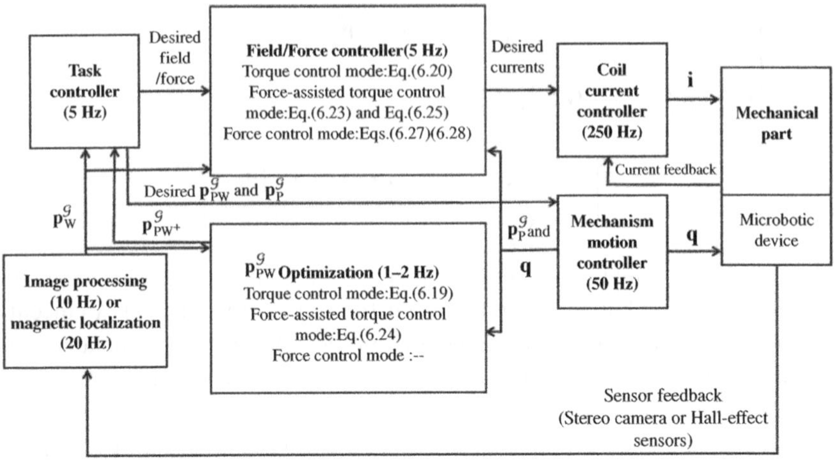

Figure 6.4 Unified control architecture of the PMCS.

where ${}^{\mathcal{G}}\mathbf{R}_{\mathcal{C}_k}\left(\mathbf{p}_{\mathrm{W}}^{\mathcal{C}_k}, \mathbf{p}_{\mathrm{P}}^{\mathcal{G}}\right) \in \mathbb{R}^{3\times3}$ denotes the rotation matrix that maps vectors from the $\{\mathcal{C}_k\}$ to $\{\mathcal{G}\}$. Unlike the magnetic field vector, the magnetic gradient matrix does not transform as a simple rotational quantity. Therefore, we first compute the unit magnetic forces in each coil's local coordinate frame as:

$$\begin{aligned}\bar{\mathbf{f}}_k^{\mathcal{C}_k}\left(\mathbf{p}_{\mathrm{W}}^{\mathcal{C}_k}, \mathbf{m}^{\mathcal{G}}\right) &= \bar{\mathbf{G}}_k^{\mathcal{C}_k}\left(\mathbf{p}_{\mathrm{W}}^{\mathcal{C}_k}\right) \cdot \mathbf{m}^{\mathcal{C}_k} \\ &= \bar{\mathbf{G}}_k^{\mathcal{C}_k}\left(\mathbf{p}_{\mathrm{W}}^{\mathcal{C}_k}\right) \cdot {}^{\mathcal{C}_k}\mathbf{R}_{\mathcal{G}}\left(\mathbf{p}_{\mathrm{W}}^{\mathcal{G}}, \mathbf{p}_{\mathrm{P}}^{\mathcal{G}}\right) \cdot \mathbf{m}^{\mathcal{G}}\end{aligned} \tag{6.12}$$

where ${}^{\mathcal{C}_k}\mathbf{R}_{\mathcal{G}}(\mathbf{p}_{\mathrm{W}}^{\mathcal{G}}, \mathbf{p}_{\mathrm{P}}^{\mathcal{G}}) = {}^{\mathcal{G}}\mathbf{R}_{\mathcal{C}_k}^{-1}\left(\mathbf{p}_{\mathrm{W}}^{\mathcal{C}_k}, \mathbf{p}_{\mathrm{P}}^{\mathcal{G}}\right)$, and $\mathbf{m}^{\mathcal{G}}$ denotes the microrobot's magnetic moment. The resulting unit magnetic forces are then mapped into the global coordinate frame as follows:

$$\begin{aligned}\bar{\mathbf{f}}_k^{\mathcal{G}}\left(\mathbf{p}_{\mathrm{W}}^{\mathcal{C}_k}, \mathbf{m}^{\mathcal{G}}\right) &= {}^{\mathcal{G}}\mathbf{R}_{\mathcal{C}_k}\left(\mathbf{p}_{\mathrm{W}}^{\mathcal{C}_k}, \mathbf{p}_{\mathrm{P}}^{\mathcal{G}}\right) \cdot \bar{\mathbf{f}}_k^{\mathcal{G}}\left(\mathbf{p}_{\mathrm{W}}^{\mathcal{C}_k}, \mathbf{m}^{\mathcal{G}}\right) \\ &= {}^{\mathcal{G}}\mathbf{R}_{\mathcal{C}_k}\left(\mathbf{p}_{\mathrm{W}}^{\mathcal{C}_k}, \mathbf{p}_{\mathrm{P}}^{\mathcal{G}}\right) \cdot \bar{\mathbf{G}}_k^{\mathcal{C}_k}\left(\mathbf{p}_{\mathrm{W}}^{\mathcal{C}_k}\right) \cdot {}^{\mathcal{C}_k}\mathbf{R}_{\mathcal{G}}\left(\mathbf{p}_{\mathrm{W}}^{\mathcal{G}}, \mathbf{p}_{\mathrm{P}}^{\mathcal{G}}\right) \cdot \mathbf{m}^{\mathcal{G}}\end{aligned} \tag{6.13}$$

Based on the magnetic field superposition principle, the resultant magnetic field and magnetic force at $\mathbf{p}_{\mathrm{W}}^{\mathcal{G}}$ can be obtained as:

$$\begin{cases}\mathbf{b}^{\mathcal{G}}\left(\mathbf{p}_{\mathrm{W}}^{\mathcal{G}}\right) = \left[\overline{\mathbf{b}}_1^{\mathcal{G}}\left(\mathbf{p}_{\mathrm{W}}^{\mathcal{G}}\right) \overline{\mathbf{b}}_2^{\mathcal{G}}\left(\mathbf{p}_{\mathrm{W}}^{\mathcal{G}}\right) \overline{\mathbf{b}}_3^{\mathcal{G}}\left(\mathbf{p}_{\mathrm{W}}^{\mathcal{G}}\right)\right] \cdot \mathbf{i} = \overline{\mathbf{B}} \cdot \mathbf{i} \\ \mathbf{f}^{\mathcal{G}}\left(\mathbf{p}_{\mathrm{W}}^{\mathcal{G}}, \mathbf{m}^{\mathcal{G}}\right) = \left[\mathbf{f}_1^{\mathcal{G}}\left(\mathbf{p}_{\mathrm{W}}^{\mathcal{G}}, \mathbf{m}^{\mathcal{G}}\right) \mathbf{f}_2^{\mathcal{G}}\left(\mathbf{p}_{\mathrm{W}}^{\mathcal{G}}, \mathbf{m}^{\mathcal{G}}\right) \mathbf{f}_3^{\mathcal{G}}\left(\mathbf{p}_{\mathrm{W}}^{\mathcal{G}}, \mathbf{m}^{\mathcal{G}}\right)\right] \cdot \mathbf{i} \\ \qquad = \overline{\mathbf{F}} \cdot \mathbf{i}\end{cases} \tag{6.14}$$

where $\mathbf{i} = [i_1 i_2 i_3]^{\mathrm{T}}$ is the vector of coil currents.

6.3 Multimode Control and Validation

This section presents a multimode control strategy for the PMCS, enabling it to accommodate different actuation conditions and various microrobot designs. Three control modes are introduced: torque control, force-assisted torque control, and force control. The torque control mode targets devices that primarily require torque output, while the force-assisted torque mode improves actuation efficiency for microrobots with low propulsion efficiency. In addition, the force control mode is specifically developed for microrobots driven mainly by magnetic forces. By exploiting the mobility of the PMCS, these modes optimize coil placement to enhance actuation capability while reducing energy consumption. The section further describes how the three modes can be incorporated into a unified control framework and explains their use in achieving accurate motion tracking and control over large workspaces.

6.3.1 Torque Control Mode

In the simplified torque-control setting, which can be treated as a field-control problem, the goal is to compute the required current vector $-\mathbf{i}$ for a specified desired magnetic field $-\mathbf{b}^{\mathcal{G}}(\mathbf{p}_{\mathrm{W}}^{\mathcal{G}})$. For clarity, $\mathbf{b}^{\mathcal{G}}(\mathbf{p}_{\mathrm{W}}^{\mathcal{G}})$ is referred to as b. A metric is derived to assess the impact of variations in $\mathbf{p}_{\mathrm{W}}^{\mathcal{G}}$ on magnetic field generation performance. The unit direction of the desired field $-\hat{\mathbf{b}}$ is defined as:

$$\hat{\mathbf{b}} = \begin{bmatrix} \cos\beta\cos\alpha & \cos\beta\sin\alpha & \sin\beta \end{bmatrix}^{\mathrm{T}} \tag{6.15}$$

where β and α denote the pitch and yaw angles of the magnetic field, respectively. From Eq. (6.14), it follows that:

$$\mathbf{b} = B\hat{\mathbf{b}} = B \cdot \begin{bmatrix} \cos\beta\cos\alpha & \cos\beta\sin\alpha & \sin\beta \end{bmatrix}^{\mathrm{T}} = \overline{\mathbf{B}} \cdot \mathbf{i} \tag{6.16}$$

In this context, B denotes the magnetic field strength in the direction $\hat{\mathbf{b}}$ when a coil current i is applied. A higher B signifies that the PMCS can produce the same field with a reduced current i, indicating that B can quantify energy consumption for direction $\hat{\mathbf{b}}$. By rearranging Eq. (6.16), it follows that $\mathbf{i} = B\bar{\mathbf{B}}^{-1}\hat{\mathbf{b}}$. Assuming the coil current is a unit vector, i.e., $\mathbf{i}^{\mathrm{T}}\mathbf{i} = 1$, the subsequent analysis proceeds as follows:

$$B^2\hat{\mathbf{b}}^{\mathrm{T}}(\bar{\mathbf{B}}^{-1})^{\mathrm{T}}\bar{\mathbf{B}}^{-1}\hat{\mathbf{b}} = 1 \tag{6.17}$$

Rearranging (6.17), we have:

$$B = \frac{1}{\sqrt{\hat{\mathbf{b}}^{\mathrm{T}}(\bar{\mathbf{B}}\,\bar{\mathbf{B}}^{\mathrm{T}})^{-1}\hat{\mathbf{b}}}} \tag{6.18}$$

The optimization goal is to maximize B for the specified field direction. For dynamic fields, such as rotating fields, N directions should be considered for B', i.e., $B' = \sum_{n=1}^{N} \frac{1}{\sqrt{\hat{\mathbf{b}}_n^{\mathrm{T}}(\bar{\mathbf{B}}\,\bar{\mathbf{B}}^{\mathrm{T}})^{-1}\hat{\mathbf{b}}_n}}$. Let B'_{P} and $B'_{\mathrm{P+}}$ denote the values of $B'$when the coil-end plate is located at the current and next positions, respectively. Then, optimizing energy consumption is equivalent to minimizing $1 - \frac{B'_{\mathrm{P+}}}{B'_{\mathrm{P}}}$. Other forms resulting in the minimization of $-B'_{p+}$ are also applicable, e.g., $B'_{\mathrm{P}} - B'_{\mathrm{P+}}$. To prevent abrupt long-range movements of the coils, the coil motion should also be penalized in the optimization, e.g., through the term $\|\mathbf{p}^{\mathcal{G}}_{\mathrm{PW+}} - \mathbf{p}^{\mathcal{G}}_{\mathrm{PW}}\|_{xy}$. Accordingly, the overall optimization problem can be formulated as follows:

$$\begin{aligned} \min_{\mathbf{p}^{\mathcal{G}}_{\mathrm{PW+}}} \quad & W_{\mathrm{b}} \cdot \|\mathbf{p}^{\mathcal{G}}_{\mathrm{PW+}} - \mathbf{p}^{\mathcal{G}}_{\mathrm{PW}}\|_{xy} + \left(1 - \frac{B'_{\mathrm{P+}}}{B'_{\mathrm{P}}}\right) \\ \text{subject to} \quad & \|\mathbf{p}^{\mathcal{G}}_{\mathrm{PW+}}\|_{xy} \leq R_s \end{aligned} \tag{6.19}$$

where

$$B'_{\mathrm{P+}} = \sum_{n=1}^{N} \frac{1}{\sqrt{\hat{\mathbf{b}}_n^{\mathrm{T}}\left(\bar{\mathbf{B}}_{\mathrm{P+}}\bar{\mathbf{B}}_{\mathrm{P+}}^{\mathrm{T}}\right)^{-1}\hat{\mathbf{b}}_n}},$$

$$B'_{\mathrm{P}} = \sum_{n=1}^{N} \frac{1}{\sqrt{\hat{\mathbf{b}}_n^{\mathrm{T}}\left(\bar{\mathbf{B}}_{\mathrm{P}}\bar{\mathbf{B}}_{\mathrm{P}}^{\mathrm{T}}\right)^{-1}\hat{\mathbf{b}}_n}}.$$

The objective is to maximize B for the desired field direction. W_b dictates the trade-off between coil movement distance and field generation performance improvement percentage. R_s ensures the microrobot remains within the feedback instrument's field of view. The optimization initializes with $W_b = 0$ to start with the optimal $\mathbf{p}^{\mathcal{G}}_{\mathrm{PW}}$. After the initialization stage, W_b is restored to its nominal value. Once the optimal $\mathbf{p}^{\mathcal{G}}_{\mathrm{PW}}$ is determined, the coil-end plate position $\mathbf{p}^{\mathcal{G}}_{\mathrm{P}}$ is calculated based on the localized microrobot position $\mathbf{p}^{\mathcal{G}}_{\mathrm{W}}$. The unit field matrix $\bar{\mathbf{B}}$ is calculated using the modeling approach described in Section 6.2, and the required coil currents are derived by solving the corresponding equations.

$$\mathbf{i} = \bar{\mathbf{B}}^{-1} \cdot \mathbf{b} \tag{6.20}$$

6.3.2 Force-assisted Torque Control Mode

Once the desired magnetic field $\mathbf{b}$ is generated to provide the primary torque actuation, the resulting magnetic force $\mathbf{f}$ (shorthand for $\mathbf{f}^{\mathcal{G}}(\mathbf{p}^{\mathcal{G}}_{\mathrm{W}}, \mathbf{m}^{\mathcal{G}})$) can be evaluated using (6.14). Assuming $\mathbf{m}^{\mathcal{G}}$ aligns with $\mathbf{b}^{\mathcal{G}}(\mathbf{p}_{\mathrm{W}}^{\mathcal{G}})$, the magnetic force can be computed using the following relationship:

$$\mathbf{m}^{\mathcal{G}} = \|\mathbf{m}^{\mathcal{G}}\| \cdot \hat{\mathbf{b}} \tag{6.21}$$

The goal is to determine the optimal future $\mathbf{p}_{\mathrm{PW}}^{\mathcal{G}}$ such that the resulting magnetic force $\mathbf{f}$ provides the greatest assistance to the intended microrobot motion. Let $\hat{\mathbf{f}}_{\mathrm{d}}$ denote the desired force direction; then, the effective force contribution can be quantified by:

$$C_f(\mathbf{f}, \hat{\mathbf{f}}_{\mathrm{d}}) = \mathbf{f}^{\mathrm{T}} \cdot \hat{\mathbf{f}}_{\mathrm{d}} - \sqrt{\|\mathbf{f}\|^2 - (\mathbf{f}^{\mathrm{T}} \cdot \hat{\mathbf{f}}_{\mathrm{d}})^2} \tag{6.22}$$

The proposed optimization algorithm aims to accommodate control strategies for microrobots driven by dynamic periodic magnetic fields, such as rotating-field actuation. Forces $\mathbf{f}_{1+} \cdots \mathbf{f}_{N+}$ are calculated corresponding to each candidate of the optimal $\mathbf{p}_{\mathrm{PW+}}^{\mathcal{G}}$ for given $\hat{\mathbf{b}}_1 \cdots \hat{\mathbf{b}}_N$

$$\begin{cases} \mathbf{i}_n = \bar{\mathbf{B}}^{-1} \cdot \hat{\mathbf{b}}_n, & n = 1, \dots, N \\ \mathbf{f}_{n+} = \bar{\mathbf{F}} \cdot \mathbf{i}_n, & n = 1, \dots, N \end{cases} \tag{6.23}$$

Then, the optimal $\mathbf{p}_{\mathrm{PW+}}^{\mathcal{G}}$ is obtained by solving:

$$\begin{aligned} \min_{\mathbf{p}_{\mathrm{PW+}}^{\mathcal{G}}} \quad & W_{\mathrm{fb}} \cdot \|\mathbf{p}_{\mathrm{PW+}}^{\mathcal{G}} - \mathbf{p}_{\mathrm{PW}}^{\mathcal{G}}\|_{xy} - \sum_{n=1}^{N} \frac{C_f(\mathbf{f}_{n+}.xy, \hat{\mathbf{f}}_{\mathrm{d}}.xy)}{N} \\ \text{subject to} \quad & \|\mathbf{p}_{\mathrm{PW+}}^{\mathcal{G}}\|_{xy} \le R_s \end{aligned} \tag{6.24}$$

In the proposed optimization formulation, the first term suppresses abrupt long-range movements of the coil-end plate, whereas the second term determines the optimal $\mathbf{p}_{\mathrm{PW+}}^{\mathcal{G}}$ that maximizes the xy plane force component along the desired xy direction. The relative weight between these terms is determined by W_{fb}. Before the actuation begins, the optimization is initialized by setting $W_{\mathrm{fb}} = 0$, ensuring that the system starts from the optimal $\mathbf{p}_{\mathrm{P}}^{\mathcal{G}}$. After initialization, W_{fb} is restored to its nominal value.

Once the optimized $\mathbf{p}_{\mathrm{PW+}}^{\mathcal{G}}$ is acquired, the required magnetic field magnitude is evaluated to determine the coil current vector $\mathbf{i}$. Since the magnetic force scales linearly with the field strength, the desired field magnitude can be computed by dividing the target z-direction magnetic force by the mean z-direction force generated under a unit field, which is expressed as $\frac{\mathbf{f}_{\mathrm{d}.z}}{\frac{1}{N}\sum_{n=1}^{N} \mathbf{f}_{n+,z}}$. Consequently, $\mathbf{i}$ can be derived using this method.

$$\mathbf{i} = \frac{\mathbf{f}_{\mathrm{d}}.z}{\frac{1}{N} \cdot \sum_{n=1}^{N} \mathbf{f}_{n+}.z} \cdot \bar{\mathbf{B}}^{-1} \cdot \hat{\mathbf{b}} \tag{6.25}$$

6.3.3 Force Control Mode

Considering (6.13), (6.14), and (6.21), the magnetic force acting on an untethered magnetic microrobot in a fluid environment can be expressed as:

$$\mathbf{f} = [{}^{\mathcal{G}}\mathbf{R}_{C_1} \cdot \bar{\mathbf{G}}_1^{C_1} \cdot {}^{C_1}\mathbf{R}_{\mathcal{G}} \cdot \bar{\mathbf{B}} \cdot \mathbf{i}, {}^{\mathcal{G}}\mathbf{R}_{C_2} \cdot \bar{\mathbf{G}}_2^{C_2} \cdot {}^{C_2}\mathbf{R}_{\mathcal{G}} \cdot \bar{\mathbf{B}} \cdot \mathbf{i},$$
$${}^{\mathcal{G}}\mathbf{R}_{C_3} \cdot \bar{\mathbf{G}}_3^{C_3} \cdot {}^{C_3}\mathbf{R}_{\mathcal{G}} \cdot \bar{\mathbf{B}} \cdot \mathbf{i}] \cdot \frac{\|\mathbf{m}^{\mathcal{G}}\|}{\|\bar{\mathbf{B}} \cdot \mathbf{i}\|}\mathbf{i} \tag{6.26}$$

where $\mathbf{f}$, ${}^{\mathcal{G}}\mathbf{R}_{C_1}$, and $\bar{\mathbf{G}}_1^{C_1}$ are abbreviations of $\mathbf{f}^{\mathcal{G}}(\mathbf{p}_{\mathrm{W}}^{\mathcal{G}})$, ${}^{\mathcal{G}}\mathbf{R}_{C_k}(\mathbf{p}_{\mathrm{W}}^{C_k}, \mathbf{p}_{\mathrm{P}}^{\mathcal{G}})$, and $\bar{\mathbf{G}}_k^{C_k}(\mathbf{p}_{\mathrm{W}}^{C_k})$, respectively. Deriving the current vector $\mathbf{i}$ for a specified desired force $\mathbf{f}_{\mathrm{d}}$ cannot be achieved through an analytical solution. Therefore, an iterative optimization approach is developed to solve this inverse problem.

In the approach outlined, the management of magnetic force along the z-axis is prioritized over the xy axes due to the persistent presence of net gravitational force F_{g} (the result of gravity minus buoyancy). Consequently, Eq. (6.27) is considered a constraint in the optimization process.

$$\mathbf{f}.z = \mathbf{f}_{\mathrm{d}}.z = F_{\mathrm{g}} + f_z \tag{6.27}$$

The net force in the z-axis is denoted by f_z. To achieve rapid convergence for real-time control, the objective for the force components along the x and y axes is to align the direction of $\mathbf{f}.xy$ with that of $\mathbf{f}_{\mathrm{d}}.xy$, without specifying the magnitude. Consequently, the optimization problem can be formulated as follows:

$$\begin{aligned} &\min_{\mathbf{i}} W_{\mathrm{f}1} \cdot \begin{cases} \|\mathrm{YAW}(\mathbf{f}) - \mathrm{YAW}(\mathbf{f}_{\mathrm{d}})\|, & \text{for } \mathbf{f}_{\mathrm{d}}.x \neq 0 \text{ or } \mathbf{f}_{\mathrm{d}}.y \neq 0 \\ \|\mathbf{f}.x + \mathbf{f}.y\|, & \text{for } \mathbf{f}_{\mathrm{d}}.x = 0 \text{ and } \mathbf{f}_{\mathrm{d}}.y = 0 \end{cases} \\ &-W_{\mathrm{f}2} \cdot (\|\mathbf{f}.x\| + \|\mathbf{f}.y\| + \|\mathbf{i}\|) \\ &\text{subject to } \begin{cases} \mathbf{f}.z = \mathbf{f}_{\mathrm{d}}.z = F_{\mathrm{g}} + f_z \\ \mathbf{i}_0 = [-1\ -1\ -1]^{\mathrm{T}} \\ [-i_{\mathrm{bd}}\ -i_{\mathrm{bd}}\ -i_{\mathrm{bd}}]^{\mathrm{T}} \leq \mathbf{i} \leq [i_{\mathrm{bd}}\ i_{\mathrm{bd}}\ i_{\mathrm{bd}}]^{\mathrm{T}} \end{cases} \end{aligned} \tag{6.28}$$

The initial value of $\mathbf{i}$ is denoted by $\mathbf{i}_0$, and i_{bd} represents the limit of coil current. The function $\mathrm{YAW}(\mathbf{f})$ calculates the yaw angle of $\mathbf{f}$, defined as $\mathrm{YAW}(\mathbf{f}) = \arctan 2(\mathbf{f}.y, \mathbf{f}.x)$. $W_{\mathrm{f}1}$ and $W_{\mathrm{f}2}$ are positive weights. In the cost function, the first and third terms are introduced to reduce the xy-plane error of the generated magnetic force and the coil current vector, respectively. The second term is formulated to enhance the magnitude of the magnetic force in the xy direction. It has been observed that solving this optimization problem using MATLAB's '*fmincon*' function typically requires on the order of tens of milliseconds, making real-time selection of the optimal $\mathbf{p}^{\mathcal{G}}_{\mathrm{PW+}}$ by evaluating Eq. (6.28) for a set of candidates challenging. Instead of selecting $\mathbf{p}^{\mathcal{G}}_{\mathrm{PW+}}$ in real time, the selection is achieved through simulation for a specific task (e.g., a 3D path). This approach yields the appropriate $\mathbf{p}^{\mathcal{G}}_{\mathrm{PW+}}$ that ensures correct convergence results for Eq. (6.28) for the given task.

6.3.4 Validation

The implementation of the three control modes within a unified framework is depicted in Figure 6.4. It is necessary to apply the relevant equations of each control mode within the field/force controller and the $\mathbf{p}^{\mathcal{G}}_{\mathrm{PW+}}$ optimization algorithm. The desired coil currents and mechanism motion (i.e., $\mathbf{p}^{\mathcal{G}}_{\mathrm{PW+}}$ and $\mathbf{p}^{\mathcal{G}}_{\mathrm{P}}$) are updated at a frequency of 5 Hz by the task controller and field/force controller to achieve the desired motion of the microrobotic device. The coil current controller and the mechanism motion controller run at 250 Hz and 50 Hz, respectively, to guarantee accurate magnetic field generation and motion execution. The computational speed of the $\mathbf{p}^{\mathcal{G}}_{\mathrm{PW+}}$ optimization algorithm depends on the number of candidate points evaluated to identify the optimal $\mathbf{p}^{\mathcal{G}}_{\mathrm{PW+}}$. The search space is discretized by defining the polar radius r_c and polar angle ϕ_c of the candidates within the polar coordinates, where $\|\mathbf{p}^{\mathcal{G}}_{\mathrm{PW+}}\|_{xy} \leq R_s$.

$$\begin{cases} \mathbf{r}_c = [0, 0.2, 0.4, 0.6, 0.8, 1] \cdot R_s \\ \phi_c = \left[\dfrac{360^\circ}{C_\phi}, \dfrac{360^\circ}{C_\phi} \cdot 2, \dots, \dfrac{360^\circ}{C_\phi} \cdot (C_\phi - 1), 360^\circ\right] \end{cases} \tag{6.29}$$

The constant C_ϕ determines the total number of $\mathbf{p}^{\mathcal{G}}_{\mathrm{PW+}}$ candidates, which is $5C_\phi + 1$. A higher value of C_ϕ yields better optimization outcomes but requires more computation time. The selection of C_ϕ should be based on the computer's computational power. Given that microrobotic devices typically operate at speeds not exceeding several millimeters per second, an optimization frequency of 1 to 2 Hz for $\mathbf{p}^{\mathcal{G}}_{\mathrm{PW+}}$ is adequate for microrobotic actuation.

6.4 System Structure Design with Parameter Optimization

This section introduces the proposed approach for determining the key structural parameters of the PMCS. The design procedure starts from a target workspace, which serves as a fundamental requirement for any robotic system. Given the axisymmetric nature of the PMCS, the desired workspace is defined as a cylindrical volume with radius R_{W} and a height of H_{W}. According to Eq. (6.6), the system kinematics can be categorized into two cases. In this work, we derive the parameter design method for $0° \leq \theta \leq 45°$, which can be readily extended to the other case.

6.4.1 The Methodology for PMCS Design

In Figure 6.5, the alignment of the workspace's symmetry axis and its bottom center with $\hat{\mathbf{z}}^{\mathcal{G}}$ and $O^{\mathcal{G}}$ is evident. The main objective of the design is to enhance both the mechanical performance and magnetic actuation capability of the PMCS, while keeping the overall system size as compact as possible. The figure also shows the plane extended by the symmetry axis of the workspace $\hat{\mathbf{z}}^{\mathcal{G}}$ and the direction of the kth sliding track ($\hat{\mathbf{n}}_a^{\mathcal{G}}$). Here, ϕ is defined as the angle of intersection between a coil branch and its corresponding sliding track. It is observed that within this plane, ϕ attains its extreme values—namely, the minimum $-\phi'$ and the maximum $-\phi''$, when the coil-end plate is located at points P′ and P″, respectively. To circumvent singularities in the parallel mechanism, it is crucial that no coil branch is positioned perpendicularly to its sliding track. Moreover, considering mechanical constraints, the coil branch should also be prevented from becoming parallel to its sliding track. Consequently, the extreme values of ϕ must adhere to specific constraints.

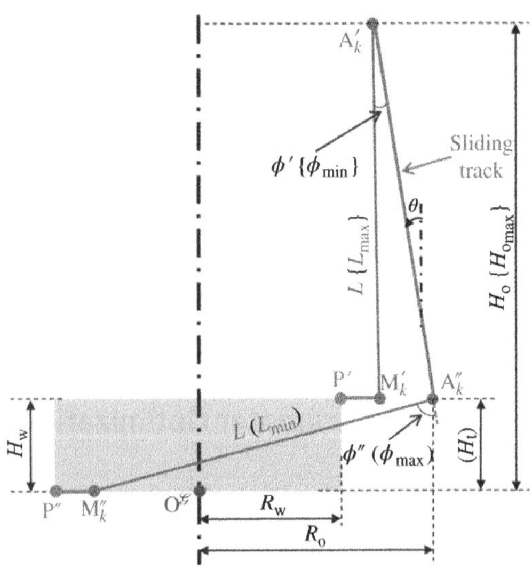

Figure 6.5 Illustration of the plane spanned by the workspace symmetry axis ($\hat{\mathbf{z}}^{\mathcal{G}}$) and the direction of the kth sliding track ($\hat{\mathbf{n}}_a^{\mathcal{G}}$).

$$\begin{cases} \phi' \geq \phi_{\min} > 0° \\ \phi'' \leq \phi_{\max} < 90° \end{cases} \tag{6.30}$$

where $\phi_{\min}$ and $\phi_{\max}$ are constants that are determined with flexibility, and it is recommended that $\phi_{\min} \in [5°, 10°]$ and $\phi_{\max} \in [80°, 85°]$.

To balance minimizing the overall system radius R_0 while maintaining sufficient space for mounting mechanical components, R_0 is selected as follows:

$$R_0 = 1.5\, R_w + L_{MP} \tag{6.31}$$

Referring to Figure 6.5, it can be seen that the horizontal coordinate of the sliding track's bottom end is R_o. To decrease the overall system height H_0, the vertical coordinate H_t at the lower end of each sliding track should be minimized. This minimum is achieved when ϕ'' attains $\phi_{\max}$, which can be expressed as:

$$H_t = \frac{R_o + R_w - L_{MP}}{\tan(\phi_{\max} - \theta)} \tag{6.32}$$

With the sliding track locations defined, the next step is to determine the remaining structural parameters, namely the coil branch length L_{AM} and the sliding track length L_{st}, which are the structural parameters yet to be resolved. To simplify the discussion, L_{AM} will henceforth be referred to as L. Recognizing the pivotal role L plays in the system's actuation performance, optimizing its value becomes essential. This optimization begins with defining the range $[L_{\min}, L_{\max}]$. As depicted in Figure 6.5, the coil branch length hits its minimum threshold, $L_{\min}$, when ϕ'' aligns with $\phi_{\max}$. This minimum length can be ascertained through a specific calculation:

$$L_{\min} = \frac{R_o + R_w - L_{MP}}{\sin(\phi_{\max} - \theta)} \tag{6.33}$$

Similarly, L attains its maximum permissible value, $L_{\max}$, when ϕ' equals $\phi_{\min}$. Utilizing the Sine theorem, $L_{\max}$ can be derived as follows:

$$L_{\max} = \frac{(R_o - R_w - L_{MP} - \Delta H \tan\theta)\sin(90° - \theta)}{\sin\phi_{\min}} \tag{6.34}$$

where

$$\Delta H = H_w - \frac{R_o + R_w - L_{MP}}{\tan(\phi_{\max} - \theta)} \tag{6.35}$$

The permissible value of H_w is contingent upon θ and can be ascertained through structural constraints: (1) $L_{\max} \geq L_{\min}$ establishes the upper limit of H_w, referred to as $H_{w\,\max}$; (2) The condition A_k' not aligning with the system's symmetry axis, i.e., $L_{\max}\sin(\theta - \phi_{\min}) < R_w + L_{MP}$, sets the lower limit for H_w, known as $H_{w\,\min}$. Consequently, $H_{w\max}$ is determined as the solution to the following:

$$L_{\max} = L_{\min} \tag{6.36}$$

and H_{wmin} would be the solution of:

$$L_{\max} \cdot \sin(\theta - \phi_{\min}) = R_{\text{w}} + L_{\text{MP}} \tag{6.37}$$

Since H_{w} must exceed zero, it follows that $\text{Max}(0, H_{\text{wmin}}) \leq H_{\text{w}} \leq H_{\text{wmax}}$. The term H_{wmax} is disregarded if $\theta = 0$, given that $L_{\max}$ surpasses $L_{\min}$ for any H_{w}. Additionally, H_{wmin} is omitted if $\theta \leq \phi_{\min}$, since A'_k would not coincide with the system's symmetry axis in this case.

6.4.2 Metrics of Actuation Performances

Performance indices are required to identify the optimal L. During the optimization, it is recommended to evaluate both the mechanical motion capability and the magnetic actuation performance.

In terms of motion actuation performance, the parallel mechanism—designed according to the proposed structural parameter selection method—is guaranteed to operate without encountering singular configurations. However, its kinematic properties must be taken into account to make an optimal selection of L. To achieve this, the Jacobian matrix $\mathbf{J}_m$ of the parallel mechanism for a given L is formulated as described. Substituting Eqs. (6.2) and (6.4) into Eq. (6.5) yields:

$$(x_{\text{P}} - x_k)^2 + (y_{\text{P}} - y_k)^2 + (z_{\text{P}} - z_k)^2 = L \tag{6.38}$$

where

$$\begin{cases} x_k = (L_{\text{h}} + q_k \sin\theta)\cos a_k - L_{\text{MP}}\cos a_k \\ y_k = (L_{\text{h}} + q_k \sin\theta)\sin a_k - L_{\text{MP}}\sin a_k \\ z_k = L_{\text{v}} - q_k\cos\theta \end{cases} \tag{6.39}$$

L_{h} and L_{v} have the following expressions:

$$\begin{cases} L_{\text{h}} = R_{\text{w}} + L_{\text{MP}} - L\sin(\theta - \phi') \\ L_{\text{v}} = H_{\text{o}} = H_{\text{w}} + L\cos(\theta - \phi') \end{cases} \tag{6.40}$$

where ϕ' is calculated by:

$$\phi' = \arcsin\left(\frac{(R_{\text{o}} - R_{\text{w}} - L_{\text{MP}} - \Delta H\tan\theta)\sin(90^\circ - \theta)}{L}\right) \tag{6.41}$$

Taking the time derivative of Eq. (6.38) results in:

$$(x_{\text{P}} - x_k)\dot{x}_{\text{P}} + (x_{\text{P}} - y_k)\dot{y}_{\text{P}} + (z_{\text{P}} - z_k)\dot{z}_{\text{P}} = D_k\dot{q}_k \tag{6.42}$$

where

$$D_k = (x_P - x_k)\sin\theta\cos a_k + (y_P - y_k)\sin\theta\sin a_k - (z_P - z_k)\cos\theta \tag{6.43}$$

Rearranging Eq. (6.42), then we have:

$$\dot{\mathbf{q}} = \mathbf{J}_m \dot{\mathbf{p}}_P^{\mathcal{G}} \tag{6.44}$$

The Jacobian matrix $\mathbf{J}_m$ reads:

$$\mathbf{J}_m = \begin{bmatrix} \frac{x_P - x_1}{D_1} & \frac{y_P - y_1}{D_1} & \frac{z_P - z_1}{D_1} \\ \frac{x_P - x_2}{D_2} & \frac{y_P - y_2}{D_2} & \frac{z_P - z_2}{D_2} \\ \frac{x_P - x_3}{D_3} & \frac{y_P - y_3}{D_3} & \frac{z_P - z_3}{D_3} \end{bmatrix} \tag{6.45}$$

Accordingly, the reciprocal of the condition number of $\mathbf{J}_m$ is selected as the motion actuation performance metric:

$$\mu_m = \frac{1}{\kappa_m} \tag{6.46}$$

The condition $\kappa_m = ||\mathbf{J}_m^{1}|| \cdot ||\mathbf{J}_m||$ is defined, with $|| \cdot ||$ denoting the second norm of the matrix. The value of μ_m ranges between [0 and 1], signifying the proximity to actuation singularity. $\mu_m = 0$ suggests that the parallel mechanism has reached a singular state, rendering it uncontrollable. Conversely, a higher μ_m value indicates a greater distance from singularities and enhanced isotropic actuation capability. Although μ_m is a local performance index, its workspace-wide statistics—namely the average value, AVG(μ_m) and the minimum value MIN(μ_m), can be used to evaluate the overall motion isotropy and the proximity to singular configurations across the entire workspace [40].

To evaluate magnetic actuation performance throughout the workspace, the condition number and its distribution over the workspace can likewise be adopted as effective metrics [27]. Unlike mechanical mechanisms, coil currents can be manipulated at very high speeds. Thus, instead of examining kinematics, the condition number of $\bar{\mathbf{B}}$ and/or $\bar{\mathbf{F}}$ in Eq. (6.14) is directly employed for performance evaluation, contingent upon the system's specific application. For instance, for torque control systems, AVG(μ_b) and the minimum MIN(μ_b) are considered.

$$\mu_b = \frac{1}{\kappa_b} \tag{6.47}$$

where $\kappa_b = ||\mathbf{B}_m^{-1}|| \cdot ||\mathbf{B}_m||$. To characterize the overall system performance, a combined metric Q is defined as:

$$
\begin{aligned}
Q = {} & W_1 \cdot \mathrm{AVG}(\mu_m) + W_2 \cdot \mathrm{MIN}(\mu_m) \\
& + W_3 \cdot \mathrm{AVG}(\mu_b) + W_4 \cdot \mathrm{AVG}(\mu_b)
\end{aligned}
\tag{6.48}
$$

W_1 to W_4 represent constant weights. Ultimately, the optimal L prevents actuation singularities, and the choice of W_1 to W_4 balances optimal mechanical and magnetic actuation. The absolute field magnitude is not required in the performance metrics, since the parallel-mobile-coil mechanism keeps the coils close to the microrobot, and the achievable field strength is mainly determined by the coil design and selection. Similarly, for pure force control applications, such as the control of floating microrobots [41], magnetic actuation performance is characterized using $\mathrm{AVG}(\mu_f)$ and $\mathrm{MIN}(\mu_f)$.

Once Q is defined, the optimal coil branch length L_{opt} that maximizes Q can be determined through offline simulations. Subsequently, the corresponding ϕ'_{opt} and ϕ''_{opt} are obtained by substituting L_{opt} into Eqs. (6.41) and (6.49), respectively.

$$
\phi''_{opt} = \arcsin\left(\frac{L_{min}\sin(180^\circ - \phi_{max})}{L_{opt}}\right) \tag{6.49}
$$

The required length of the sliding track L_{st} is calculated as follows:

$$
\begin{aligned}
L_{st} = {} & \frac{(R_o - R_w - L_{MP} - \Delta H \tan\theta)\sin(90^\circ + \theta - \phi'_{opt})}{\sin\phi'_{opt}} \\
& + \frac{\Delta H}{\cos\theta} - \frac{L_{min}\sin(\phi_{max} - \phi''_{opt})}{\sin\phi''_{opt}}
\end{aligned}
\tag{6.50}
$$

Now, the sliding track parameters have been determined.

6.5 Validation of Microrobotic Navigation Control

This section verifies the system's ability to generate the expected field and force by applying algorithms in Section 6.2. It then executes three navigation tasks: semi-automated control of a miniature helical robot within a 3D lumen, as well as fully automated navigation using two closed-loop controllers for torque-driven and force-driven microrobots, demonstrating high-resolution motion control over a large workspace.

6.5.1 Semi-automated Navigation Control of Miniature Helical Robot in 3D Lumen

To assess the system's 3D motion navigation and control capabilities, a miniature helical robot ($\Phi 4 \times 10$ mm) depicted in Figure 6.6 is maneuvered within a 3D lumen structure with a diameter of 8 mm. The experimental setup in Figure 6.6 features a lumen with a cuboid volume of $200 \times 200 \times 60$ mm^3. The xy position of the coil-end plate mirrors that of the helical robot, with $||\mathbf{p}_{\mathrm{PW}}^{\mathcal{G}}|$ maintained at 50 mm. For magnetic flux densities of 9, 5, and 2 mT, the maximum achievable rotation frequencies are 6 Hz, 11 Hz, and 23 Hz, respectively. A small permanent magnet ($\Phi 1 \times 2$ mm) is embedded in the robot's head. In a 2 mT magnetic field, the robot's step-out frequency is determined to be 8 Hz, leading to the selection of 2 mT and 5 Hz for the rotating magnetic field to drive the helical robot. Precision tracking of the robot is achieved by the system through automatic updates of the coil-end plate's 3D position via a stereo camera. The robot's center is extracted via thresholding, followed by triangulation to determine its 3D position. During the experiment, manual keyboard inputs adjust the magnetic field's pitch and yaw angles to control the robot's swimming pose, enabling movement within the 3D lumen. The experiment verifies effective manipulation of the helical robot inside the lumen, and the navigation trajectory shown in Figure 6.7 is consistent with the physical experimental setup.

6.5.2 Automated Control of Rolling Magnetic Microrobot

Rolling microrobots are prevalent in microrobotics [42]. In this study, the system is employed to autonomously steer a microscopic magnet, as depicted in Figure 6.8. This micro-magnet moves on the acrylic tank's substrate, leveraging friction and fluid forces. To establish a low Reynolds number setting, 96 wt% glycerol is utilized at a depth of 45 mm. Traditional triangulation's localization errors due to air-fluid interface refraction are mitigated by a refraction-corrected method for 3D microrobot localization [43].

In this control scheme, the coil-end plate's xy position is automatically adjusted to follow the microrobot's motion, with $|\mathbf{p}_{\mathrm{PW}}|$ maintained at 55 mm. A rotating magnetic field with a magnitude of 2 mT is generated in planes perpendicular to the xy plane to actuate the microrobot. The dynamic model of the rolling microrobot is given by:

$$\begin{cases} \dot{x}_{\mathrm{rm}} = a_0 f \cos(\alpha) + d_x \\ \dot{y}_{\mathrm{rm}} = a_0 f \sin(\alpha) + d_y \end{cases} \tag{6.51}$$

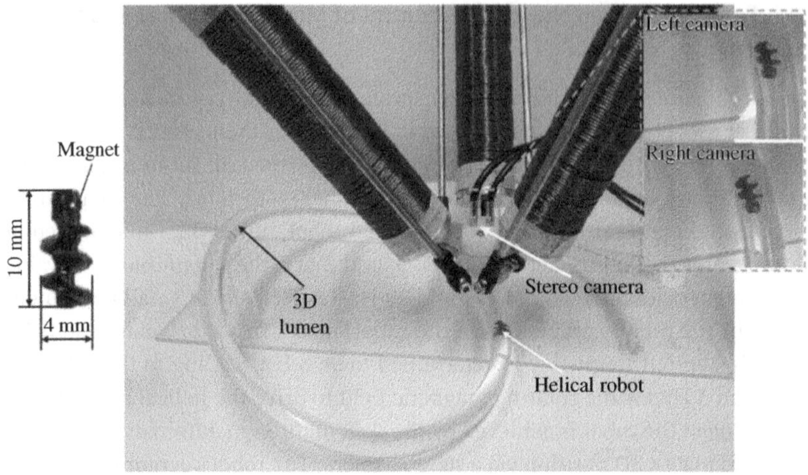

Figure 6.6 Semi-automated navigation control of a miniature helical robot inside a 3D lumen filled with 96 wt% glycerol. A photograph and dimensions of the miniature robot. Experimental setup. The lumen spans a 3D cuboid workspace of 200 × 200 × 60 mm. Rotating magnetic fields (strength: 2 mT; frequency: 5 Hz) are applied to actuate the robot. The pitch and yaw angles of the fields are manually adjusted to regulate the robot's swimming posture.

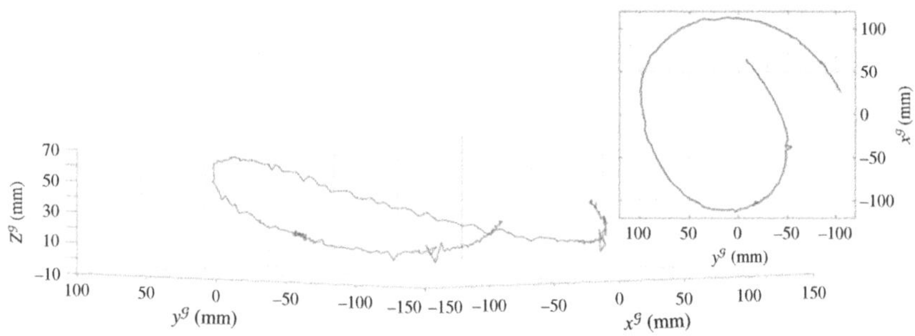

Figure 6.7 Robot navigation trajectory, with the inset providing the corresponding *xy*-plane view.

The system is composed of two elements: a baseline model and a lumped disturbance $\mathbf{d}_{\mathrm{rm}}^{\mathcal{G}} = [d_x \; d_y]^{\mathrm{T}}$. The disturbance $\mathbf{d}_{\mathrm{rm}}^{\mathcal{G}}$ arises from various factors including imprecise robot manufacturing, flawed working surfaces, and erroneously generated fields. The parameter a_0, estimated experimentally as 1.5, represents a constant. The variables f and α denote the rotation frequency and yaw angle of the rotating magnetic field, respectively. To mitigate these disturbances, a sliding

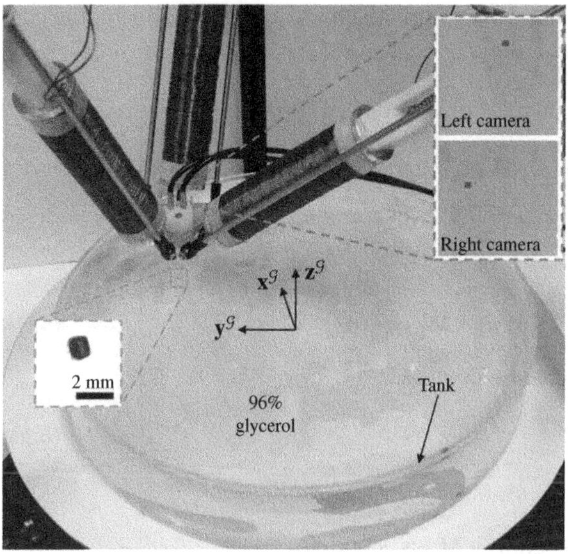

Figure 6.8 Experimental setup, where the microrobot rolls on the substrate of an acrylic tank filled with 96 wt% glycerol. The dashed rectangles indicate the stereo-camera feedback images and the close-up photographs of the rolling microrobot, respectively.

mode controller (SMC) is developed for automated motion regulation, as it is well known for strong robustness against external uncertainties and disturbances.

1) SMC Design: To facilitate controller development, the dynamic model is first reformulated into a state-space representation as:

$$\mathbf{p}_{\mathrm{rm}}^{\mathcal{G}} = a_0\mathbf{u} + \mathbf{d}_{\mathrm{rm}}^{\mathcal{G}} \tag{6.52}$$

The position vector is denoted as $\mathbf{p}_{\mathrm{rm}}^{\mathcal{G}} = [x_{\mathrm{rm}}\ y_{\mathrm{rm}}]^{\mathrm{T}}$, while the control input vector is represented by $\mathbf{u} = [u_x\ u_y]^{\mathrm{T}} = [f\cos(\alpha)\ f\sin(\alpha)]^{\mathrm{T}}$. The sliding surface $\mathbf{s} = \mathbf{0}$ delineates the desired system state. To achieve tracking of the desired trajectory $-\mathbf{p}_{\mathrm{d}}^{\mathcal{G}} = [x_{\mathrm{d}}\ y_{\mathrm{d}}]^{\mathrm{T}}$, the sliding surface is selected as follows:

$$\mathbf{s} = \mathbf{p}_{\mathrm{d}}^{\mathcal{G}} - \mathbf{p}_{\mathrm{rm}}^{\mathcal{G}} \tag{6.53}$$

To guarantee convergence of the system states to the sliding surface, the SMC reaching law is defined as:

$$\dot{\mathbf{s}} = -(M_1 + M_2|\mathbf{s}|) \cdot \mathrm{Sgn}(\mathbf{s}) = -M_1 \cdot \mathrm{Sgn}(\mathbf{s}) - M_2 \cdot \mathbf{s} \tag{6.54}$$

In this context, M_1 and M_2 are constant positive gain parameters, and $\mathrm{Sgn}(\cdot)$ denotes the sign function, which returns the sign of its argument. By integrating

Eqs. (6.52) through (6.54) and disregarding the disturbance component, the control input $\mathbf{u}$ is formulated as follows:

$$\mathbf{u} = \frac{1}{a_0}\left[\dot{\mathbf{p}}_{\mathrm{d}}^{\mathcal{G}} + M_1 \cdot \mathrm{Sgn}(\mathbf{p}_{\mathrm{d}}^{\mathcal{G}} - \mathbf{p}_{\mathrm{rm}}^{\mathcal{G}}) + M_2 \cdot (\mathbf{p}_{\mathrm{d}}^{\mathcal{G}} - \mathbf{p}_{\mathrm{rm}}^{\mathcal{G}})\right] \tag{6.55}$$

By substituting the control law (6.55) into the system model (6.52), we get:

$$\dot{\mathbf{p}}_{\mathrm{rm}}^{\mathcal{G}} = \dot{\mathbf{p}}_{\mathrm{d}}^{\mathcal{G}} + M_1 \cdot \mathrm{Sgn}(\mathbf{s}) + M_2 \cdot \mathbf{s} + \mathbf{d}_{\mathrm{rm}}^{\mathcal{G}} \tag{6.56}$$

By Eq. (6.53), we have:

$$\dot{\mathbf{s}} = \dot{\mathbf{p}}_{\mathrm{d}}^{\mathcal{G}} - \dot{\mathbf{p}}_{\mathrm{rm}}^{\mathcal{G}} = -M_1 \cdot \mathrm{Sgn}(\mathbf{s}) - M_2 \cdot \mathbf{s} - \mathbf{d}_{\mathrm{rm}}^{\mathcal{G}} \tag{6.57}$$

Select the Lyapunov candidate function as:

$$V = \frac{1}{2}\mathbf{s}^T\mathbf{s} \tag{6.58}$$

Then,

$$\begin{aligned} \dot{V} &= \mathbf{s}^{\mathrm{T}}\dot{\mathbf{s}} = -M_1 \cdot \mathbf{s}^{\mathrm{T}} \cdot \mathrm{Sgn}(\mathbf{s}) - M_2 \cdot \mathbf{s}^{\mathrm{T}}\mathbf{s} - \mathbf{s}^{\mathrm{T}}\mathbf{d}_{\mathrm{rm}}^{\mathcal{G}} \\ &\leq -M\_1 \cdot \|\mathbf{s}\| - \mathbf{s}^{\mathrm{T}} d_{\mathrm{rm}}^{\mathcal{G}} \\ &= \|\mathbf{s}\| \cdot \left[-M_1 - \mathrm{Sgn}^{\mathrm{T}}(\mathbf{s})\mathbf{d}_{\mathrm{rm}}^{\mathcal{G}}\right] \end{aligned} \tag{6.59}$$

Consequently, if M_1 is chosen to exceed each component of $\mathbf{d}_{\mathrm{rm}}^{\mathcal{G}}$, meaning M_1 surpasses the disturbances in both the x and y directions, then $\dot{V}$ becomes negative definite, leading to $\mathbf{s}$ converging toward $\mathbf{0}$. This ensures that the closed-loop system achieves exponential stability. Ultimately, the input field frequency and yaw angle are computed as follows:

$$\begin{cases} f = \sqrt{u_x^2 + u_y^2} \\ \alpha = \arctan 2\left(\dfrac{u_y}{u_x}\right) \end{cases} \tag{6.60}$$

The experimental determination of the disturbance bound involves the microrobot's navigation through the workspace, with its yaw angle adjusted manually. By contrasting the experimental position data against the nominal position derived from the nominal model, the bound of $\mathbf{d}_{\mathrm{rm}}^{\mathcal{G}}$ can be established. This process may be executed either manually or through the design of a disturbance observer, such as an extended state observer (ESO) [44], for Eq. (6.51).

2) Trajectory Tracking Experiments: The rolling microrobot is commanded to follow an "∞"-shaped trajectory described by:

$$\begin{cases} x_{\mathrm{d}} = -100\sin(0.005\pi t) \\ y_{\mathrm{d}} = -80\sin(0.01\pi t) \end{cases} \tag{6.61}$$

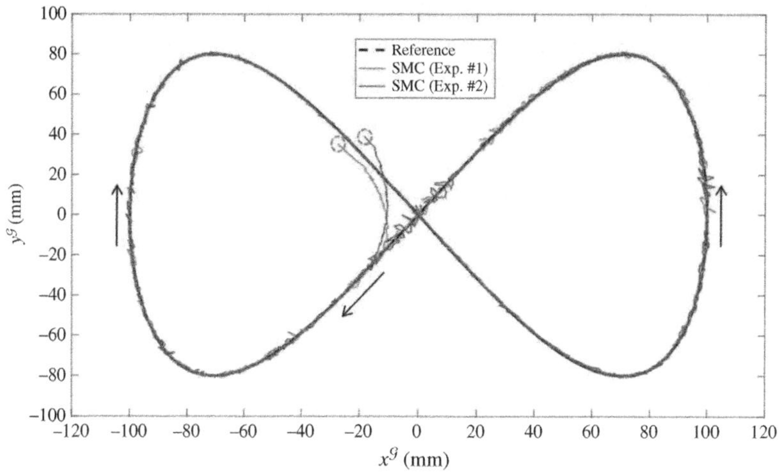

(a)

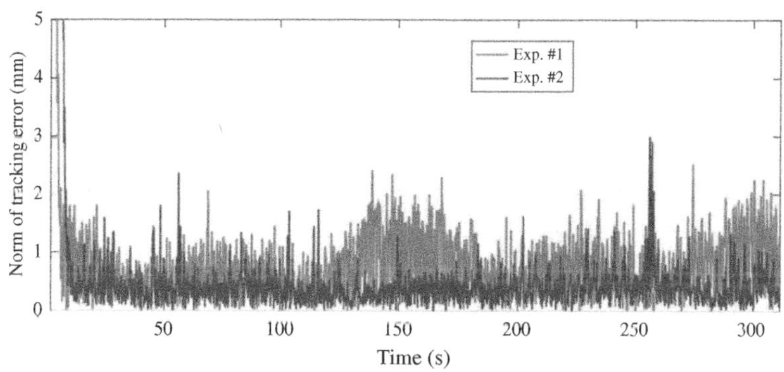

(b)

Figure 6.9 (a) The trajectory tracking results using the SMC in two experiments (Exp. #1 and Exp. #2). Arrows show the motion direction of the microrobot. The dashed circles mark the starting points. (b) Norms of the tracking errors in the two experiments. The observed variations are mainly attributed to uncertainties arising from the magnetic field generated by the mobile coils and the microrobot's operating environment.

The workspace diameter significantly exceeds that of the microrobot, with a ratio surpassing 200, underscoring the system's high control precision and expansive operational area. The SMC, characterized by $M_1 = 0.5$ and $M_2 = 0.05$, is implemented to manage these dynamics.

As shown in Figure 6.9a, the experimental results validate the SMC's effectiveness in achieving stable and accurate trajectory tracking. Specifically, in the first experiment (Exp. #1), the xy plane's tracking errors average at 0.87 mm, peaking at 2.45 mm. To substantiate the control's robustness, the experiment is repeated, yielding consistent results. Furthermore, the findings from another experiment (Exp. #2), presented in Figure 6.9a,b, mirror the accuracy observed in the initial trial.

6.5.3 Automated Control of a Force-driven Floating Microrobot

Three-coil magnetic actuation systems can achieve pure force control with stationary coils. This capability is now showcased for a floating microrobot using the optimally designed PMCS. As seen in Figure 6.10, the microrobot is affixed to a larger plastic disk. Immersed in a 96% glycerol solution, it is buoyed to the surface, with its magnetic moment confined to the $\mathbf{z}^{\mathcal{G}}$-axis ($\mathbf{m}_{\text{fm}}^{\mathcal{G}} = [0\ 0\ 1]^{\text{T}}$). Throughout the experiment, $|\mathbf{p}_{\text{PW}}^{\mathcal{G}}|$ is held at 50 mm. The setup is illustrated in Figure 6.11, where the goal is to automate the microrobot's circular trajectory (diameter: 160 mm).

$$\begin{cases} x_d = 80\cos(0.005\pi t) \\ y_d = 80\sin(0.005\pi t) \end{cases} \tag{6.62}$$

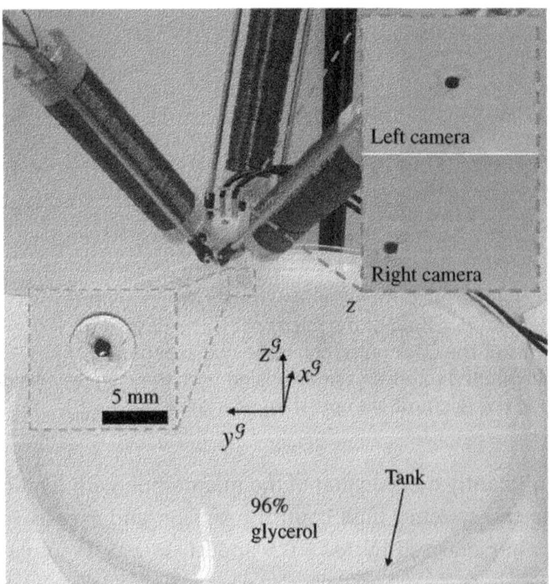

Figure 6.10 The experimental scenario. The dashed rectangles indicate the feedback images and the close-up photograph of the floating microrobot, respectively.

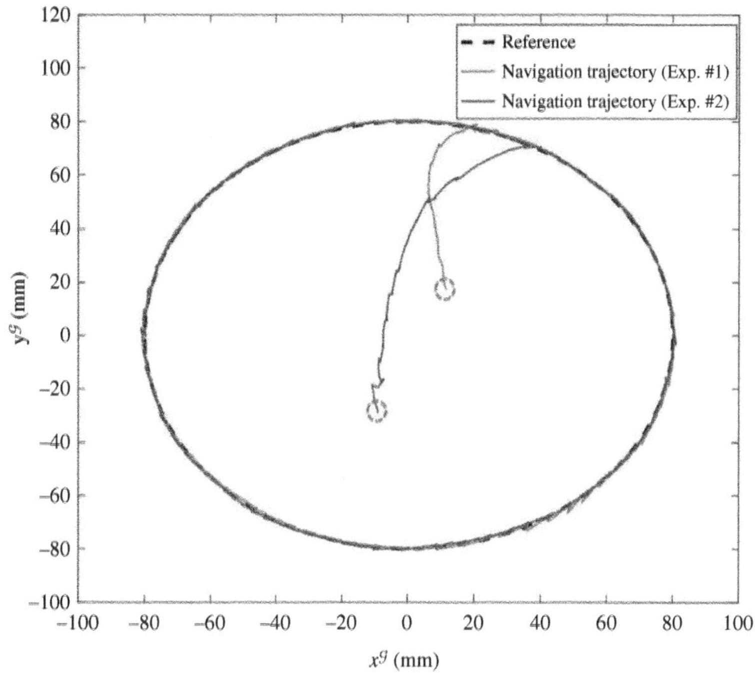

Figure 6.11 Experimental results of the trajectory tracking control in two experiments (Exp. #1 and Exp. #2). The small dashed circles mark the start points.

To achieve this goal, a proportional controller is employed to compute the desired force as:

$$\mathbf{f}^{\mathcal{G}}(\mathbf{p}_{\mathrm{W}}^{\mathcal{G}}, \mathbf{m}_{\mathrm{fm}}^{\mathcal{G}}) = K_{\mathrm{p}} \cdot \begin{bmatrix} x_d - x_{\mathrm{fm}} \\ y_d - y_{\mathrm{fm}} \\ 0 \end{bmatrix} \tag{6.63}$$

Upon simulation verification that $\bar{\mathbf{F}}$ is free of singularities along the intended path, Eq. (6.14) can be inverted to compute the required $\mathbf{i}$, as follows:

$$\mathbf{i} = \bar{\mathbf{F}}^{-1} \cdot \mathbf{f}^{\mathcal{G}}(\mathbf{p}_{\mathrm{W}}^{\mathcal{G}}, \mathbf{m}^{\mathcal{G}}) \tag{6.64}$$

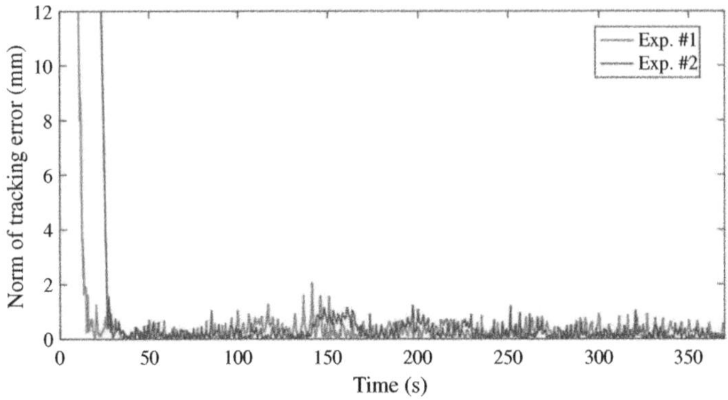

Figure 6.12 Time evolution of the trajectory-tracking error norm.

Given the redundancy of three coils in the 2D control scenarios, a nonlinear optimization approach is employed to refine coil currents and limit their maximum values. This method is formulated to enhance the optimization process.

$$\begin{aligned} \underset{\mathbf{i}}{\text{Minimize}} \quad & \|\mathbf{i}\|^2 \\ \text{subject to} \quad & \bar{\mathbf{F}} \cdot \mathbf{i} = \mathbf{f}^{\mathcal{G}}(\mathbf{p}_{\mathrm{W}}^{\mathcal{G}}, \mathbf{m}_{\mathrm{fm}}^{\mathcal{G}}) \\ & -i_1 \leq i_k \leq i_1, \quad k = 1, 2, 3. \end{aligned} \tag{6.65}$$

where i_1 denotes the upper limit of the coil current. The optimization then yields an optimal $\mathbf{i}$ that satisfies this current constraint. Following parameter tuning, K_{p} is set to 0.001 and i_1 is set to 3 A. The trajectory tracking outcomes from two experiments (Exp. #1 and Exp. #2), depicted in Figures 6.11 and 6.12, the results confirm that the proposed control approach enables the PMCS to realize precise and robust automated microrobotic motion control. As illustrated in Figure 6.12, after the controller converges, the maximum tracking error along the circular trajectory remains below 2 mm. The constraint in Eq. (6.65) enforces an upper limit on the coil currents, crucial for preventing overheating or potential hazards. If Eq. (6.64) were directly applied as the controller, the coil currents would saturate during the convergence phase. This experiment further demonstrates the effectiveness and accuracy of force-based microrobotic navigation control using the developed PMCS.

The results from three microrobotic navigation experiments demonstrate that the system can effectively actuate and control different types of microrobots within the workspace, thereby validating the accuracy of the proposed modeling and control algorithms. Since the optimization metrics are based on these modeling formulations, the corresponding performance metric values are also verified to be reliable.

6.6 Discussion and Conclusion

This chapter presents an overview of the multimode control strategies for the PMCS and introduces a unified control framework that broadens its applicability to different microrobot types and actuation scenarios. In addition, it investigates the optimal system design and navigation control of the PMCS. To streamline the redesign process of such systems, a computational approach is proposed, facilitating the creation of PMCS with optimized parameters based on specific workspace requirements. Utilizing this method, a prototype PMCS with dimensions of 230 × 100 mm^3 has been developed. Its effectiveness was demonstrated through the control of diverse microrobot types. Specifically, three magnetic microrobots actuated by either torque or force were successfully navigated under semi-automated and fully automated configurations. Notably, the implementation of automated controllers enables the system to achieve high-precision and large-scale navigation control.

Future research aims to explore the expansion of PMCS coil numbers beyond three to enhance magnetic control capabilities for concurrent magnetic field and force control.

References

1 C. Heunis, J. Sikorski, and S. Misra. Flexible instruments for endovascular interventions: improved magnetic steering, actuation, and image-guided surgical instruments. *IEEE Robotics and Automation Magazine*, 25 (3): 71–82, 2018.

2 Metin Sitti, et al. Biomedical applications of untethered mobile milli/microrobots. *Proceedings of the IEEE*, 103 (2): 205–224, 2015.

3 P. E. Dupont, et al. A decade retrospective of medical robotics research from 2010 to 2020. *Science Robotics*, 6 (60): eabi8017, 2021.

4 D. Jang, J. Jeong, H. Song, and S. K. Chung. Targeted drug delivery technology using untethered microrobots: a review. *Journal of Micromechanics and Microengineering*, 29 (5): 053002, 2019.

5 L. Yang, Y. Zhang, Q. Wang, K.-F. Chan, and L. Zhang. Automated control of magnetic spore-based microrobot using fluorescence imaging for targeted delivery with cellular resolution. *IEEE Transactions on Automation Science and Engineering*, 17 (1): 490–501, 2020.

6 X. Wang, et al. Intracellular manipulation and measurement with multipole magnetic tweezers. *Science Robotics*, 4 (28): eaav6180, 2019.

7 O. Felfoul, et al. Magneto-aerotactic bacteria deliver drug-containing nanoliposomes to tumour hypoxic regions. *Nature Nanotechnology*, 11 (11): 941, 2016.

8 J. Li, et al. Development of a magnetic microrobot for carrying and delivering targeted cells. *Science Robotics*, 3 (19): eaat8829, 2018.

9 J. W. Martin, et al. Enabling the future of colonoscopy with intelligent and autonomous magnetic manipulation. *Nature Machine Intelligence*, 2 (10): 595–606, 2020.

10 C. Chautems and B. J. Nelson. The tethered magnet: force and 5-DOF pose control for cardiac ablation. In *Proceedings of the IEEE International Conference on Robotics and Automation (ICRA)*, pages 4837–4842, 2017.

11 A.-M. Singeap. Capsule endoscopy: the road ahead. *World Journal of Gastroenterology*, 22 (1): 369, 2016.

12 J. J. Abbott, E. Diller, and A. J. Petruska. Magnetic methods in robotics. *Annual Review of Control, Robotics, and Autonomous Systems*, 3 (1): 57–90, 2020.

13 L. Zheng, et al. 3D navigation control of untethered magnetic micro-robot in centimeter-scale workspace based on field-of-view tracking scheme. *IEEE Transactions on Robotics*, 38 (3): 1583–1598, 2022.

14 A. Oulmas, N. Andreff, and S. Regnier. 3D closed-loop swimming at low Reynolds numbers. *International Journal of Robotics Research*, 37 (11): 1359–1375, 2018.

15 T. Xu, C. Huang, Z. Lai, and X. Wu. Independent control strategy of multiple magnetic flexible millirobots for position control and path following. *IEEE Transactions on Robotics*, 38 (5): 2875–2887, 2022.

16 L. Yang, J. Jiang, X. Gao, Q. Wang, Q. Dou, and L. Zhang. Autonomous environment-adaptive microrobot swarm navigation enabled by deep learning-based real-time distribution planning. *Nature Machine Intelligence*, 4 (5): 480–493, 2022.

17 L. Yang, J. Yu, S. Yang, B. Wang, B. J. Nelson, and L. Zhang. A survey on swarm microrobotics. *IEEE Transactions on Robotics*, 38 (3): 1531–1551, 2022.

18 J. Edelmann, A. J. Petruska, and B. J. Nelson. Magnetic control of continuum devices. *International Journal of Robotics Research*, 36 (1): 68–85, 2017.

19 A. W. Mahoney and J. J. Abbott. Five-degree-of-freedom manipulation of an untethered magnetic device in fluid using a single permanent magnet with application in stomach capsule endoscopy. *International Journal of Robotics Research*, 35 (1–3): 129–147, 2016.

20 S. E. Wright, A. W. Mahoney, K. M. Popek, and J. J. Abbott. The spherical-actuator-magnet manipulator: a permanent-magnet robotic end-effector. *IEEE Transactions on Robotics*, 33 (5): 1013–1024, 2017.

21 J. Sikorski, I. Dawson, A. Denasi, E. E. Hekman, and S. Misra. Introducing BigMag: a novel system for 3D magnetic actuation of flexible surgical manipulators. In *Proceedings of the IEEE International Conference on Robotics and Automation (ICRA)*, pages 3594–3599, 2017.

22 B. Veron, A. Hubert, J. Abadie, and N. Andreff. Dealing with redundancy of a multiple mobile coil magnetic manipulator: a 3RPR magnetic parallel kinematics manipulator. In *Advances in Robot Kinematics*, pages 201–208, Cham, Switzerland, 2018. Springer.

23 X. Du, M. Zhang, J. Yu, L. Yang, P. W. Y. Chiu, and L. Zhang. Design and real-time optimization for a magnetic actuation system with enhanced flexibility. *IEEE/ASME Transactions on Mechatronics*, 26 (3): 1524–1535, 2021.

24 L. Yang, X. Du, E. Yu, D. Jin, and L. Zhang. DeltaMag: an electromagnetic manipulation system with parallel mobile coils. In *Proceedings of the IEEE International Conference on Robotics and Automation (ICRA)*, pages 9814–9820, 2019.

25 A. Pourkand and J. J. Abbott. A critical analysis of eight-electromagnet manipulation systems: the role of electromagnet configuration on strength, isotropy, and access. *IEEE Robotics and Automation Letters*, 3 (4): 2957–2962, 2018.

26 L. Manamanchaiyaporn, T. Xu, and X. Wu. An optimal design of an electromagnetic actuation system towards a large homogeneous magnetic field and accessible workspace for magnetic manipulation. *Energies*, 13 (4): 911, 2020.

27 R. Chen, D. Folio, and A. Ferreira. Mathematical approach for the design configuration of magnetic system with multiple electromagnets. *Robotics and Autonomous Systems*, 135: 103674, 2021.

28 C. M. Heunis, Y. P. Wotte, J. Sikorski, G. P. Furtado, and S. Misra. The ARMM system-autonomous steering of magnetically-actuated catheters: towards endovascular applications. *IEEE Robotics and Automation Letters*, 5 (2): 705–712, 2020.

29 Z. Yang, L. Yang, M. Zhang, Q. Wang, C. H. Simon, and L. Zhang. Magnetic control of a steerable guidewire under ultrasound guidance using mobile electromagnets. *IEEE Robotics and Automation Letters*, 6 (2): 1280–1287, 2021.

30 J. Sikorski, A. Denasi, G. Bucchi, S. Scheggi, and S. Misra. Vision-based 3D control of magnetically actuated catheter using BigMag—an array of mobile electromagnetic coils. *IEEE/ASME Transactions on Mechatronics*, 24 (2): 505–516, 2019.

31 L. Yang and L. Zhang. Large-workspace and high-resolution magnetic microrobot navigation using global-local path planning and eye-in-hand visual servoing. In *Proceedings of the IEEE International Conference on Automation Science and Engineering (CASE)*, pages 876–881, 2020.

32 M. Larbi, E.-H. Guechi, A. Chah, A. Maidi, and K. Belharet. Sliding mode observer of a two-agent microrobotic system. In *Proceedings of the 5th International Conference on Advanced Systems and Emergent Technologies (ICASET)*, pages 268–273, 2022.

33 J. J. Jiang, L. Yang, and L. Zhang. Closed-loop control of a Helmholtz coil system for accurate actuation of magnetic microrobot swarms. *IEEE Robotics and Automation Letters*, 6 (2): 827–834, 2021.

34 L. Sadelli, M. Fruchard, and A. Ferreira. 2D observer-based control of a vascular microrobot. *IEEE Transactions on Automatic Control*, 62 (5): 2194–2206, 2017.

35 X. Huo, B. Lian, P. Wang, Y. Song, and T. Sun. Dynamic identification of a tracking parallel mechanism. *Mechanical Systems and Signal Processing*, 155: 104091, 2021.

36 H. Shen, Q. Meng, J. Li, J. Deng, and G. Wu. Kinematic sensitivity, parameter identification and calibration of a non-fully symmetric parallel delta robot. *Mechanical Systems and Signal Processing*, 161: 104311, 2021.

37 A. J. Petruska, J. Edelmann, and B. J. Nelson. Model-based calibration for magnetic manipulation. *IEEE Transactions on Magnetics*, 53 (7): 1–6, 2017.

38 M. Richter, V. K. Venkiteswaran, and S. Misra. Multi-point orientation control of discretely-magnetized continuum manipulators. *IEEE Robotics and Automation Letters*, 6 (2): 3607–3614, 2021.

39 R. Yu, et al. Modeling electromagnetic navigation systems for medical applications using random forests and artificial neural networks. In *Proceedings of the IEEE International Conference on Robotics and Automation (ICRA)*, pages 9251–9256, 2020.

40 J. P. Merlet. Jacobian, manipulability, condition number, and accuracy of parallel robots. *Journal of Mechanical Design*, 128 (1): 199–206, 2006.

41 A. Barbot, H. Tan, M. Power, F. Seichepine, and G.-Z. Yang. Floating magnetic microrobots for fiber functionalization. *Science Robotics*, 4 (34): eaax8336, 2019.

42 R. Pieters, H.-W. Tung, S. Charreyron, D. F. Sargent, and B. J. Nelson. RodBot: a rolling microrobot for micromanipulation. In *Proceedings of the IEEE International Conference on Robotics and Automation (ICRA)*, pages 4042–4047, 2015.

43 Z. Yang, L. Yang, and L. Zhang. Eye-in-hand 3D visual servoing of helical swimmers using parallel mobile coils. In *Proceedings of the IEEE International Conference on Robotics and Automation (ICRA)*, pages 9223–9229, 2020.

44 L. Yang, Q. Wang, and L. Zhang. Model-free trajectory tracking control of two-particle magnetic microrobot. *IEEE Transactions on Nanotechnology*, 17 (4): 697–700, 2018.

7

Autonomous Microrobot Navigation Using Parallel-mobile-coil System

This chapter proposes a novel navigation framework, using a home-designed eye-in-hand mobile-coil system. And a double-loop motion controller is designed to make the microrobot accurately follow the computed path and intelligently avoid collisions. Moreover, to enable robust tracking with pulsatile flow, an iterative statistics-based strategy is proposed to extract the position and orientation of the robot from the ultrasound (US) imaging.

7.1 Introduction

Microrobots exhibit considerable promise in biomedical applications [1–3] and micromanipulation tasks [4], owing to their capacity to navigate at low Reynolds numbers and traverse confined spaces. In contrast to their larger counterparts, integrating onboard components such as processors or power sources into microbots remains challenging; hence, alternative methods like optical, chemical, and biological actuation have been explored [5]. Among the various techniques available, the magnetic field stands out due to its ability to penetrate deeply into tissues and its high level of safety [6]. Furthermore, it offers a range of field types (including rotating and oscillating fields) for versatile propulsion mechanisms and boasts excellent controllability. Consequently, magnetic microrobots have garnered significant attention in recent research endeavors [7–10].

Despite significant advancements in the construction, functional enhancement, and mobility of magnetic microrobots, the realm of autonomous navigation remains an area of ongoing development. In many open settings, these microrobots can traverse a predefined route through closed-loop control mechanisms. However, when confronted with complex environments, the aforementioned method falls short in achieving autonomous guidance of microrobots to their intended destination, as it necessitates the definition of viable, obstacle-circumventing pathways tailored to the specific environmental conditions.

The majority of current autonomous navigation techniques for magnetic microrobots leverage well-established macro-scale planning and control algorithms, albeit with substantial customization and adaptation. Li and colleagues [11] introduced a method grounded in artificial potential fields (APF) to navigate a magnetic Janus microsphere within maze-inspired microstructures. Xie et al. [12] created a system for autonomous movement of a snake-like microrobot swarm, utilizing the DA* path planner. Folio and Ferreira [13] suggested a multi-objective optimal path planning algorithm, rooted in Pareto optimality, to ensure magnetic resonance-based navigation of a microrobot through channels. Liu et al. [14] exhibited an informed optimal rapidly exploring random tree (Informed RRT) path planning strategy, enabling a scaled-up helical swimmer to propel through congested environments. Yang et al. [15] proposed an obstacle-avoidance algorithm employing particle swarm optimization (PSO) to direct the motion of magnetic spores.

Recent research endeavors have witnessed substantial advancements, yet the majority of navigation studies have been confined to relatively compact workspaces. The primary limitations encompass: (1) the restricted workspace of magnetic manipulation setups, such as multiaxis Helmholtz coils, and the inherent challenge in scaling up stationary-coil systems due to the rapid decline in magnetic field strength with increasing distance; and (2) the confined imaging scope of visual modalities, for instance, fixed optical cameras, which stems from the inherent trade-off between achieving a broad field of view and maintaining high imaging resolution. Nevertheless, there is a pressing demand for large-workspace steering in various potential applications, including clinical procedures like intravascular therapy and industrial tasks such as circuit repair [16]. Consequently, it is of great significance to devise a cost-effective approach for enabling autonomous navigation of microrobots within expansive workspaces.

This chapter introduces an innovative electromagnetic actuation system that distinguishes itself from existing systems [17–25] through the integration of parallel-mobile coils and an "eye-in-hand" vision setup. In this context, "eye-in-hand" signifies a configuration where the camera is mounted onto and moves in tandem with the parallel mechanism. This design enables both the coils and the imaging device to come closer to the controlled agent, thereby effectively expanding the magnetic field generation area and imaging range without necessitating higher energy consumption or compromising on imaging accuracy. Subsequently, a navigation framework is proposed, characterized by global accessibility and local adaptability, aimed at achieving autonomous control of magnetic microrobots within large workspaces. This framework contains:

1. Three-step Motion Planner: A motion planner rooted in system architecture is devised to determine global trajectories utilizing a camera with constrained visual range. Initially, a feature-centric image mosaicking technique is introduced for reconstructing the environment. Subsequently, a correlation-based

matching method is incorporated to facilitate self-registration of relative positional data. Ultimately, a refined A* search algorithm is tailored to devise a seamless path traversing the central corridor of the workspace.

2. Double-loop Motion Controller: A motion controller is introduced for precise trajectory tracking and local motion adjustments, tailored for large-workspace applications. This controller ensures synchronized operation between the mechanism's movement and the microrobot's steering. To address the challenges posed by motion uncertainties in microrobot modeling, a disturbance observer (DOB) is designed to estimate and compensate for generalized disturbances. Furthermore, a hybrid control scheme, combining model predictive control (MPC) and fuzzy logic mapping (FLM), is presented to provide adaptive motion control based on real-time environmental conditions.

To achieve precise in vivo control of miniature magnetic robots, medical imaging feedback is indispensable [26]. Magnetic resonance imaging (MRI), known for its high tissue contrast, has been employed for concurrent tracking and actuation of microrobots [27, 28]. However, MRI's capability is confined to generating quasistatic magnetic fields, making it challenging to integrate a magnetic manipulation system for dynamic fields (such as rotating fields essential for helical robot actuation) due to MRI's intense magnetic environment. Consequently, MRI-based magnetic actuation is limited to gradient-induced pulling [29]. Computed tomography (CT), another prevalent medical imaging technique, offers sufficient imaging depth but is concerned with relatively low resolution (approximately 1 mm) and potentially harmful long term radiation exposure when used for microrobot tracking. Recently, optical coherence tomography (OCT) has been utilized for microrobot tracking in mice [30], yet its limited imaging depth (<1 cm) hinders its application in humans. For imaging magnetic miniature robots in the submillimeter to centimeter range, US imaging has proven reliable for precise localization [31–33]. It boasts deep imaging depth and compatibility with magnetic fields. Stationary US-guided motion control has been achieved for magnetic microparticles [34], helical robots [17, 18], and nanoparticle swarms [19]. Nevertheless, the restricted imaging area of a US probe confines the robot's workspace to a few millimeters, inadequate for human applications. Hence, three-dimensional (3D) mobility of the US probe is essential for large-workspace imaging. Additionally, large-workspace magnetic control necessitates either large stationary coils or mobile coils/magnets due to the rapid decline of magnetic fields [20]. Heunis et al. employed two robot arms to independently manipulate the US probe and the coil [21], enabling long-distance endovascular navigation of magnetic catheters. However, this approach is unsuitable for miniature magnetic robots driven by dynamic fields, such as helical [22], rolling [23, 24], and undulating robots [25, 35], as a single coil cannot generate 3D rapidly changing fields. Therefore, long-distance US-guided navigation of untethered magnetic robots necessitates three or more mobile coils and a cooperative mobile US probe. This objective poses significant challenges: (1) sophisticated system design is required for concurrent motion control of multiple coils and the US probe; (2) robust algorithms are necessary for stable position and orientation tracking of untethered miniature robots in pulsatile flow using noisy two-dimensional (2D) US feedback; and (3) intricate algorithms are demanded for precise field control leveraging multiple mobile coils.

In this chapter, we introduced an innovative approach for navigating untethered miniature robots, guided by US, within endovascular environments characterized by pulsatile blood flow over long distances. As depicted in Figure 7.1, our design integrates three mobile coils with a parallel mechanism, enabling them to traverse 3D space and produce dynamic fields essential for actuating miniature magnetic robots (such as helical and rolling robots) across extensive workspaces. A US probe, attached to the "coil-end plate," moves in unison with the coils to

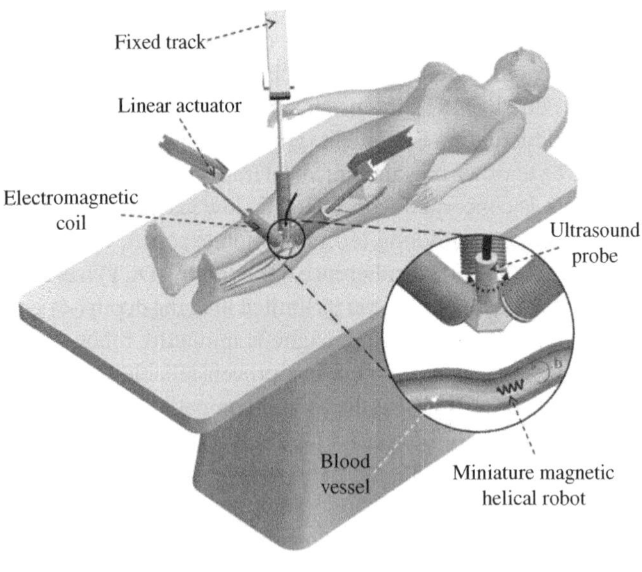

Figure 7.1 Conceptual illustration of the navigation of a miniature magnetic helical robot in blood vessels based on mobile US tracking and magnetic control. Actuated by the parallel mechanism and the probe rotation motor (the probe rotation action is indicated in the enlarged inset), the eye-in-hand US probe can adjust its position and rotation angle to track the robot in a large 3D workspace. Simultaneously, according to the tracked robot position and orientation, the three electromagnetic coils generate demanded rotating magnetic fields **b** for helical propulsion.

provide concurrent US feedback. This probe possesses four motion degrees-of-freedom (DOFs): three translational and one rotational, forming the basis for our proposed methodologies to achieve real-time tracking of the position and orientation of the untethered magnetic robots. Building upon this tracked information, we updated the required 3D actuation magnetic field and adjusted the position/rotation of the US probe in real time, facilitating US-guided endovascular navigation over extended distances. Following the implementation of the approach, we demonstrated the successful 3D endovascular navigation of a miniature helical robot against pulsatile flow within a silicone phantom of constant diameter. Additionally, we showcase the navigation capabilities of the helical robot in a bifurcation phantom and a life-sized replica of human arteries.

7.2 Design of Motion Planner and Motion Controller

7.2.1 Three-step Motion Planner

The aim is to devise a motion planning system capable of producing optimal trajectories. Here, contemplate a scenario featuring a channel-maze environment, where an intimate understanding of the terrain proves instrumental in determining a path that is both obstacle-circumventing and target-reaching. The motion planning approach encompasses three sequential stages, tailored to our magnetic actuation mechanism. Initially, utilize image mosaic techniques to reconstruct the comprehensive map from fragmented submaps, a process that involves aligning internal features and imaging coordinates. Subsequently, develop an image registration method to ascertain the positional relationship between the magnetic actuation system and the reconstructed map, achieving this without resorting to additional markers. Lastly, introduce a refined version of the A* search algorithm, tailored to compute efficient and smooth paths for the microrobot.

1. **Environment Reconstruction:** The system incorporates a close-focusing endoscopy for capturing real-time images. Given the disparity between the extensive workspace, measured in hundreds of millimeters, and the limited view field of just tens of millimeters, gaining a comprehensive understanding of the entire environment is challenging. To address this, a parallel mechanism is employed to manipulate the camera and scan the surroundings initially, resulting in the acquisition of a series of submaps, as illustrated in Figure 7.2a. Numerous approaches have been suggested for solving map recovery challenges. Considering the unique properties of wall boundaries, we have devised a tailored image mosaic strategy. Furthermore, the reconstruction algorithm possesses flexibility and can be adapted to accommodate varying environmental conditions. The image distortions are rectified, and the corrected images are converted to grayscale format. Subsequently, corner points exhibiting distinctive boundary attributes are identified by

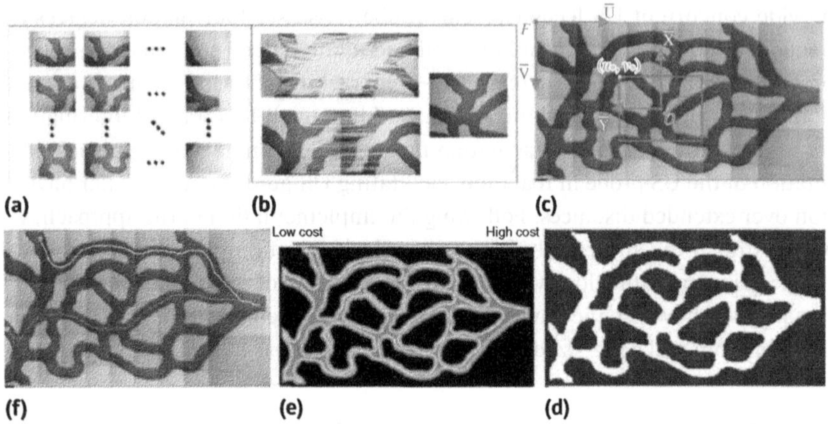

Figure 7.2 Working procedure of the motion planner. (a) A group of submaps acquired from a channel maze. (b) The upper image shows the initial feature detection and matching results. The lower image shows the final matching after outlier removal. The right image is the reference image captured at the home position. (c) Result of map reconstruction by image mosaic. The highlighted block is the registered reference image. *O* is the world coordinate frame and *F* is the map coordinate frame. (d) The gridded map after preprocessing, where light regions are available channels and dark regions are walls. (e) Path priority diagram of a microrobot moving in the channel maze: lighter regions have lower cost and higher priority, darker regions have higher cost and lower priority, and the darkest regions indicate unavailable areas. (f) Two different planned paths from the same start point to different destinations in a channel maze.

computing the minimum eigenvalue ($\lambda_{\min}$) of a structure tensor and assessing them against a predefined threshold (ξ). The submaps undergo convolution using gradient operators, enabling the extraction of the histogram of oriented gradients feature (HOGF) for each identified corner point. To enhance the resilience of feature detection against variations in lighting and shadowing, a normalization step is incorporated. The specific procedures are outlined in Algorithm 1 in the Appendix.

Feature matching is conducted between every adjacent image pair using normalized HOGF. The outliers are removed by two rounds. First, match pairs that dissatisfy the imaging coordinates are kicked (e.g., with match line angles far away from 0 or 180). Second, match pairs that unmatch estimation geometry transformation using the maximum likelihood estimator sample consensus (MLESAC) algorithm are excluded. The matching results are shown in Figure 7.2b. By the computed transformation matrices, submaps can be stitched in sequence. Finally, median filtering is adopted to eliminate noise, and the entire map is shown in Figure 7.2c.

2. **Position Registration:** To align the coordinate system of the reconstructed map with the actuation system, a reference photograph is captured when

the parallel mechanism occupies its initial, or home, position (depicted in Figure 7.2b), at which point the center of the end plate precisely overlays the center of the baseplate. Given the self-locating capability of the parallel mechanism, our system's relative positional registration process necessitates no additional markers. For the purpose of registering the positions between coordinate frames F and O, further elaboration on this method is outlined in Algorithm 2 in Appendix. The resultant alignment, expressed in pixel coordinates relative to frame F, is denoted as (u_*, v_*) (visualized in Figure 7.2c).

3. **Path Planning:** The A* search algorithm stands out as a prominent path planning technique, incorporating elements of heuristic searching akin to greedy best-first-search and the shortest path strategy employed by Dijkstra's algorithm [36]. Initially, binarization was applied to differentiate between the channel area and the walls. Subsequently, image dilation is utilized to broaden the wall representation, thereby incorporating an additional safety margin. Following this, the map is partitioned into uniform grid cells, as illustrated in Figure 7.2d. However, paths generated by the conventional A* algorithm tend to hug the boundaries, where disturbances are pronounced and detrimental to the movement of microrobots. To devise a trajectory that navigates through the central portion of channels, the cost function $\mathcal{F}$ for the current node N_p is formulated as follows:

$$\mathcal{F}(N_p) = \mathcal{G}(N_p) + \mathcal{H}(N_p) + \mathcal{W}(N_p) + \mathcal{T}(N_p) \quad (7.1)$$

where $\mathcal{G}$ represents the precise expenditure function measuring the cost from the initial node to the current node. $\mathcal{H}$ denotes the heuristic estimation of the cost from the current node to the target node. Additionally, $\mathcal{W}$ stands for the weighted cost function associated with each grid in the channel, where grids closer to the wall incur higher weights, as illustrated in Figure 7.2e. Furthermore, $\mathcal{T}$ functions as a process evaluation metric, assessing the proximity of the current node to both the starting and goal nodes. Both $\mathcal{W}$ and $\mathcal{T}$ are considered as factors introducing disturbances, with $\mathcal{W}$ ensuring the planned path maintains a safe distance from walls and $\mathcal{T}$ aiming to minimize excessive turning variations to a certain extent. The search procedure is detailed in Algorithm 3 in the Appendix.

Ultimately, an algorithm for path smoothing is incorporated to alleviate abrupt angle changes along the trajectory. The sequence of points comprising the initial path can be denoted as a set X, containing m points: $X = \{X_1, X_2, \dots, X_m\}$. Let us designate the refined path sequence as another set of m points, $Y = \{Y_1, Y_2, \dots, Y_m\}$. Subsequently, the smoothed path S is obtained by addressing the subsequent optimization challenge:

$$\mathcal{S} = \arg\min_{\mathcal{Y}} \lambda_1 \sum_{i=2}^{m-1} \left\| \mathrm{X}_i \mathrm{Y}_i \right\|^2 + \lambda_2 \sum_{i=2}^{m-1} \left\| \mathrm{Y}_i \mathrm{Y}_{i+1} \right\|^2 \quad (7.2)$$

where λ_1 and λ_2 are parameters deciding the smoothness. The path planning results are shown in Figure 7.2f.

Currently, the nodes of the path list are represented in the frame of reference F. A calibration-derived mapping C exists between image distance and physical distance, and this mapping is contingent upon the working height. By employing position registration techniques, the transformation required for the planned path is determined.

$$\begin{bmatrix} x_{S_j} \\ y_{S_j} \\ 1 \end{bmatrix} = \begin{bmatrix} 0 & -1 & C\left(v_* + \frac{N}{2}\right) \\ -1 & 0 & C\left(u_* + \frac{M}{2}\right) \\ 0 & 0 & 1 \end{bmatrix} \begin{bmatrix} C\left(u_{S_j}^F\right) \\ C\left(v_{S_j}^F\right) \\ 1 \end{bmatrix} \tag{7.3}$$

In the frame of reference F, $(u_{S_j}^F, v_{S_j}^F)$ denotes the pixel coordinates of point S_j. Meanwhile, (x_{S_j}, y_{S_j}) represents the physical coordinates of S_j in the reference frame O. The camera captures images with a resolution of $M \times N$ pixels. Throughout the remainder of this section, a consistent notation convention will be adopted: points are designated by uppercase standard letters, with their coordinates expressed as lowercase italic letters x and y (or u and v) followed by right subscripts indicating the corresponding point and a right superscript denoting the reference coordinate frame (with O being the default).

The number of points is limited by the size of the grid; therefore, the interpolation method is included to produce points with an arbitrary number. The calculated path is then represented by a set of n points as $R = \{R_1, R_2, \dots, R_j, \dots, r_j\}$.

7.2.2 Double-loop Motion Controller

A dual-loop control algorithm has been devised for the system, comprising an inner loop for vision feedback (regulating the motion of the control mechanism) and an outer loop for microrobot steering (regulating the generation of the control magnetic field). In the current implementation of the inner loop, the parallel mechanism is commanded to track and maintain a suitable working height above the magnetic microrobot, ensuring isotropic imaging and stable tracking. Potential future enhancements include optimizing the relative position between the coils and the microrobot. In the outer loop, the DOB estimates and compensates for generalized disturbances, addressing the complexities of hydrodynamics and boundary effects. Meanwhile, flexible trajectory tracking is achieved through a hybrid

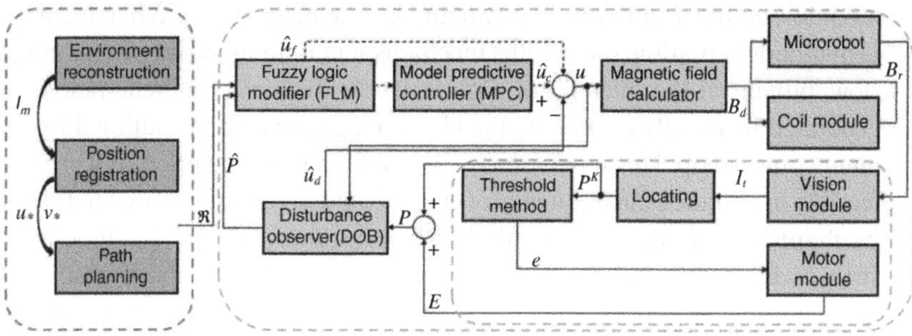

Figure 7.3 Block diagram of the overall autonomous navigation framework of the microrobot using mobile-coil system.

approach combining FLM and MPC, which enhances the microrobot's decision-making priority in dynamic environments. The comprehensive autonomous navigation framework is illustrated in Figure 7.3.

1. **Mechanism Motion Loop:** The region beneath the endplate constitutes the area where the effective magnetic field is generated. For the purpose of navigating the microrobot across its entire operational space, it is imperative to employ a parallel mechanism capable of automatically tracking the microrobot. At each point in time, an image denoted as $I_t(t) \in R^{N \times M \times 3}$ is captured. To ensure reliable target tracking in an environment with numerous distractions, a mean-shift algorithm is utilized. The tracked center, labeled as P, has coordinates $(u_P^K(t), v_P^K(t))$ within the set K. Subsequently, a thresholding technique is iteratively applied, as described below:

$$\Delta e_x(t) = \begin{cases} 0 & \text{if } \left| v_P^K(t) - \dfrac{N}{2} \right| \le \delta \\ -C\left(v_P^K(t) - \dfrac{N}{2} \right) & \text{if } \left| v_P^K(t) - \dfrac{N}{2} \right| > \delta \end{cases}$$
$$\Delta e_y(t) = \begin{cases} 0 & \text{if } \left| u_P^K(t) - \dfrac{M}{2} \right| \le \delta \\ -C\left(u_P^K(t) - \dfrac{M}{2} \right) & \text{if } \left| u_P^K(t) - \dfrac{M}{2} \right| > \delta \end{cases} \tag{7.4}$$

Here, δ represents the predefined threshold, dictating the precision with which the mechanism tracks the microrobot. The vector $\mathbf{e}(t) = [\Delta e_x(t)\ \Delta e_y(t)]^T$ quantifies the endplate center's displacement over one operational cycle. Additionally, the desired endpoint coordinate $E(t)$ is determined through superposition and subsequently transmitted to the motor module, serving as the control command for the mechanism. When it comes to magnetic actuation, the parallel mechanism maintains a continuous, albeit slow, pursuit of the microrobot. Concurrently, dynamic magnetic fields are generated by rapidly energizing the electromagnetic coils, as outlined in the outer control loop. This separation of control strategies is justified, as it is more feasible to enhance the frequency of current adjustments compared to that of mechanical movements, particularly when dealing with heavy components such as the coils in our system. Furthermore, the microrobot's positional data can be transformed from the L coordinate system to the O coordinate system by:

$$\begin{cases} x_P(t) = x_E(t) - C\left(v_P^K(t) - \dfrac{N}{2}\right) \\ y_P(t) = y_E(t) - C\left(u_P^K(t) - \dfrac{M}{2}\right) \end{cases} \tag{7.5}$$

where $(x_E(t), y_E(t))$ is the coordinate of the endplate center $E(t)$; $(x_P(t), y_P(t))$ is the coordinate of the microrobot position $P(t)$.

2. **Microrobot Steering Loop:** To guarantee precise and adaptable path tracking by the microrobot, its steering system incorporates three core components: a DOB that provides dependable feed-forward control to mitigate the effects of disturbances and ensures smooth state feedback, taking into account imaging noise [37]; a flexible lane management module for dynamic path adjustments in real-time to prevent potential collisions; and a MPC system for accurate path tracking, leveraging its optimization and prediction capabilities.
   1. Problem Statement: In this chapter, we utilize a cylindrical microrobot that propels itself along a surface through fluidic drag imbalance, an effective means of locomotion for such robots. The dynamics of this tumbling microrobot are significantly influenced by fluid properties and proximity to the substrate, introducing model uncertainties. Additionally, external factors such as walls and dynamic fluid flows further impact its motion performance, serving as external disturbances. To streamline the modeling process, the microrobot's motion is delineated as follows:

$$\begin{cases} \dot{q}_x(t) = h_x(t) + d_x(t) \\ h_x(t) = a_0 f_m(t)\cos(\alpha_m(t)) \\ \dot{q}_y(t) = h_y(t) + d_y(t) \\ h_y(t) = a_0 f_m(t)\sin(\alpha_m(t)) \end{cases} \tag{7.6}$$

Here, the velocities of the microrobot in the x and y directions are denoted as $\dot{q}_x(t)$ and $\dot{q}_y(t)$, respectively. The coefficient a_0 represents a constant factor that depends on both the structural characteristics of the robot and the properties of its operating environment. The rotating magnetic field that influences the microrobot's motion is characterized by its frequency $f_m(t)$ and yaw angle $\alpha_m(t)$. Meanwhile, $d_x(t)$ and $d_y(t)$ encapsulate combined disturbances, encompassing both model uncertainties and external factors. The equation reveals that manipulation of the magnetic fields allows for the regulation of the microrobot's locomotion.

2. Disturbance Observer: For the purpose of assessing the motion states and generalized disturbances, we introduce an augmented state variable $z(t)$ and its corresponding estimated counterpart $\widehat{z}(t)$, defined as follows:

$$\begin{aligned} [z_1(t)\, z_2(t)\, z_3(t)\, z_4(t)]^T &= [q_x(t)\, d_x(t)\, q_y(t)\, d_y(t)]^T \\ [\widehat{z}_1(t)\, \widehat{z}_2(t)\, \widehat{z}_3(t)\, \widehat{z}_4(t)]^T &= \left[\widehat{q_x}(t)\, \widehat{d_x}(t)\, \widehat{q_y}(t)\, \widehat{d_y}(t)\right]^T \end{aligned} \tag{7.7}$$

According to Eq. (7.6), a DOB is designed as:

$$\begin{cases} \dot{\widehat{z}}_1(t) = \widehat{z}_2(t) + h_x(t) + \beta_1 (x_P(t) - \widehat{z}_1(t)) \\ \dot{\widehat{z}}_2(t) = \beta_2 (x_P(t) - \widehat{z}_1(t)) \\ \dot{\widehat{z}}_3(t) = \widehat{z}_3(t) + h_y(t) + \beta_3 (y_P(t) - \widehat{z}_3(t)) \\ \dot{\widehat{z}}_4(4) = \beta_4 (y_P(t) - \widehat{z}_3(t)) \end{cases} \tag{7.8}$$

where $\beta_1, \beta_2, \beta_3$, and β_4 are the observer gains to be tuned. The estimation error $r(t)$ equals $\widehat{z}(t) - z(t)$, which obeys the following first order equation:

$$\dot{\mathbf{r}}(t) = \mathbf{\Phi}\mathbf{r}(t) + \mathbf{\Psi}(t) \tag{7.9}$$

where

$$\mathbf{\Phi} = \begin{bmatrix} -\beta_1 & 1 & 0 & 0 \\ -\beta_2 & 0 & 0 & 0 \\ 0 & 0 & -\beta_3 & 1 \\ 0 & 0 & -\beta_4 & 0 \end{bmatrix}, \mathbf{\Psi}(t) = \begin{bmatrix} 0 \\ -\dot{d}_x(t) \\ 0 \\ -\dot{d}_y(t) \end{bmatrix} \tag{7.10}$$

Assume that the derivatives of the lumped disturbances are bounded, which is reasonable for our system. Theoretically, if β_1, β_2, β_3, and β_4 are appropriately chosen such that Φ matrix is Hurwitz, then the estimation error is bounded. The output of the DOB is generalized disturbances $\widehat{\mathbf{u}_d}(t) = [\widehat{z}_2(t)\, \widehat{z}_4(t)]^T$ and the estimate position $\hat{P} = (\widehat{z}_1(t), \widehat{z}_3(t))$.

3. Fuzzy Logic Modifier: Assuming the constructed map is precise and the predetermined route is devoid of obstacles, satisfactory tracking in unchanging environments can be ensured. Nevertheless, such completely static environments are uncommon in reality, and even minor alterations

in the surroundings can disrupt navigation. By incorporating virtual sights, the microrobot is endowed with the capability to observe and reason, thereby enhancing the intelligence of its controller. While in motion, the microrobot continuously scans its surroundings, capturing obstacle information within three distinct sight sectors, as illustrated in Figure 7.4. Here, s_f denotes the central axis of the forward sight sector, aligned with the vector pointing from the microrobot's current position to the intended tracking point. Meanwhile, s_l and s_r represent the central axes of the left-front and right-front sight sectors, respectively, each deviating from s_f by angles ϕ and $-\phi$. The angle ψ signifies the width of each sight sector's field of view. The distances between the microrobot and obstacles along each sight vector are calculated using real-time images, with the microrobot's position accurately accounted for. If all calculated distances exceed a predefined safety threshold, the microrobot is deemed to be in a safe state ("*sa*"), prompting it to follow a precise trajectory controlled by MPC. Conversely, if any distance falls below this threshold, indicating an unsafe state ("*us*"), the microrobot adjusts its propulsion forces based on FLM.

The fuzzy logic controller (FLC) takes as its inputs the distances measured by three sight sectors: the left sight (LS), front sight (FS), and right sight (RS). The outputs of this controller are the adjusted angle (AA) and adjusted velocity (AV), represented mathematically by ζ_α and ζ_v, respectively. The FLC employs three membership functions for each input variable, as depicted in Figure 7.5a. For the AA output, seven membership functions are utilized, while for the AV output, three membership

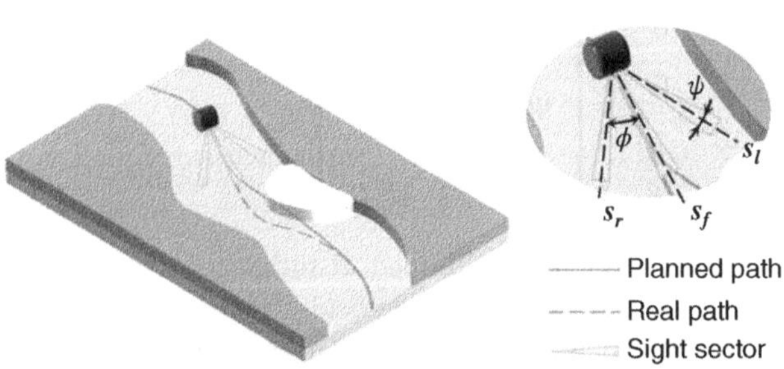

Figure 7.4 Virtual sights of the microrobot during actuation. Dark shaded blocks represent walls, and light shaded blocks denote new obstacles detected after map scanning. Solid lines denote planned paths, and dashed lines denote real paths. s_l, s_f, and s_r are the centering lines of the virtual sights; ϕ is the deflection angle; ψ is the viewing angle.

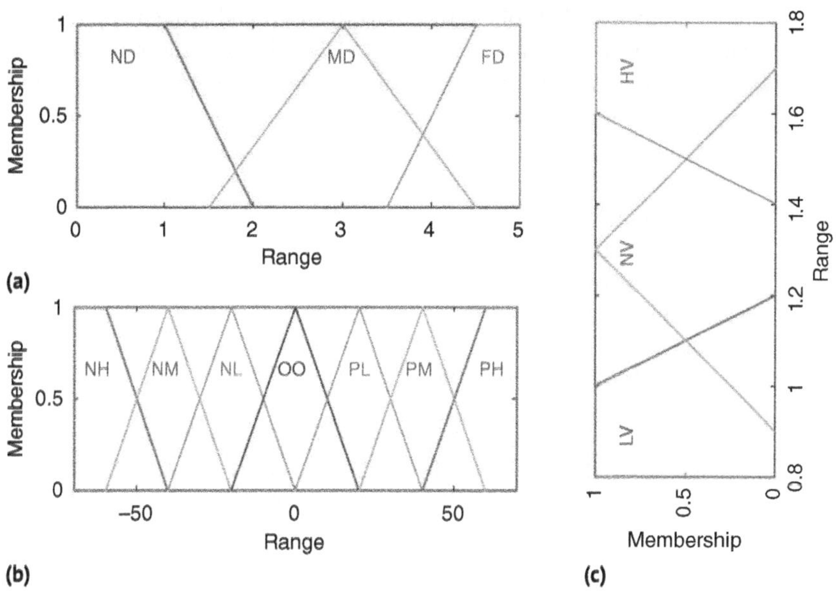

Figure 7.5 Membership functions of input and output variables. (a) Membership functions for LS, FS and RS: ND (near distance); MD (middle distance); FD (far distance). (b) Membership functions for AA: NH (negative high); NM (negative medium); NL (negative low); OO (zero); PL (positive low); PM (positive medium); PH (positive high). (c) Membership functions for AV: LV (low velocity); NV(normal velocity); HV (high velocity).

functions are applied, as shown in Figure 7.5b,c. The linguistic variables associated with these membership functions are specified as follows:

$$\begin{aligned} &\mathcal{L}_{\text{LS}} = \mathcal{L}_{\text{FS}} = \mathcal{L}_{\text{RS}} = \{\text{ND, MD, FD}\} \\ &\mathcal{L}_{\text{AA}} = \{\text{NH, NM, NL, OO, PL, PM, PH}\} \\ &\mathcal{L}_{\text{AV}} = \{\text{LV, NV, FV}\} \end{aligned} \tag{7.11}$$

A group of 27 fuzzy rules is set. The combination of the deflection angle, viewing angle, membership functions, and fuzzy rules determines the transient smoothness and obstacle-avoidance performance. These are determined by simulations. The maximum-minimum composition technique is used for inference and the center-of-area method is adopted for defuzzi-fication [38]. Further, the output of the FLM can be derived as

$$\widehat{\mathbf{u}_f} = \frac{\upsilon_a \zeta_\upsilon}{\left\| \mathbf{s}_f \right\|} \text{Rotz}\left(\mathbf{s}_f, \zeta_\alpha\right) \tag{7.12}$$

where υ_a is the average velocity of trajectory tracking.

4. Model Predictive Controller: MPC, an advanced and refined approach to process control, determines the input signal through the minimization of a cost function. This function evaluates both the current output and the predicted future output of the system, thereby optimizing its performance [39]. Within this framework, the system's aggregated disturbances are compensated for by the feed-forward mechanism, allowing the controller to focus solely on the nominal model.

Suppose the prediction horizon is H, which contains H steps. A part of the predicted position sequence $\hat{\mathbf{q}}_o(H|t)$ at sampling time t caused by the previous input is

$$\hat{\mathbf{q}}_o(H \mid t) = [\hat{\mathbf{q}}_o(t+1 \mid t)\hat{\mathbf{q}}_o(t+2 \mid t) \ldots \hat{\mathbf{q}}_o(t+H \mid t)]^T \tag{7.13}$$

where $\hat{\mathbf{q}}_o\,(t+k \mid t) = \left[\widehat{q_{ox}}\,(t+k \mid t)\,\widehat{q_{oy}}\,(t+k \mid t)\right]^T$ is kth predicted free response in the position sequence without control action. Similarly, the other part of the predicted position sequence $\hat{\mathbf{q}}_i(H|t)$ at sampling time t arising from optimized input increment equals

$$\hat{\mathbf{q}}_i(H \mid t) = [\hat{\mathbf{q}}_i(t+1 \mid t)\hat{\mathbf{q}}_i(t+2 \mid t) \ldots \hat{\mathbf{q}}_i(t+H \mid t)]^T \tag{7.14}$$

where $\hat{\mathbf{q}}_i\,(t+k \mid t) = \left[\widehat{q_{ix}}\,(t+k \mid t)\,\widehat{q_{iy}}\,(t+k \mid t)\right]^T$ is kth predicted free response in the position sequence without control action. Then the total predicted position sequence $\hat{\mathbf{q}}(H|t)$ at sampling time t contains both Eqs. (7.13) and (7.14), which can be further derived by

$$\begin{aligned}\hat{\mathbf{q}}\,(H \mid t) &= \hat{\mathbf{q}}_o\,(H \mid t) + \hat{\mathbf{q}}_i\,(H \mid t)\\ &= \hat{\mathbf{p}}\,(H \mid t) + \mathbf{M}\,(H \mid t) + \mathbf{G}\Delta\mathbf{u}\,(H \mid t)\end{aligned} \tag{7.15}$$

where $\hat{\mathbf{p}}\,(H \mid t) = \left[\hat{\mathbf{p}}(t+1 \mid t)^T \hat{\mathbf{p}}(t+2 \mid t)^T \ldots \hat{\mathbf{p}}(t+H \mid t)^T\right]^T$ and $\hat{\mathbf{p}}\,(t+1 \mid t) = \hat{\mathbf{p}}\,(t+2 \mid t) = \ldots = \hat{\mathbf{p}}\,(t+H \mid t) = [x_{\hat{p}}\,(t)\,y_{\hat{p}}\,(t)]^T$ is the estimated position from the DOB, which is resulted from the previous input; $\mathbf{M}(H|t)$ is the unvaried control output calculated by Eq. (7.6), which is related to the input at last sampling time $\mathbf{u}(t - \triangle(t))$ and sampling time interval $\triangle(t)$; G is the matrix standing for the dynamic properties of the system; $\triangle\mathbf{u}(H|t)$ is the predicted control increment sequence; the latter two items equal to

$$\begin{aligned}\Delta\mathbf{u}\,(H \mid t) &= [\Delta\mathbf{u}\,(t+1 \mid t)\,\Delta\mathbf{u}\,(t+2 \mid t) \ldots \Delta\mathbf{u}\,(t+H \mid t)]^T\\ \mathrm{G} &= \begin{bmatrix}\mathbf{g}_1 & \mathbf{g}_2 & \mathbf{g}_3 & \cdots & \mathbf{g}_H\end{bmatrix}\end{aligned} \tag{7.16}$$

where $\Delta \mathbf{u}(t+k \mid t) = [u_x(t+k \mid t)\, u_y(t+k \mid t)]^T$ is the kth control increment in the increment signal sequence; $\mathbf{g}_1 = \begin{bmatrix} g_1 & g_2 & \dots & g_H \end{bmatrix}^T$, $\mathbf{g}_2 = \begin{bmatrix} 0 & g_1 & \dots & g_{H-1} \end{bmatrix}^T$, ..., $g_H = \begin{bmatrix} 0 & 0 & \dots & g_1 \end{bmatrix}^T$ are unit step response coefficients. The cost function of our system is defined as

$$\begin{aligned} \mathbf{J}(t) = {} & [\mathbf{r}(H \mid t) - \hat{\mathbf{q}}(H \mid t)]^T \mathbf{L} [\mathbf{r}(H \mid t) - \hat{\mathbf{q}}(H \mid t)] \\ & + \Delta \mathbf{u}(H \mid t)^T \mathbf{R} \Delta \mathbf{u}(H \mid t) \end{aligned} \tag{7.17}$$

where **L** and **R** are weighted matrices of the control effect and the control input, respectively; $\mathbf{r}(H|t)$ is the desired trajectory sequence in the receding horizon at sampling time t

$$\mathbf{r}(H \mid t) = [\mathbf{r}(t+1 \mid t)\, \mathbf{r}(t+2 \mid t) \dots \mathbf{r}(t+H \mid t)]^T \tag{7.18}$$

where $\mathbf{r}(t+k \mid t) = \begin{bmatrix} x_{R_{t/\Delta t+k}} & y_{R_{t/\Delta t+k}} \end{bmatrix}^T$ is the kth target point in the trajectory sequence, which is also the $(t/\Delta t + k)$th point of the path set $\mathcal{R}$. Substituting Eq. (7.15) into (7.17), the optimal solution of the cost function with respect to input increment sequence can be solved as

$$\Delta \mathbf{u}^*(t) - (\mathbf{G}^T \mathbf{L} \mathbf{G} \mid \mathbf{R})^{-1} \mathbf{G}^T \mathbf{L} [\mathbf{r}(H \mid t) - \hat{\mathbf{q}}(H \mid t)] \tag{7.19}$$

Then the control signal $\hat{\mathbf{u}}_c(t)$ computed by MPC is updated by

$$\hat{\mathbf{u}}_c(t) = \hat{\mathbf{u}}_c(t - \Delta t) + \mathbf{s}^T \Delta \mathbf{u}^*(t) \tag{7.20}$$

In the context where $s \in R^H$, it serves as a selection vector characterized by having its first element equal to one and all other elements set to zeros. The process involves selecting solely the nearest control signal from the optimized input signal sequence, and this selection procedure is iteratively applied at each sampling instant.

5. Magnetic Field Calculator: The microrobot operates under the remote guidance of dynamic magnetic fields, necessitating the utilization of a magnetic field calculator to manage the input field. The system's adjusted control signal is formulated as described.

$$\mathbf{u}(t) = [u_x(t)\, u_y(t)]^T = \begin{cases} \hat{\mathbf{u}}_c(t) - \widehat{\mathbf{u}_d}(t) & \text{if}\ 'sa' \\ \widehat{\mathbf{u}_f}(t) - \widehat{\mathbf{u}_d}(t) & \text{if}\ 'us' \end{cases} \tag{7.21}$$

An additional step is essential for determining the orientation and frequency of the magnetic field

$$\begin{cases} f_m(t) = \text{sat}\left(\frac{1}{a_0}\sqrt{u_x(t)^2 + u_y(t)^2}, f_{\max}\right) \\ \alpha_m(t) = \text{atan2}\left(u_y(t), u_x(t)\right) - \pi/2 \end{cases} \tag{7.22}$$

where $f_{\max}$ represents the maximum frequency of the microrobot's operation, and assuming A_m denotes the magnitude of the rotating magnetic field, the targeted time-varying magnetic field vector, $\mathbf{B}_d(t)$, can be subsequently deduced by

$$\mathbf{B}_d(t) = \begin{bmatrix} A_m \sin(\alpha_m(t)) \sin(2\pi t f_m(t)) \\ A_m \cos(\alpha_m(t)) \sin(2\pi t f_m(t)) \\ A_m \cos(2\pi t f_m(t)) \end{bmatrix} \tag{7.23}$$

which is sent to the coil module as the magnetic field control instruction, and $\mathbf{B}_r(t)$ can be produced by powering appropriate currents into different coils.

7.3 Simulation and Experimental Validation

1. **System Overview:** Illustrated in Figure 7.6 is the prototype of the magnetic actuation system. This system employs a parallel mechanism powered by three NEMA17 step motors, which are regulated by an Arduino Mega 2560 microcontroller. The electromagnetic coils receive power from three ESCON 70/10 servo-amplifiers (manufactured by Maxon Inc.), while signals are transmitted via an 826 multifunctional I/O card from Sensory Inc. For visual feedback, an HT-660 endoscope is mounted on the endplate. The test setup includes a laser-cut PMMA vascular-like channel maze, measuring 220 mm in length and 120 mm in width. The microrobot used in our experiments is a cylindrical permanent magnet, with dimensions of 450 μm in radius and 900 μm in length. This maze is submerged in pure glycerol. Experimental evaluations indicate that the system's average positioning error is 0.89 mm, a figure that could be optimized through more refined system manufacturing. Workspace analysis confirms that the system is devoid of field generation singularities within its operational range. Additionally, mechanical vibrations are negligible in the context of the microrobot actuation tasks discussed in this study.

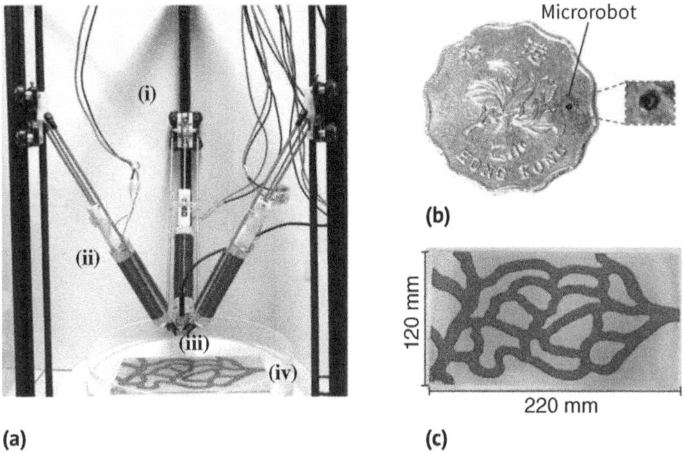

Figure 7.6 Experimental robotic platform. (a) The overall structure of DeltaMag prototype. (i) is the slide block, (ii) is the electromagnetic coil, (iii) is the camera, (iv) is the glycerol solution. (b) The microrobot compared with a Hong Kong coin. (c) The PMMA channel maze.

2. **Simulations of Controller:** Prior to implementation, simulations are undertaken to optimize controller parameters and verify the efficacy of the algorithms. Initially, the DOB and MPC are individually tuned. Subsequently, the entire motion controller undergoes comprehensive simulation testing.

   During the simulation phase for DOB, the control frequency and kinematic model of the microrobot were maintained consistent with those of the actual system. A portion of the results is presented in Figure 7.7, where all eigenvalues of Φ were configured to be equal. In this simulation scenario, the magnetic field was maintained at zero, and a square-wave disturbance was introduced in the x-direction. Additionally, a random noise signal was incorporated to emulate image noise. The findings underscore the effectiveness of the proposed DOB, which is capable of estimating both the aggregated disturbance and positional information. The observer gains play a pivotal role in determining the performance of the DOB, necessitating a balance. Smaller eigenvalues result in a quicker convergence rate and enhanced accuracy in position estimation, but compromise the noise-filtering capability. Conversely, smaller eigenvalues also expedite the convergence rate for disturbance estimation, albeit accompanied by more pronounced fluctuations. Similar trends were observed in the disturbance and position estimations for the y-direction.

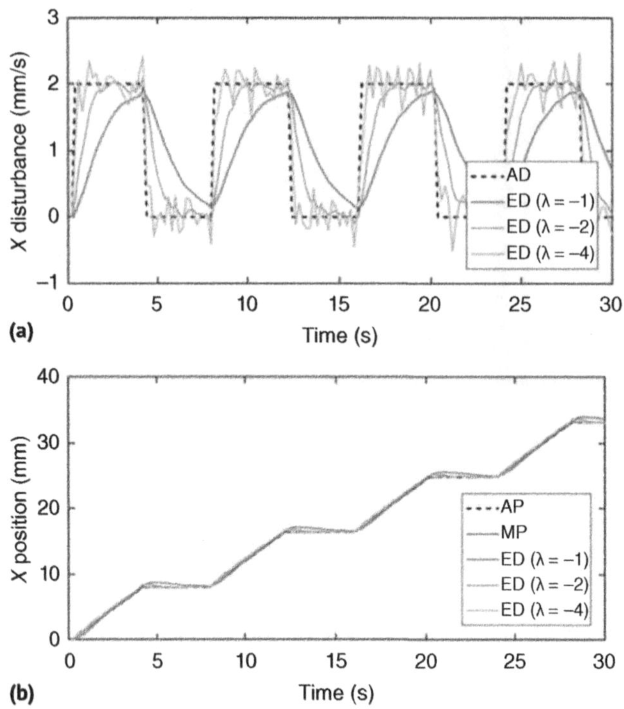

Figure 7.7 Simulation results of the DOB. (a) Disturbances in *x* direction. AD is actual disturbance and ED are estimated disturbances under different observer parameters. (b) Positions in *x* direction. AP is actual position, MP is measured position, and EP are estimated positions under different observer parameters.

Subsequently, simulations were conducted for the MPC system. A prediction horizon of five was selected. The microrobot was actuated using rotating magnetic fields of low frequency (0–5 Hz) and constant intensity (1 mT). Consequently, the optimization of control input weights was of lesser significance compared to the control effect. Both a finely tuned proportional-integral (PI) controller and the MPC were designed to track a star-shaped trajectory. The results of these simulations are depicted in Figure 7.8. When compared to the PI controller, the proposed MPC strategy exhibited superior tracking performance, particularly at the trajectory's inflection points (such as the corners of the star-shaped path). This enhanced performance is attributed to the predictive capability of the MPC.

Ultimately, simulations encompassing the integrated system of the DOB, formation learning module, and MPC were carried out. Prior to the simulations, a series of points were planned using a modified A^* search algorithm based on a provided map, where the channel regions denoted accessible

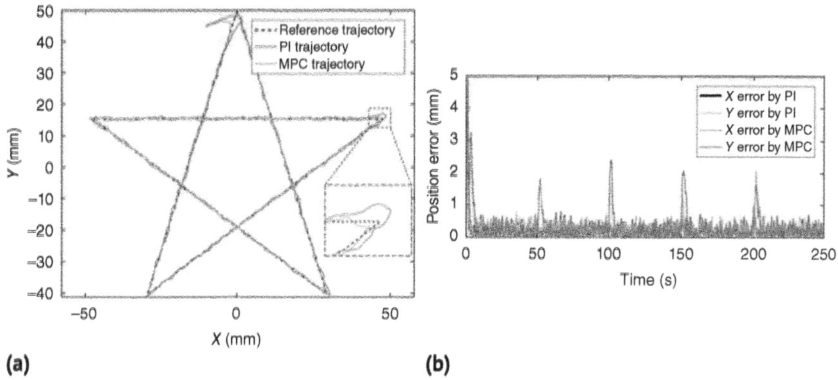

Figure 7.8 Simulation results of the MPC. (a) Path tracking results of a star-shaped trajectory via the designed MPC and PI controller. (b) Tracking errors of different controllers in *x* and *y* directions.

pathways and the non-channel regions denoted obstacles or walls. Subsequently, additional obstacles were introduced after path planning. Throughout the simulation, obstacles within the visual sectors were detected promptly based on the updated map. Concurrently, triangular-shaped disturbances in both the *x* and *y* directions were imposed to simulate unwanted perturbations. Additionally, random noise signals were integrated to emulate imaging noise. With the selected DOB and MPC in place, the parameters of the FLM were meticulously adjusted. The results of these simulations are displayed in Figure 7.9. The motion controller has satisfying performance: accurate trajectory tracking is performed when the state is "*sa*;" effective path adjustment is executed when the state is "*us*;" undesired disturbances can be estimated and compensated; the transition between two modes is smooth.

3. **Navigation Experiment of Microrobot:** The tracking performance under different parameters is first characterized. Considering the hardware constraints, the maximum frequency (f_{max}) was established at 5 Hz, resulting in an average microrobot motion speed of approximately 6.2 mm/s. To ensure stable motion tracking of the microrobot, the coil motion speed was set to 15 mm/s. A circular reference path with a 50 mm radius was designed for testing purposes, with each experimental group replicated three times to ensure consistency. Figure 7.10a presents the mean absolute deviations (MAD) for tracking speeds ranging from 1 to 5 mm/s, while maintaining a constant control distance of 45 mm. To further investigate the impact of control distance on tracking performance, the control distance was varied from 35 to 55 mm, with the tracking speed fixed at 3 mm/s. The corresponding MAD values are plotted in Figure 7.10b. The results indicate that both high tracking speeds

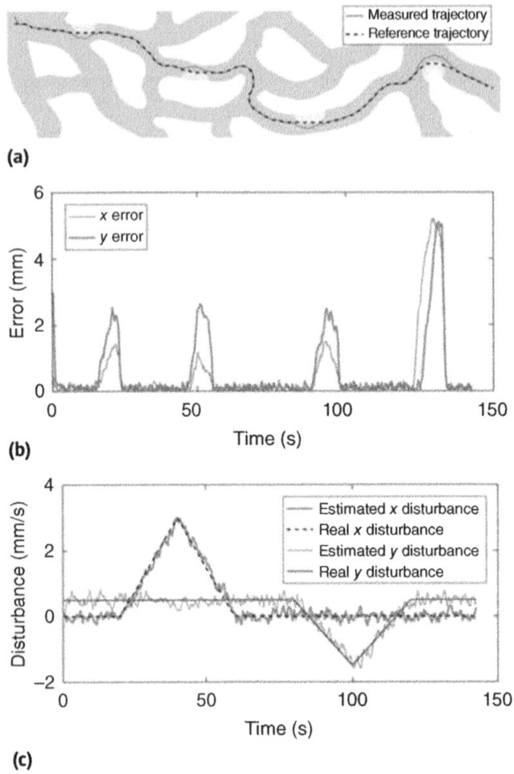

Figure 7.9 Simulations of the microrobot steering loop. (a) The trajectory tracking results of a microrobot inside a channel maze, showing the original map, the changed map with additional obstacles, the reference trajectory (preplanned), and the measured trajectory (adjusted). (b) The plot of tracking errors, where four peaks correspond to the four added obstacles. (c) The comparison between real disturbances and estimated disturbances.

and short control distances can adversely affect tracking performance, which is caused by reaching motion restriction and increased disturbance of magnetic force effect.

The system's navigation process encompasses two distinct phases. Initially, the motion planning phase involves utilizing a parallel mechanism to manipulate a camera for a comprehensive environmental scan. This scan is then used to reconstruct a comprehensive map from individual submaps. Subsequently, self-localization techniques are employed to determine the relative positioning between the actuation system and the channel maze. By selecting start and target points on the constructed map, feasible navigation pathways are computed, as illustrated in Figure 7.2f. To validate the efficacy of the motion planner, the parallel mechanism is controlled to follow the

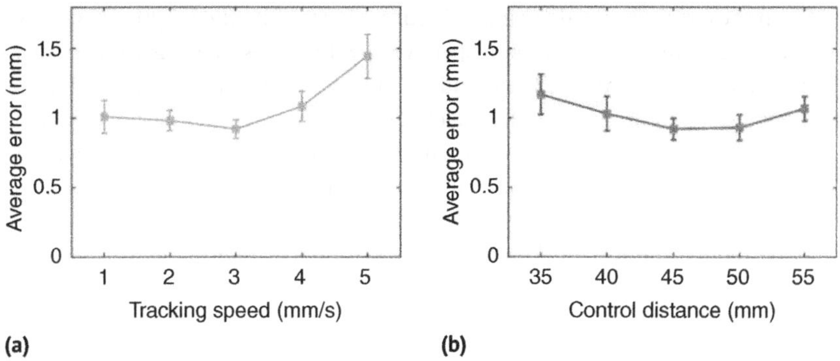

Figure 7.10 Trajectory tracking errors under (a) different motion speeds and (b) different control distances.

pre-determined path in a point-by-point manner. For each intended position of the endplate center E, the x and y coordinates align precisely with those specified in the planned path, while the z coordinate is adjusted to a preset height. The results demonstrate that the central point of the captured images consistently remains within the channel region, transitioning smoothly from the initial to the designated target positions.

With the planned path, the microrobot is controlled to perform trajectory tracking according to the proposed motion controller. Two different trajectories are selected. The control distance and tracking speed are set as 45 mm and 2.5 mm/s, respectively. The merged snapshots during tracking and results are shown in Figure 7.11.

To assess the capacity for dynamic path adjustment, additional clay obstacles were manually placed over the pre-scanned and planned path. The experimental outcomes reveal that the microrobot possesses the ability to deviate from its original route in proximity to these obstacles, effectively navigating around them. Notably, the microrobot maintains its ability to traverse through the narrowed channel despite these deviations, as demonstrated in Figure 7.9.

4. **Discussion:** By leveraging feature-based environment reconstruction and correlation-based positioning, the system achieves a comprehensive understanding of the surrounding environment, thereby overcoming the constraints imposed by the imaging device's field of view. Furthermore, the introduced modified A^* algorithm is capable of generating seamless pathways within the central channel area for microrobots to follow. It is worth noting that each component within the proposed navigation architecture can be independently adapted to accommodate varying scenarios and specific requirements.

The two trajectories depicted in Figure 7.11 exhibit total lengths of approximately 235 mm and 250 mm, respectively, with corresponding average tracking errors of roughly 1.13 mm and 1.21 mm. These results indicate that the double-loop motion controller effectively achieves precise trajectory tracking for the microrobot within a substantial workspace. The observed increase in tracking errors compared to simulations is primarily attributed to modeling inaccuracies and imperfections in system implementation.

Moreover, the results demonstrate that the DOB accurately estimates the microrobot's position, suggesting effective estimation and compensation of

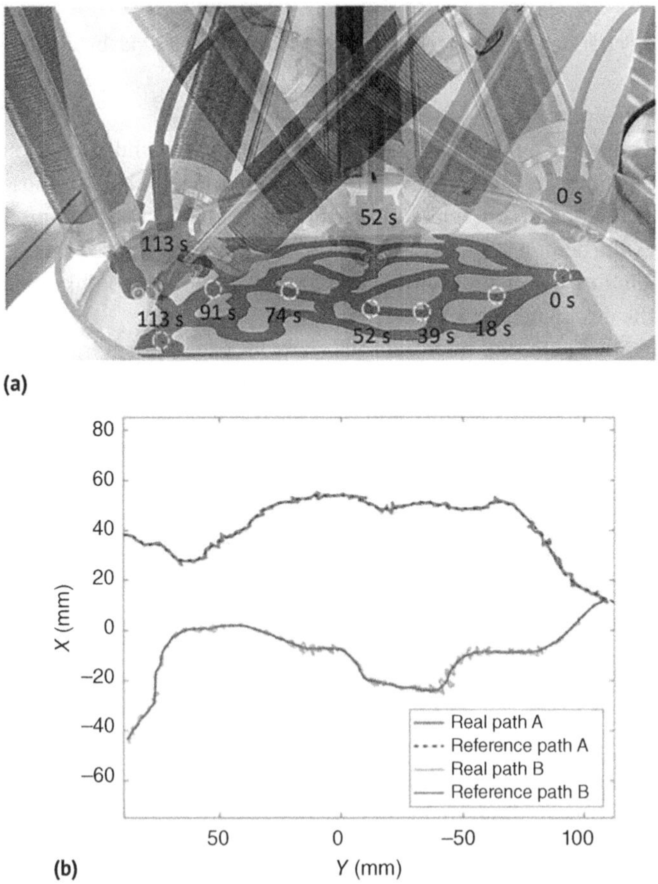

Figure 7.11 Experimental results. (a) Trajectory tracking process. (b) Comparison between reference paths and real trajectories from similar starting points to different destinations.

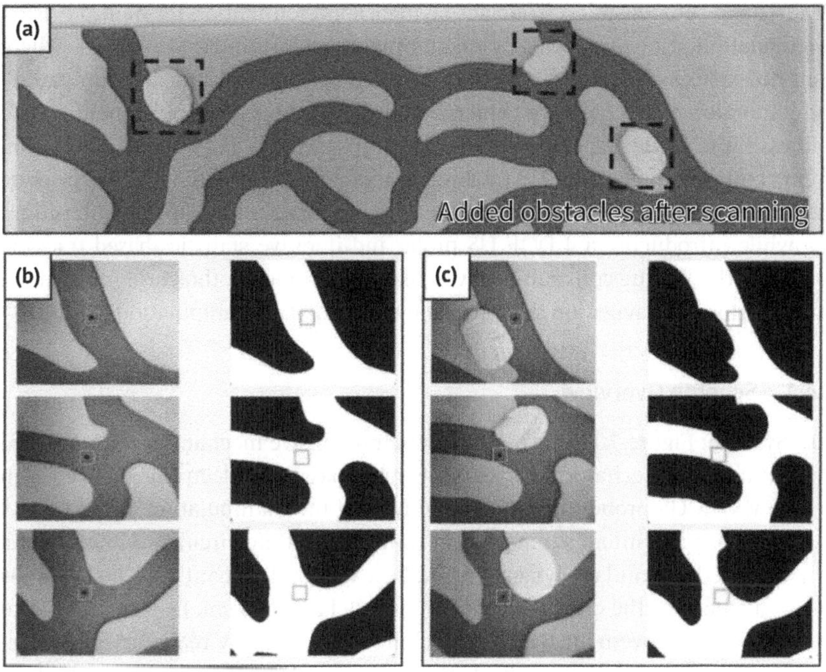

Figure 7.12 Demonstration of obstacle avoidance. (a) The changed environment with added obstacles marked by dashed blocks. (b) Screenshots of tracking in an invariant environment. In each group, the left figure is the real-time image and the right figure is the processed image with microrobot eliminated. (c) Screenshots of tracking in a changed environment.

the aggregated disturbances (Figure 7.12). The integration of the FLM and MPC endows the navigation system with both broad accessibility and localized adaptability. Experimental observations reveal that, in the absence of the FLM, the microrobot may become ensnared by additional obstacles, continuously rubbing against them. It is important to note that our algorithm is designed to handle scenarios with a limited degree of variability, such as slightly narrowed channels. In cases where new obstacles obstruct the channel, replanning of viable paths becomes necessary.

7.4 Control Scheme and Environment Registration

The mobile-coil system and dual-loop control framework proposed in Sections 7.2 and 7.3 provide a robust solution for autonomous navigation of magnetic microrobots in large, unstructured workspaces, with its three-step motion planner and disturbance compensation strategies laying a solid foundation for high-precision manipulation. However, when transitioning to endovascular scenarios—where confined spaces, pulsatile flows, and the need for medical imaging compatibility dominate—new challenges emerge that demand specialized adaptations. To address these clinical-specific requirements, this section extends the mobile-coil architecture by integrating US tracking, a modality better suited for deep-tissue visualization. This evolution retains the core advantage of dynamic coil positioning while introducing a 4-DOF US probe and iterative statistics-based tracking to handle the unique constraints of vascular environments, thus bridging general large-workspace navigation and precision endovascular manipulation.

7.4.1 Scheme Overview

1. **System:** Figure 7.13 illustrates the comprehensive mechanical and electrical setups. The mechanical component encompasses an electromagnetic system (EMS), a US probe integrated directly with the manipulation device, a life-size artery phantom submerged in water, and a fluid circulation mechanism designed to simulate dynamic blood flow within the phantom. The US probe is attached to the end-effector of the parallel mechanism, facilitating its synchronized movement with the tracking robot, thereby realizing the eye-in-hand arrangement. To replicate the pulsatile nature of human arterial blood

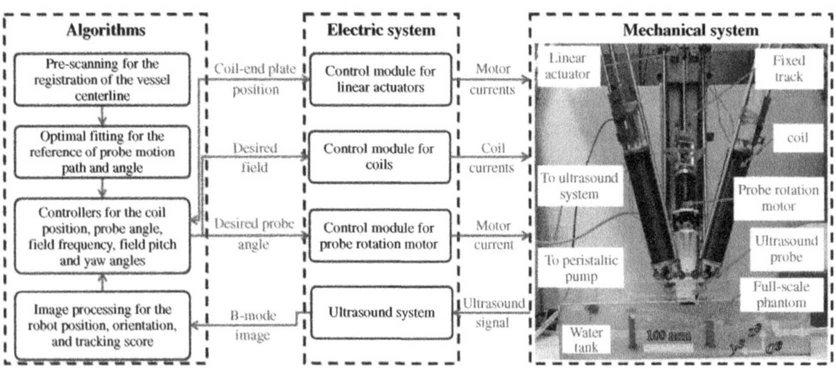

Figure 7.13 System and algorithm overview. In the mechanical system, three mobile coils can generate dynamic fields (e.g., rotating fields) in a large workspace for magnetic actuation. The eye-in-hand US probe with controllable rotation enables the position and orientation US tracking. A full-scale phantom of human iliac and aorta arteries with pulsatile flow of blood-mimicking fluid is used for experimental demonstration. The electric system provides related control for linear actuators, coils, and probe rotation. To realize the long-distance endovascular navigation with magnetic control and automated mobile US tracking, a series of vessel registration, robot tracking and system control algorithms are designed. $\{\mathcal{G} : O^{\mathcal{G}} - x^{\mathcal{G}}y^{\mathcal{G}}z^{\mathcal{G}}\}$ is the global coordinate frame.

flow, a peristaltic pump is utilized. This pulsatile flow exhibits periodic fluctuations and encompasses a laminar-to-turbulent transition phase. Furthermore, to emulate the physical properties of blood, we adopt a fluid mixture specified in Kumar et al.'s study [40], consisting of 75 wt% water, 25 wt% glycerol, and 6 wt% cellulose particles with a diameter of 20 μm.

The EMS discussed in this study represents an advanced iteration inspired by the parallel mobile coil concept introduced in our prior research [41]. It features larger coils and an expanded cylindrical workspace measuring $\Phi 500 \times 150\,m$. In this setup, each coil's termini are connected to a linear actuator and a coil-end plate, respectively, using ball joints. This configuration ensures that the coil-end plate maintains a consistent orientation while its 3D position is adjustable through linear actuator movements facilitated by a ball-screw mechanism. Upon receiving the desired position input for the coil-end plate, a microcontroller (specifically, the myRIO from National Instruments Inc.) computes the inverse kinematics of the system and issues pulse-width modulation (PWM) signals to manage the positions of three linear actuators. Simultaneously, real-time positional data of the coil-end plate is relayed back to the host computer for field control purposes. The coil control module's role is to produce the required rotating magnetic fields. Depending on the instantaneous current demands of the three coils, the myRIO microcontroller translates analog input signals into corresponding PWMs for the full-bridge circuits that regulate coil currents. The US probe utilized here (Model: 16HL7 by Terason Inc.) offers an imaging width of 26 mm and a depth range of 20–70 mm, with a resolution exceeding 100 μm, making it adequate for tracking miniature magnetic robots. To facilitate robot orientation tracking, the US probe cannot be mounted directly onto the coil-end plate as per Yang et al.'s method [42]. Instead, we have engineered a module enabling the probe to rotate around the z-axis, with rotation actuated by a stepper motor. The microcontroller (myRIO) dynamically adjusts the rotation angle based on the desired probe orientation. Using the US signals echoed back to the transducer, the US system (Model: uSmart 3300 NexGen by Terason Inc.) computes and generates US images captured by the host computer at a frequency of 30 Hz. Building upon this established system, we have developed pertinent image processing and control algorithms, which are elaborated upon below.

2. **Algorithms:** In the outlined plan, three core algorithmic components are involved. Initially, given the inherent limitation of 2D imaging by US probes, a preoperative scanning procedure, such as CT or US, is indispensable for marking the centerline of blood vessels [43, 44]. The approach employs US scanning for this purpose, where precise manipulation of the US probe's position and rotation facilitates the identification of the centerline. Subsequently,

introduce an efficient methodology for autonomous tracking of the position and orientation of miniature magnetic robots. This methodology comprises two algorithms: one for tracking in the 2D US image plane and another for tracking along the axis perpendicular to the US image. Lastly, during prolonged navigation, the movement of the probe results in positional and orientational shifts of the three coils. Furthermore, as the robot navigates within complex 3D endovascular settings, its position and orientation also undergo changes. Therefore, it is imperative to dynamically adjust the magnetic field parameters in real-time, accounting for both of these variables.

7.4.2 US Scanning for Vessel Centerline Registration

Figure 7.14a demonstrates the fundamental approach for registering the vessel centerline. Initially, the rotation angle θ of the probe is adjusted to ensure that the US image plane aligns perpendicularly with the vessel's cross section. Subsequently, a rectangle is manually delineated using mouse input to encompass the vessel area. The center of this rectangle is designated as the vessel center, and its position within the global coordinate frame $\mathcal{G}$ is determined through computational means.

$$\mathbf{p}_{\mathrm{vc}}^{\mathcal{G}} = \begin{bmatrix} \mathbf{p}_{P}^{\mathcal{G}} \cdot x + \mathbf{p}_{\mathrm{vc}}^{\mathcal{U}} \cdot x \cdot \cos(\theta) \\ \mathbf{p}_{P}^{\mathcal{G}} \cdot y + \mathbf{p}_{\mathrm{vc}}^{\mathcal{U}} \cdot x \cdot \sin(\theta) \\ \mathbf{p}_{P}^{\mathcal{G}} \cdot z - L_{\mathrm{PU}} - \mathbf{p}_{\mathrm{vc}}^{\mathcal{U}} \cdot z \end{bmatrix} \tag{7.24}$$

In the context of our system, $\mathbf{p}_{P}^{\mathcal{G}}$ represents the central location of the coil-end plate in the global coordinate system, while $\mathbf{p}_{\mathrm{vc}}^{\mathcal{U}}$ signifies the vessel center's position within the US image frame, as exemplified in Figure 7.14b. Additionally, $\mathbf{p}_{P}^{\mathcal{G}}.x$ denotes the x-coordinate of $\mathbf{p}_{P}^{\mathcal{G}}$. As illustrated in Figure 7.14a, LPU stands for the fixed z-distance separating the coil-end plate's position ($\mathbf{p}_{P}^{\mathcal{G}}$) from the origin of the image frame ($O^{\mathcal{U}}$), with a predefined value of 64 mm for our setup. Figure 7.14b displays a case study involving the identification of the vessel center in a 3D silicone phantom. For this specific phantom, the US imaging depth was configured to 60 mm. After each vessel center identification step, adjustments are made to the coil-end plate's position to facilitate a 2 mm probe movement perpendicular to the image plane, ensuring that the entire vessel remains within the imaging plane's field of view. By iteratively applying this process, we are able to register the vessel centerline, as exemplified in Figure 7.14c. This figure depicts the US imaging planes for three distinct locations of the 3D phantom, each characterized by unique probe positions and orientations. The registration outcomes indicate that the dimensions of the 3D silicone phantom span approximately 200, 350, and 40 mm along the x, y, and z axes, respectively, resulting in an overall length

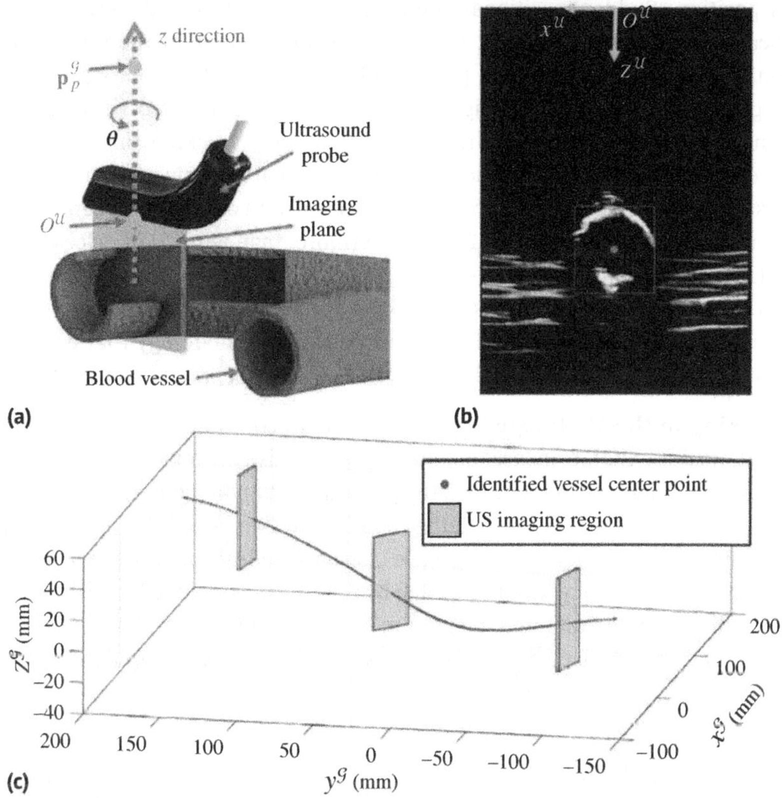

Figure 7.14 (a) The principle illustration of the vessel centerline registration. The probe rotation angle θ is controlled to be perpendicular to the vessel, and the *xyz* position of the probe is controlled via the coil-end plate position $\mathbf{p}_p^g$ control to maintain the entire vessel inside the US imaging plane. (b) One example during the process of the vessel centerline registration, in which the oulined rectangle and its center point mark the identified vessel and its center, respectively. $\{\mathcal{U} : O^u - x^u z^u\}$ denotes the coordinate frame of the US image. (c) The US image planes for three registration positions are also plotted in the figure.

of around 500 mm. To mitigate occasional identification inaccuracies, we employ optimal fitting of the identified centerline using Matlab's Curve Fitting Toolbox.

7.4.3 Robot Position and Orientation Tracking with a Mobile US Probe

A reliable and resilient tracking approach is crucial for guiding miniature robots within complex 3D endovascular settings using mobile US feedback. The unpredictability stemming from fluctuating fluid dynamics and varying vessel wall interactions poses significant challenges in accurately modeling and anticipating the robot's position. Sudden positional shifts due to pulsatile blood flow further complicate matters. Consequently, real-time, precise tracking of the robot's 3D position is imperative to facilitate prompt adjustments in the probe's position, enabling stable long-distance navigation. Additionally, given that field adjustments necessitate knowledge of the robot's orientation, tracking this aspect is equally essential. However, this tracking endeavor is particularly arduous due to the substantial noise interference from the vessel wall and other US-reflective elements, as exemplified in Figure 7.15b. Furthermore, managing 2D US feedback necessitates concurrent control over both the probe's positional and rotational movements to achieve comprehensive tracking of the robot's 3D position and 2D orientation. To address these challenges, introduce a methodology that encompasses tracking within the image plane as well as tracking perpendicular to it.

1. **Tracking in the US Image Plane:** To maintain continuous tracking of the robot, it is essential to ensure that the entire robot is encompassed within the US imaging field. To achieve this, the probe's rotation angle is adjusted such that the image plane aligns with the robot's direction of motion, as depicted in Figure 7.15a. In our study, we have chosen a helical robot, known for its high motion efficiency when propelled by rotating magnetic fields. This robot has a diameter of 6 mm and a length of 12 mm, and incorporates a cylindrical, radially magnetized magnet ($\Phi 2 \times 6$ mm) for magnetic manipulation [45].

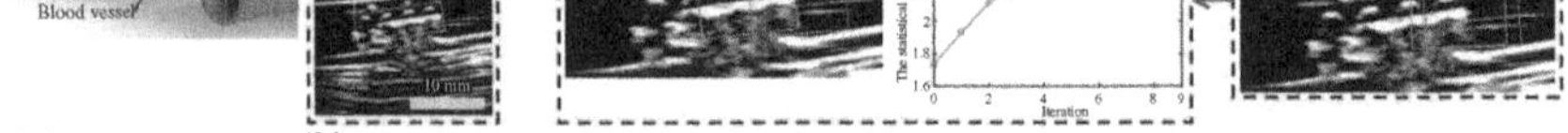

(a) (b)

Figure 7.15 (a) The principle illustration of US tracking of a miniature helical robot during the real-time endovascular navigation. The probe rotation angle θ is controlled to align with the vertical tangent plane of the vessel. In the imaging plane, the probe position is adjusted via the coil-end plate position $\mathbf{p}_p^{\mathcal{G}}$ to keep the robot inside the US imaging region. For the direction perpendicular to the imaging plane, the probe position moves a distance l_{pcom} to compensate for the deviation between the registered centerline and the real one. (b) The proposed iterative statistics-based method for robot tracking in the US image plane. $\{\mathcal{U} : o^{\mathcal{U}} - x^{\mathcal{U}}z^{\mathcal{U}}\}$ denotes the coordinate frame of the US image. The two dashed rectangles mark the robot and its US image, respectively. In (1), the solid rectangle marks the robot tracking region in the previous tracking frame. In (2), a masking strip between the dashed and solid rectangles is overlaid to isolate the tracking region. In (4), the solid rectangle marks the updated robot tracking region after the first iteration. In (5), the solid rectangle marks the updated robot tracking region after the ninth iteration.

Given that the robot operates at the vessel's base, where the vessel walls act as constraints on the robot's orientation, we postulate that the robot's central axis is confined to the vertical tangent plane of the vessel's centerline. As illustrated in Figure 7.15a, the probe's rotation angle θ is adjusted to align the image plane with this tangent plane, which is determined through optimal fitting of the identified vessel centerline.

Figure 7.15b depicts the fundamental concept of our proposed tracking methodology within the US image plane. Notably, the helical robot lacks distinctive features that distinguish it from other objects or noise, rendering template-based tracking approaches, such as MeanShift, ineffective. Instead, we observe that the cylindrical magnet embedded within the robot exhibits a robust US signal due to its high reflectivity. Consequently, this magnet is designated as the primary tracking target, with additional contributions from surrounding robot components. Taking these considerations into account, we introduce an iterative, statistics-driven method comprising five stages, as outlined in Figure 7.15b. The solid rectangle denotes the previous tracking area of the robot. Due to pulsatile blood flow, the robot's current location significantly diverges from this initial tracking region. Our method iteratively updates the tracking region based on statistical analysis of US data from the previous region until convergence with the robot's current position. Initially, to streamline computations, we isolate a region of interest (ROI) encompassing the vessel for image analysis. The z-coordinate of the ROI's centroid aligns with that of the registered vessel center, and the ROI's height is adaptable to accommodate variations in vessel diameters. To mitigate the impact of the vessel wall, which shares similar features with the robot, the second step involves overlaying a masking strip (situated between the dashed and solid rectangles) around the tracking area, thereby isolating the robot tracking zone. Subsequently, in the third step, the ROI undergoes binarization, with connected white regions adjacent to the wall being excluded. This isolation ensures the elimination of most vessel wall components without compromising the integrity of the robot within the tracking area. Lastly, to transition from the previous tracking area to the current robot location, the fourth step involves establishing statistical metrics for all N white pixels within the tracking zone.

$$U_i = \{i, \mathbf{p}_i^{\mathcal{U}} \cdot x, \mathbf{p}_i^{\mathcal{U}} \cdot z, a_i, \beta_i \mid i = 1, 2, \cdots, N\} \tag{7.25}$$

where U_i represents the parameter set of the ith unit: the identifier i, the coordinates of its center-of-mass $\mathbf{p}_i^{\mathcal{U}} \cdot x$ and $\mathbf{p}_i^{\mathcal{U}} \cdot z$, area a_i, and the orientation of its minimum bounding rectangle in the image plane β_i. Utilizing the statistics, in order to make the tracking region contain maximized white area with the largest white unit (i.e., the magnet) locating near the center of the tracking region, the optimization problem for the tracking region center $\mathbf{p}_{tc}^{\mathcal{U}}$ is formulated as

$$\max_{\mathbf{p}_{tc}^{\mathcal{U}}} \left\{ \frac{1}{\sum_{i=1}^{N} a_i} \sum_{i=1}^{N} \frac{a_i}{d_i^2 + C_0} \right\} \tag{7.26}$$

where C_0 is a small positive constant to avoid zero denominator and d_i is the distance between the ith unit and the tracking region center.

$$d_i = \sqrt{(\mathbf{p}_i^{\mathcal{U}} \cdot x - \mathbf{p}_{tc}^{\mathcal{U}} \cdot x)^2 + (\mathbf{p}_i^{\mathcal{U}} \cdot z - \mathbf{p}_{tc}^{\mathcal{U}} \cdot z)^2} \tag{7.27}$$

To maximize the statistical metric in the curly bracket of Eq. (7.26), in each iteration, $\mathbf{p}_{tc}^{\mathcal{U}}$ is updated by

$$\mathbf{p}_{tc}^{\mathcal{U}} \cdot x = \frac{1}{\text{Sum}} \sum_{i=1}^{N} \{(a_i - W \times d_i)\, \mathbf{p}_i^{\mathcal{U}} \cdot x\} \tag{7.28}$$

$$\mathbf{p}_{tc}^{\mathcal{U}} \cdot z = \frac{1}{\text{Sum}} \sum_{i=1}^{N} \{(a_i - W \times d_i)\, \mathbf{p}_i^{\mathcal{U}} \cdot z\} \tag{7.29}$$

where W is a constant weight, and the function Sum has the following expression:

$$\text{Sum} = \sum_{i=1}^{N} \{a_i - W \times d_i\} \tag{7.30}$$

Due to the potential oscillation in the $\mathbf{p}_{tc}^{\mathcal{U}}$ update induced by US noise signals located near the border of the tracking area, the algorithm incorporates a factor related to the unit's distance from the tracking region's center, denoted as $-W \times d_i$. Increasing the weight W enhances the suppression of oscillations but concurrently retards the convergence rate. Through an iterative tuning process, we have determined an optimal value of W as 5 for our specific application. Regarding the orientation update, recognizing that the robot and the embedded magnet share the same orientation, the tracking orientation β in the image plane is adjusted to align with the orientation of the largest unit.

$$\beta = \beta_m \tag{7.31}$$

$$m = \text{MaxID}([a_1, a_2, \cdots, a_N]) \tag{7.32}$$

The function MaxID(·) determines the index corresponding to the maximum value within a vector. The fourth step depicted in Figure 7.15b illustrates the updated tracking region (indicated by the solid rectangle) following the initial iteration. By iteratively applying this statistical update process, the tracking region gradually converges toward the current location of the robot, with the magnet positioned approximately at its center, as shown in the fifth step of the figure. The statistical metric defined in Eq. (7.26) attains its maximum value after eight iterations. Given that the statistical metric nearly attains its maximum after four iterations, in real-time tracking applications, four iterations are performed for each frame, enabling a tracking frequency of 20 frames per second (FPS).

2. **Tracking Perpendicular to the US Image Plane:** It is expected that deviations in the registration of the vessel centerline may lead to discrepancies between the US imaging plane and the robot's position during real-time navigation. To address this challenge, introduce a method for assessing the quality of robot tracking. Based on this assessment, adjustments are made to the probe's position perpendicular to the image plane, aiming to counteract the registration errors and ensure that the robot's center is accurately tracked in the direction perpendicular to the imaging plane. To quantify the extent to which the imaging plane diverges from the robot's position, employ a metric termed "tracking score," which is defined as the cumulative area occupied by white units within the defined tracking region.

$$S_{\text{track}} = \sum_{i=1}^{N} a_i \tag{7.33}$$

During the initialization phase, the perpendicular compensation distance (l_{pcom}) is manually tuned to optimize the tracking score to its maximum initial value, denoted as $S_{\text{track-ini}}$. Throughout the navigation process, the normalized tracking score $S'_{\text{track}} = \frac{S_{\text{track}}}{S_{\text{track-ini}}}$ is continuously monitored. If S'_{track} falls below a predefined threshold value S'_{thr}, l_{pcom} is adjusted according to Algorithm 1 to restore the tracking quality. The adjustment of l_{pcom} is carried out in increments of 1 mm, corresponding to half the diameter of the magnet. Within the algorithm, actions that directly improve tracking quality take precedence when subsequent l_{pcom} adjustments are required. Figure 7.16 illustrates an example of the tracking quality restoration during navigation within a phantom. At 433.6 s, the normalized tracking score drops below the threshold of 0.5, prompting Algorithm 1 to adjust l_{pcom} from −3 to −4 mm, leading to the recovery of tracking quality, as evidenced by the insets in Figure 7.16b. By employing techniques for tracking the robot both within and perpendicular to the US imaging plane, stable tracking of the robot

Algorithm 1: The proposed strategy for controlling the l_{pcom}. The function Length() outputs the length of the vector. The function Mod(a, b) outputs the remainder of a divided by b. L_{pscan} is a constant vector that stores the candidates for the l_{pcom} adjustment. k_{temp} stores the prior adjustment index of L_{pscan}.

Initialization: $S_{track-ini}$ $S_{track-thr} = 0.5$, $l_{pcom} = 0$, $L_{pscan} = [-1, 0, 1, 0]$, $k_{temp} = 1$, $j = 0$.
while Navigation in do
  $S'_{track} = \frac{S_{track}}{S_{track-ini}}$
  if $S_{track} < S_{track-thr}$ **then**
    $l_{pcom} = l_{pcom} + L_{pscan}(k_{temp} + j)$
    $p_G^P.x = p_G^P.x + l_{pcom} \cdot cos(\theta)$
    $p_G^P.y = p_G^P.y + l_{pcom} \cdot sin(\theta)$
  $S'_{track} = \frac{S_{track}}{S_{track-ini}}$
  if $S_{track} < S_{track-thr}$ **then**
    $j = j + 1$
    $j = Mod(k_{temp} + j, Length(L_{pscan})) - k_{temp}$
  else
  $k_{temp} = k_{temp} + j$
  $j = 0$
  end if
  end if
end while

is achieved, enabling the real-time acquisition of its position $\mathbf{p}_R^{\mathcal{G}}$ and orientation $\mathbf{o}_R^{\mathcal{G}}$, obtained as:

$$\mathbf{p}_R^{\mathcal{G}} = \begin{bmatrix} \mathbf{p}_P^{\mathcal{G}} \cdot x + \mathbf{p}_{\mathrm{tc}}^{\mathcal{U}} \cdot x \cdot \cos(\theta) \\ \mathbf{p}_P^{\mathcal{G}} \cdot y + \mathbf{p}_{\mathrm{tc}}^{\mathcal{U}} \cdot x \cdot \sin(\theta) \\ \mathbf{p}_P^{\mathcal{G}} \cdot z - L_{\mathrm{PU}} - \mathbf{p}_{\mathrm{tc}}^{\mathcal{U}} \cdot z \end{bmatrix} \tag{7.34}$$

$$\mathbf{o}_R^{\mathcal{G}} = [\beta\theta]^T \tag{7.35}$$

where β and θ refer to the pitch and yaw angles of the robot central axis.

7.4.4 Field Control with Three Mobile Coils

The helical robot operates via rotating fields, with the planes of rotation aligned orthogonally to its central axis. Our system employs control algorithms capable of dynamically modulating coil currents based on the relative spatial configuration between the coils and the helical robot, ensuring the generation of the desired field intensity at the robot's location. The operational procedure entails the following steps: (1) The operator specifies the target magnitude and frequency for the rotating field. (2) Utilizing a dedicated tracking methodology, the position and orientation of the helical robot are ascertained, while the positions of the three coils are derived through kinematic analysis of the parallel mechanism. (3) The global coordinates of the helical robot are converted into the local coordinate systems of each respective coil. (4) For each coil, the field vector at the robot's position is computed. (5) The necessary coil currents for each coil are determined to produce the desired field. Steps 1–5 are iterated at a frequency of 30 Hz, enabling dynamic adjustments to the coil currents in response to fluctuations in the distance between the coils and the helical robot.

Detailed algorithms are described below.

The corresponding field vector with respect to time at the robot position can be modeled by

$$\mathbf{b}\left(\mathbf{p}_R^{\mathcal{G}}\right) = \left[\mathbf{b}^{\mathcal{G}}\left(\mathbf{p}_R^{\mathcal{G}}\right).x \quad \mathbf{b}\left(\mathbf{p}_R^{\mathcal{G}}\right).y \quad \mathbf{b}\left(\mathbf{p}_R^{\mathcal{G}}\right).z\right]^T$$
$$= \begin{bmatrix} B\left[\cos(\beta)\cos(\theta)\cos(2\pi\omega t) + \sin(\theta)\sin(2\pi\omega t)\right] \\ B\left[-\cos(\beta)\sin(\theta)\cos(2\pi\omega t) + \cos(\theta)\sin(2\pi\omega t)\right] \\ B\sin(\beta)\cos(2\pi\omega t) \end{bmatrix} \tag{7.36}$$

Given that ω represents the field frequency, B denotes the field strength, and t signifies time, the control of the magnetic field necessitates a sequence of coordinate transformations due to the variability in the coils' positions and orientations. Assuming that the coil cores remain unsaturated, i.e., they operate within their linear operational range, the magnetic field generated by the mth coil at the position of the robot, as expressed in the coordinate frame C_m of the mth coil, can be computed by

$$\mathbf{b}_m^{C_m}\left(p_R^{C_m}\right) = i_m \cdot \overline{\mathbf{b}}_m^{C_m}\left(\mathbf{p}_R^{C_m}\right) = i_m \cdot \overline{\mathcal{M}}\left(\mathbf{p}_R^{C_m}\right), m = 1, 2, 3 \tag{7.37}$$

In this context, i_m denotes the electric current flowing through the mth coil, while $\bar{\mathcal{M}}(\mathrm{P}_R^{C_m})$ signifies the unit field distribution pattern of that coil when energized by a current of 1 A. The quantity $\mathbf{p}_R^{C_m}$ is derived through a coordinate transformation process, transitioning from the global coordinate system G to the specific coordinate system C_m associated with the mth coil [41].

$$\mathbf{p}_R^{C_m} = {}^{C_m}\mathcal{T}_{\mathcal{G}}\left(\mathbf{p}_R^{\mathcal{G}}, \mathbf{p}_P^{\mathcal{G}}\right) \tag{7.38}$$

The microcontroller within the control module provides feedback for the global position $\mathbf{p}_P^{\mathcal{G}}$. When an update to $\mathbf{p}_P^{\mathcal{G}}$ is commanded, the positions of the three linear actuators are subsequently calculated by

$$\mathbf{q} = [q_1 q_2 q_3]^T = \mathcal{F}^{-1}\left(\mathbf{p}_P^{\mathcal{G}}\right) \tag{7.39}$$

Here, $F(\mathbf{q})$ represents the forward kinematics of the parallel mechanism. Following a coordinate rotation transformation, the unit magnetic fields $\overline{\mathbf{b}}^{C_m}\left(\mathbf{p}_R^{C_m}\right)$ corresponding to the magnetic fields generated by a current of 1 A of the three coils, as expressed in the global coordinate frame, are derived by

$$\overline{\mathbf{b}}_m^{\mathcal{G}}\left(\mathbf{p}_R^{\mathcal{G}}\right) = {}^{\mathcal{G}}\mathcal{R}_{C_m}\left(\mathbf{p}_R^{C_m}, \mathbf{p}_P^{\mathcal{G}}\right) \cdot \overline{\mathbf{b}}^{C_m}\left(\mathbf{p}_R^{C_m}\right) \tag{7.40}$$

By the linear superposition law of magnetic fields, the total field at $\mathbf{p}_R^{\mathcal{G}}$ is obtained by

$$\mathbf{b}^{\mathcal{G}}\left(\mathbf{p}_R^{\mathcal{G}}\right) = \left[\overline{\mathbf{b}}_1^{\mathcal{G}}\left(\mathbf{p}_R^{\mathcal{G}}\right) \overline{\mathbf{b}}_2^{\mathcal{G}}\left(\mathbf{p}_R^{\mathcal{G}}\right) \overline{\mathbf{b}}_3^{\mathcal{G}}\left(\mathbf{p}_R^{\mathcal{G}}\right)\right] \cdot \mathbf{i} = \overline{\mathbf{B}} \cdot \mathbf{i}. \tag{7.41}$$

where $\mathbf{i} = [i_1, i_2, i_3]^T$ is the current vector. Finally, the required coil currents to generate the desired field are calculated by

$$\mathbf{i} = \overline{\mathbf{B}}^{-1} \cdot \mathbf{b}^{\mathcal{G}}\left(\mathbf{p}_R^{\mathcal{G}}\right) \tag{7.42}$$

Based on the robot tracking result, Eqs. (7.37–7.42) are executed in real time (50 Hz) to generate the demanded rotating field.

7.5 Experimental Navigation in Flowing Conditions

During navigation, the probe's position is autonomously regulated to traverse along the pre-registered vessel centerline, ensuring the robot remains entirely within the imaging field at all times. The perpendicular displacement of the probe relative to the imaging plane, denoted as l_{pcom}, is dynamically adjusted using the methodology outlined in Section 7.4. Concurrently, the probe's rotational orientation is automatically managed in alignment with the vertical tangent plane of the registered vessel centerline. The orientation of the magnetic field, specifically the pitch and yaw angles, is autonomously controlled based on the tracked orientation of the robot. The magnetic field strength is maintained at 5 mT, allowing the operator to solely adjust the field frequency to dictate the robot's motion direction (advancing or retracting) and velocity.

7.5.1 Navigation in a Phantom with Constant Diameter

Initiate a series of navigational experiments involving a helical robot within a 3D silicone phantom, characterized by a uniform diameter of 8 mm. Both the phantom itself and the resultant centerline registration are displayed in the supplementary video, supporting documentation, and Figure 7.14c, respectively. For the purpose of real-time tracking, the algorithm is configured to iterate four times per tracking frame. To validate the robustness of our proposed tracking technique, we introduce a deviation at the distal end of the phantom, causing its actual position to diverge from the pre-registered one. Additionally, we assess the tracking robustness under pulsatile flow conditions by instantaneously introducing an 80 mL/min pulsatile flow—comparable to that observed in the popliteal artery—during the navigation phase. Figure 7.16a illustrates the navigational outcomes, encompassing the tracked position and orientation of the robot, along with the US imaging plane at three distinct locations. The comprehensive navigational procedure is further detailed in the supplementary video and accompanying documentation. Notably, the robot demonstrates successful navigation within the phantom, even in the presence of pulsatile flow.

Despite the abrupt positional shifts induced by the flow, the robot maintains stable tracking with considerable robustness. Figure 7.16b demonstrates that the normalized tracking score, a metric indicative of tracking quality, remains above

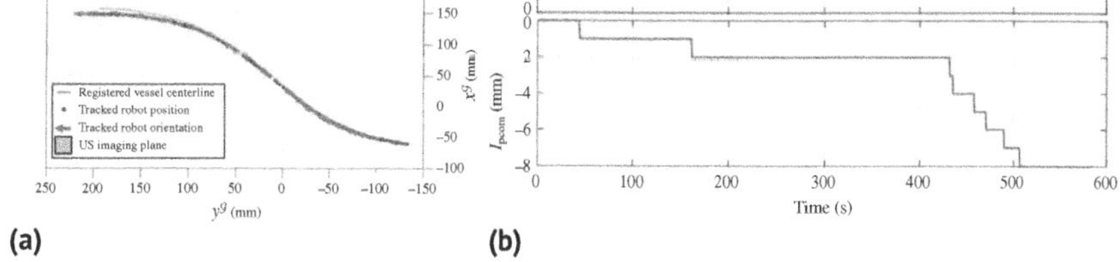

Figure 7.16 Navigation results of the first experiment. (a) The 3D view (top figure) and *xy* view (bottom figure) of the registered vessel centerlines and the tracked robot position and orientation during the navigation. Three US imaging planes with distinct positions and rotation angles are also shown. (b) Variations of the normalized tracking score S'_{track} and the compensation distance perpendicular to the imaging plane l_{pcom}, where the dashed line indicates the tracking score threshold $S'_{\text{track-thr}}$.

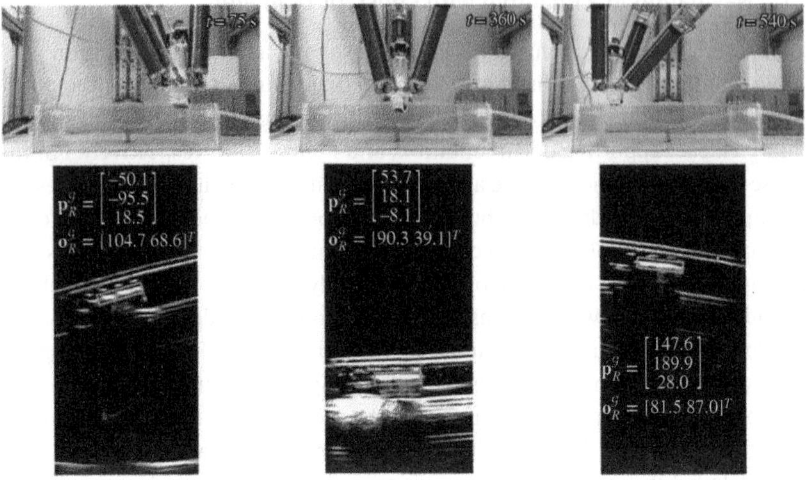

Figure 7.17 The system configurations and corresponding US tracking results at three time instants during the first navigation experiment: 75, 360, and 540 s. The tracked robot region is marked by rectangles. It is demonstrated that, despite different noise levels, probe depths, and robot orientations, the robot position and orientation can be tracked stably, and their values are labeled in the figure.

the threshold when employing the proposed l_{pcom} adjustment approach. Compensation for deviations, ranging from 0 to −8 mm, between the registered and actual vessel centerlines is evident in the robot tracking results presented in Figure 7.16a. Figure 7.17 illustrates three instances of US tracking, showcasing that despite varying noise levels from the phantom walls, blood-mimicking fluid, probe depths, and robot orientations, the robot can still be tracked accurately and stably. The tracking method exhibits resilience against sudden and significant positional changes due to pulsatile flow, enabling navigation in dynamic endovascular settings. Given that the helical propulsion force scales with rotating frequency, our system and method facilitate the navigation of miniature helical robots up steep slopes.

7.5.2 Navigation in a Bifurcation Phantom

The second experimental setup underscores the orientational control capabilities of our system by directing the robot through a bifurcation. Figure 7.18a depicts a silicone phantom designed to replicate the bifurcation found in the human iliac artery, with diameters varying between 4 and 6 mm, and subjected to a pulsatile flow at a rate of 80 mL/min. Utilizing a similar approach to the first experiment, we registered the centerlines of both branches, which intersect at an angle of approximately 30°. As the robot approaches the bifurcation, the rotation axis of the magnetic field is adjusted to guide the robot into the desired branch. The robot's

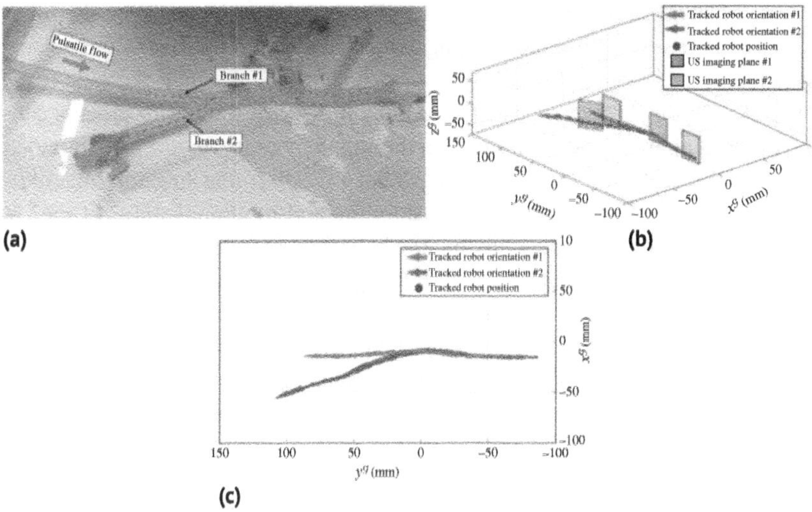

Figure 7.18 Navigation results of the bifurcation experiment. (a) Silicone phantom replicating the bifurcation of the human iliac artery. (b) Tracking and navigation result with pulsatile flow in one branch. (c) Tracking and navigation result with pulsatile flows in both branches.

position is consistently tracked within the imaging plane, based on US image tracking results. Throughout the navigation, the frequency of the rotating magnetic field is manually regulated. It is worth noting that a change in boundary conditions occurs as the robot enters the bifurcation. However, our tracking strategy effectively mitigates this disturbance, ensuring precise navigation along the intended trajectory. The tracking outcomes are presented in Figure 7.18b,c. To further validate the steering feasibility in scenarios where fluid flows exist in both branches, we conducted an additional experiment using the same phantom but with flows in both branches.

7.5.3 Navigation in a Full-scale Phantom of Human Artery

In this experimental segment, we exhibit the navigational capabilities of our system within a life-size phantom mimicking the human iliac and aorta arteries, as illustrated in Figure 7.13. This phantom replicates the aorta and its bifurcating iliac arteries, as depicted in Figure 7.19a. We have registered the vessel centerlines for both iliac artery branches ♯1 and ♯3, which are graphically represented in Figure 7.19b. The varying diameters and abrupt turns of the phantom elevate the complexity of navigation compared to our previous experiment. The experimental protocols and control methodologies employed here are consistent with those used in the initial experiment. Manual regulation of the field frequency is utilized to maneuver the robot within the two artery branches, against a pulsatile flow rate of 80 mL/min. By automatically adjusting the probe position, the robot remains consistently within the imaging plane, enabling both forward and reverse navigation.

Figure 7.19b displays the navigational outcomes, with the entire process captured in the supporting video provided in the supplementary materials. This video showcases the real-time tracking of the robot's position and orientation. Additionally, it demonstrates the updated positioning and rotational status of the US imaging plane. The results indicate that the helical robot is accurately tracked and efficiently controlled as it navigates through the life-size phantom. The phantom's varying diameters necessitate dynamic l_{pcom} adjustments to ensure high-quality tracking. It is worth noting that the phantom's protrusions and pulsatile flow patterns pose challenges in predicting robot motion through analytical modeling. However, since the proposed tracking method does not rely on such modeling, the robot remains accurately tracked despite sudden stops or thrust motions induced by these factors. In conclusion, the robust US tracking and magnetic field control techniques employed validate the effectiveness of the proposed scheme for long-distance endovascular navigation of the miniature helical robot within a 3D, full-scale human artery phantom.

Additionally, we conducted an experiment using a US probe operating at a lower frequency of 4 MHz, which offers reduced imaging resolution, to further assess the robustness of our tracking method. Under identical experimental conditions and with the same helical robot, the axial resolution of the US imaging is mathematically equivalent to half the spatial pulse length of the US waves, as per established literature. Consequently, the resolution of the 4 MHz probe is expected to be half that of the 8 MHz probe, translating to a value exceeding 200 μm. This conclusion is reinforced by the fact that both probes produce images of the same axial dimension (816 pixels), yet the actual imaging height differs, being 12 cm for the 4 MHz probe and 6 cm for the 8 MHz probe, thereby halving the resolution in the former case. To improve resolution, options include increasing the US frequency or developing microrobots constructed from materials exhibiting higher ultrasonic impedance. Furthermore, an additional experiment was designed to verify the capacity of our proposed system and methods to successfully navigate even smaller microrobots.

7.6 Discussion and Conclusion

This chapter presents a methodology for autonomous navigation of magnetic microrobots within an extensive and scalable workspace, characterized by both global accessibility and local adaptability. The employed eye-in-hand mobile-coil system facilitates the generation of magnetic fields across a large workspace with minimal energy consumption, while simultaneously enabling

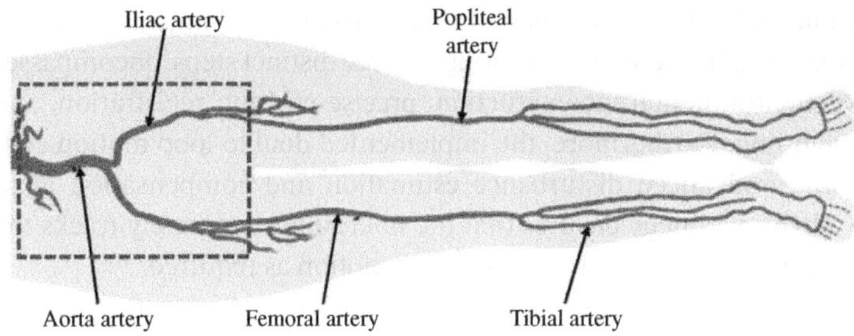

(a)

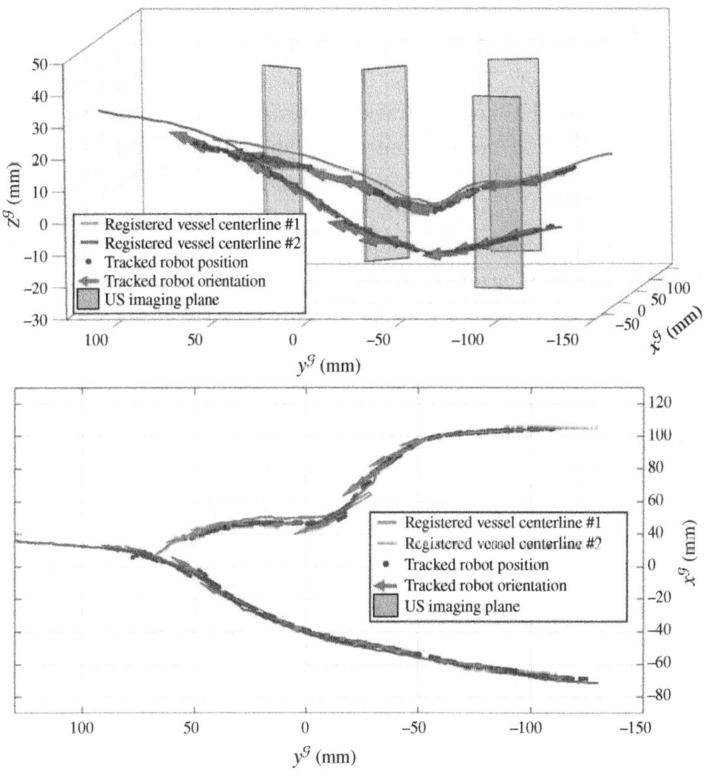

(b)

Figure 7.19 Navigation results of the second experiment in the full-scale phantom of human arteries. (a) The illustration of the human lower-extremity arteries, in which the dashed rectangle marks the region of the full-scale phantom including the iliac and aorta arteries. (b) The 3D view (top figure) and *xy* view (bottom figure) of the registered vessel centerlines and the tracked robot position and orientation during the navigation. Four US imaging planes with distinct positions and rotation angles are also shown.

image acquisition without the need for additional driving mechanisms. The proposed motion planning framework, consisting of three distinct steps, encompasses comprehensive environmental reconstruction, precise position registration, and optimal path planning. Furthermore, the implemented double-loop motion controller incorporates advanced disturbance estimation and compensation techniques, while a hybrid strategy ensures that the microrobot accurately tracks the desired trajectory and dynamically adjusts its locomotion as required.

Besides, introduce an innovative approach for navigating miniature magnetic robots within extensive endovascular environments against pulsatile flow, utilizing US imaging guidance. In comparison to other medical imaging techniques, US imaging offers sufficient resolution and depth for imaging miniature robots without the risk of irradiation. To address its limitations, such as the confined 3D imaging plane and the challenge of tracking miniature robots, we propose a four-degree-of-freedom (4-DOF) probe control system and an iterative statistics-based tracking algorithm. By providing real-time tracking of the robot's position and orientation, three mobile coils are employed to generate the required rotating magnetic fields for 3D navigation. Experiments conducted in a full-scale phantom mimicking the iliac and aorta arteries, under pulsatile flow conditions, demonstrate the robustness of our tracking method and the efficacy of our navigation scheme. A significant constraint of the existing system lies in the restricted DOFs of the US probe, which is currently limited to rotation around the z-axis. When the angle between the US array and the vessel centerline in the x–z plane exceeds a threshold, such as 40°, our tracking algorithm may fail due to a notable decline in imaging quality. This degradation arises from heightened US wave reflection on the phantom surface at larger angles. To address this issue in future iterations, the incorporation of an additional rotational DOFs for the probe could allow for better control of the angle between the vessel centerline and the US array plane. Furthermore, another notable limitation of the current system is its absence of force feedback functionality, which impedes the ability to regulate the interaction force between the US probe and the phantom. Therefore, the implementation of force feedback is crucial for future ex vivo experiments.

Appendix A Additional Algorithms

Algorithm 1: Extract Normalized HOGF at Corner Points of a Submap

---

Require: grayscale submap I_s; detection size T; block size E; cell size F; number of bins B

Ensure: corner point list $\mathcal{L}_p$, normalized HOGF list $\mathcal{L}_h$

initialization;
for each pixel of I_s **do**
    compute $\lambda_{\min}$ according to (24);
    if $\lambda_{\min} > \xi$ **then**
        $\mathbf{H}_b \leftarrow [\ ]$; $\mathbf{H}_n \leftarrow [\ ]$
        add the current pixel of I_s to $\mathcal{L}_p$
        for each cell of the block **do**
            $\mathbf{H}_c \leftarrow \mathbf{0}^B$
            for each pixel of the current cell **do**
                compute gradient features according to (25);
                $n \leftarrow \text{ceil}\left(\frac{B\,G_\theta}{2\pi}\right)$
                $\mathbf{H}_c(n) \leftarrow \mathbf{H}_c(n) + G_m$
            end for
            $\mathbf{H}_b \leftarrow [\mathbf{H}_b;\ \mathbf{H}_c]$
        end for
        $\mathbf{H}_n \leftarrow \frac{\mathbf{H}_b}{\sqrt{\|\mathbf{H}_b\|^2 + \varepsilon^2}}$ ▷ normalized HOGF $\mathbf{H}_n$
        add $\mathbf{H}_n$ to $\mathcal{L}_h$
    end if
end for

Algorithm 2: Registration Between the Map and the System

Require: grayscale reconstructed map I_m; grayscale reference image I_r; coarse interval s_c

Ensure: matched pixel (u_*, v_*)

```
initialization V_max ← 0
for each window image I_w of I_m (with searching interval of s_c pixels) do
    V_n ← Σ_(u,v) I_r(u,v) I_w(u,v) / sqrt( Σ_(u,v) I_r^2(s,t) Σ_(u,v) I_w^2(s,t) )    ▷ NCC value
    if V_n > V_max then
        (u_*, v_*) ← the origin pixel of I_w
        V_max ← V_n
    end if
end for
repeat the above process within a localized region around the best match with
searching interval of one pixel;
```

where the NCC value in the third step is

$$V_n \leftarrow \frac{\sum_{(u,v)} I_r(u,v) I_w(u,v)}{\sqrt{\sum_{(u,v)} I_r^2(s,t) \sum_{(u,v)} I_w^2(s,t)}}$$

Algorithm 3: Path Searching

Require: start point N_s; target point N_g; grided map
Ensure: path list $\mathcal{X}$

initialization;
compute $f(N_s)$ according to (17); ▷ cost value f
add N_s to $\mathcal{L}_o$; ▷ open list $\mathcal{L}_o$
while $\mathcal{L}_o$ not empty **and** N_g not in $\mathcal{L}_o$ **do**
    $N_c \leftarrow$ find the node with minimum f in $\mathcal{L}_o$
    remove N_c from $\mathcal{L}_o$; ▷ current node N_c
    add N_c to $\mathcal{L}_c$; ▷ close list $\mathcal{L}_c$
    $\mathcal{L}_n \leftarrow$ all nodes in the neighbor region of N_c, exclude those disabled and already in $\mathcal{L}_c$; ▷ neighbor list $\mathcal{L}_n$
    for all N_n in $\mathcal{L}_n$ **do**
        compute $f'(N_n)$ according to (17); ▷ successor node N_n, temporary cost value f'
        if N_n not in $\mathcal{L}_o$ **then**
            $P_a(N_n) \leftarrow N_c$; ▷ parent node $P_a(N_n)$
            $f(N_n) \leftarrow f'(N_n)$
            add N_n to $\mathcal{L}_o$ and sort $\mathcal{L}_o$ by f
        else
            if $f'(N_n) < f(N_n)$ **then**
                $P_a(N_n) \leftarrow N_c$
                $f(N_n) \leftarrow f'(N_n)$
            end if
        end if
    end for
end while
if N_g in $\mathcal{L}_o$ **then**
    successively trace the node from N_g by its parent node until reaching N_s
    $\mathcal{X} \leftarrow$ traced node sequence from N_s to N_g
else
    path planning is failed
end if

References

1 M. Sitti, H. Ceylan, W. Hu, J. Giltinan, M. Turan, S. Yim, and E. Diller. Biomedical applications of untethered mobile milli/microrobots. *Proceedings of the IEEE*, 103 (2): 205–224, 2015.

2 J. Li, B. Esteban-Fernández de Ávila, W. Gao, L. Zhang, and J. Wang. Micro/nanorobots for biomedicine: delivery, surgery, sensing, and detoxification. *Science Robotics*, 2 (4): eaam6431, 2017.

3 M. Medina-Sánchez and O. G. Schmidt. Medical microbots need better imaging and control. *Nature*, 545 (7655): 406–408, 2017.

4 Z. Zhang, X. Wang, J. Liu, C. Dai, and Y. Sun. Robotic micromanipulation: fundamentals and applications. *Annual Review of Control, Robotics, and Autonomous Systems*, 2 (1): 181–203, 2019.

5 S. Palagi and P. Fischer. Bioinspired microrobots. *Nature Reviews Materials*, 3 (6): 113–124, 2018.

6 X. Du, et al. Reconfiguration, camouflage, and color-shifting for bioinspired adaptive hydrogel-based millirobots. *Advanced Functional Materials*, 30 (10): 1909202, 2020.

7 X. Yan, et al. Multifunctional biohybrid magnetite microrobots for imaging-guided therapy. *Science Robotics*, 2 (12): eaaq1155, 2017.

8 J. Yu, B. Wang, X. Du, Q. Wang, and L. Zhang. Ultra-extensible ribbon-like magnetic microswarm. *Nature Communications*, 9 (1): 3260, 2018.

9 S. Lee, S. Lee, S. Kim, C.-H. Yoon, H.-J. Park, J.-y. Kim, and H. Choi. Fabrication and characterization of a magnetic drilling actuator for navigation in a three-dimensional phantom vascular network. *Scientific Reports*, 8 (1): 3691, 2018.

10 H. Xie, X. Fan, M. Sun, Z. Lin, Q. He, and L. Sun. Programmable generation and motion control of a snakelike magnetic microrobot swarm. *IEEE/ASME Transactions on Mechatronics*, 24 (3): 902–912, 2019.

11 T. Li, et al. Autonomous collision-free navigation of microvehicles in complex and dynamically changing environments. *ACS Nano*, 11 (9): 9268–9275, 2017.

12 H. Xie, et al. Reconfigurable magnetic microrobot swarm: multimode transformation, locomotion, and manipulation. *Science Robotics*, 4 (28): eaav8006, 2019.

13 D. Folio and A. Ferreira. Two-dimensional robust magnetic resonance navigation of a ferromagnetic microrobot using Pareto optimality. *IEEE Transactions on Robotics*, 33 (3): 583–593, 2017.

14 J. Liu, T. Xu, S. X. Yang, and X. Wu. Navigation and visual feedback control for magnetically driven helical miniature swimmers. *IEEE Transactions on Industrial Informatics*, 16 (1): 477–487, 2019.

15 L. Yang, Y. Zhang, Q. Wang, K.-F. Chan, and L. Zhang. Automated control of magnetic spore-based microrobot using fluorescence imaging for targeted

delivery with cellular resolution. *IEEE Transactions on Automation Science and Engineering*, 17 (1): 490–501, 2019.

16 D. Jin, J. Yu, K. Yuan, and L. Zhang. Mimicking the structure and function of ant bridges in a reconfigurable microswarm for electronic applications. *ACS Nano*, 13 (5): 5999–6007, 2019.

17 I. S. Khalil, A. Adel, D. Mahdy, M. M. Micheal, M. Mansour, N. Hamdi, and S. Misra. Magnetic localization and control of helical robots for clearing superficial blood clots. *APL Bioengineering*, 3 (2): 026104, 2019.

18 D. Mahdy, R. Reda, N. Hamdi, and I. S. Khalil. Ultrasound-guided minimally invasive grinding for clearing blood clots: promises and challenges. *IEEE Instrumentation & Measurement Magazine*, 21 (2): 10–14, 2018.

19 Q. Wang, et al. Ultrasound doppler-guided real-time navigation of a magnetic microswarm for active endovascular delivery. *Science Advances*, 7 (9): eabe5914, 2021.

20 J. J. Abbott, E. Diller, and A. J. Petruska. Magnetic methods in robotics. *Annual Review of Control, Robotics, and Autonomous Systems* 3 (1): 57–90,2020

21 C. M. Heunis, Y. P. Wotte, J. Sikorski, G. P. Furtado, and S. Misra. The ARMM system-autonomous steering of magnetically-actuated catheters: towards endovascular applications. *IEEE Robotics and Automation Letters*, 5 (2): 705–712, 2020.

22 J. Liu, X. Wu, C. Huang, L. Manamanchalyaporn, W. Shang, X. Yan, and T. Xu. 3-D autonomous manipulation system of helical microswimmers with online compensation update. *IEEE Transactions on Automation Science and Engineering*, 18 (3): 1380–1391, 2020.

23 X. Yang, et al. An agglutinate magnetic spray transforms inanimate objects into millirobots for biomedical applications. *Science Robotics*, 5 (48): eabc8191, 2020.

24 L. Yang and L. Zhang. Motion control in magnetic microrobotics: from individual and multiple robots to swarms. *Annual Review of Control, Robotics, and Autonomous Systems*, 4 (1): 509–534, 2021.

25 T. Xu, J. Zhang, M. Salehizadeh, O. Onaizah, and E. Diller. Millimeter-scale flexible robots with programmable three-dimensional magnetization and motions. *Science Robotics*, 4 (29): eaav4494, 2019.

26 B. Wang, et al. Endoscopy-assisted magnetic navigation of biohybrid soft microrobots with rapid endoluminal delivery and imaging. *Science Robotics*, 6 (52): eabd2813, 2021.

27 V. Lalande, et al. *In vivo* demonstration of magnetic guidewire steerability in a MRI system with additional gradient coils. *Medical Physics*, 42 (2): 969–976, 2015.

28 O. Erin, H. B. Gilbert, A. F. Tabak, and M. Sitti. Elevation and azimuth rotational actuation of an untethered millirobot by MRI gradient coils. *IEEE Transactions on Robotics*, 35 (6): 1323–1337, 2019.

29 O. Erin, M. Boyvat, M. E. Tiryaki, M. Phelan, and M. Sitti. Magnetic resonance imaging system–driven medical robotics. *Advanced Intelligent Systems*, 2 (2): 1900110, 2020.

30 D. Li, D. Dong, W. Lam, L. Xing, T. Wei, and D. Sun. Automated *in vivo* navigation of magnetic-driven microrobots using OCT imaging feedback. *IEEE Transactions on Biomedical Engineering*, 67 (8): 2349–2358, 2019.

31 O. Couture, V. Hingot, B. Heiles, P. Muleki-Seya, and M. Tanter. Ultrasound localization microscopy and super-resolution: a state of the art. *IEEE Transactions on Ultrasonics, Ferroelectrics, and Frequency Control*, 65 (8): 1304–1320, 2018.

32 Q. Chen, F.-W. Liu, Z. Xiao, N. Sharma, S. K. Cho, and K. Kim. Ultrasound tracking of the acoustically actuated microswimmer. *IEEE Transactions on Biomedical Engineering*, 66 (11): 3231–3237, 2019.

33 S. Scheggi, K. K. T. Chandrasekar, C. Yoon, B. Sawaryn, G. Van De Steeg, D. H. Gracias, and S. Misra. Magnetic motion control and planning of untethered soft grippers using ultrasound image feedback. In *2017 IEEE International Conference on Robotics and Automation (ICRA)*, pages 6156–6161, 2017, IEEE.

34 Q. Wang and L. Zhang. Ultrasound imaging and tracking of micro/nanorobots: from individual to collectives. *IEEE Open Journal of Nanotechnology*, 1: 6–17, 2020.

35 L. Manamanchaiyaporn, T. Xu, and X. Wu. Magnetic soft robot with the triangular head–tail morphology inspired by lateral undulation. *IEEE/ASME Transactions on Mechatronics*, 25 (6): 2688–2699, 2020.

36 F. Duchoň, A. Babinec, M. Kajan, P. Beňo, M. Florek, T. Fico, and L. Jurišica. Path planning with modified A-star algorithm for a mobile robot. *Procedia Engineering*, 96: 59–69, 2014.

37 W.-H. Chen, J. Yang, L. Guo, and S. Li. Disturbance-observer-based control and related methods—an overview. *IEEE Transactions on Industrial Electronics*, 63 (2): 1083–1095, 2015.

38 A. Elkhateb, N. Abd Rahim, J. Selvaraj, and M. N. Uddin. Fuzzy-logic-controller-based SEPIC converter for maximum power point tracking. *IEEE Transactions on Industry Applications*, 50 (4): 2349–2358, 2014.

39 L. M. W. K. S. Holkar. An overview of model predictive control. *International Journal of Control and Automation*, 3 (4): 47–63, 2010.

40 V. Kumar, R. Liu, R. R. Kinnick, A. Gregory, A. Alizad, M. Belohlavek, and M. Fatemi. Unambiguous identification and visualization of an acoustically active catheter by ultrasound imaging in real time: theory, algorithm, and phantom experiments. *IEEE Transactions on Biomedical Engineering*, 65 (7): 1468–1475, 2017.

41 Y. Huo, L. Yang, T. Xu, and D. Sun. Design, control, and clinical applications of magnetic actuation systems: challenges and opportunities. *Advanced Intelligent Systems*, 7: 2400403, 2024.

42 Z. Yang, L. Yang, M. Zhang, Q. Wang, S. C. H. Yu, and L. Zhang. Magnetic control of a steerable guidewire under ultrasound guidance using mobile electromagnets. *IEEE Robotics and Automation Letters*, 6 (2): 1280–1287, 2021.

43 S. E. Salcudean, H. Moradi, D. G. Black, and N. Navab. Robot-assisted medical imaging: a review. *Proceedings of the IEEE*, 110 (7): 951–967, 2022.

44 Z. Jiang, M. Grimm, M. Zhou, Y. Hu, J. Esteban, and N. Navab. Automatic force-based probe positioning for precise robotic ultrasound acquisition. *IEEE Transactions on Industrial Electronics*, 68 (11): 11200–11211, 2020.

45 J. Leclerc, H. Zhao, D. Bao, and A. T. Becker. *In vitro* design investigation of a rotating helical magnetic swimmer for combined 3-D navigation and blood clot removal. *IEEE Transactions on Robotics*, 36 (3): 975–982, 2020.

8

Magnetically Controlled Guidewire Robotic System for Vascular Intervention

In this chapter, we present a magnetically controlled guidewire robotic system (MCGRS) capable of active steering and propulsion, supported by comprehensive modeling and trajectory planning techniques. Additionally, we introduce an innovative multistage vascular embolization approach designed for hard-to-reach vessels, which involves deploying untethered, swimming shape-memory magnetic microrobots (SMMs) from a primary catheter to navigate toward vessel bifurcations.

8.1 Introduction

Cardiovascular and cerebrovascular diseases constitute a major component of the global disease burden because of their high morbidity and mortality [1]. In China, epidemiological estimates based on the seventh national census suggest that more than 330 million individuals are affected by cardio-cerebrovascular diseases. Compared with open surgery [2], vascular interventional surgery (VIS) has become a mainstream clinical option for a broad spectrum of cardio-cerebrovascular conditions. VIS is characterized by minimal tissue trauma, favorable procedural success, accelerated postoperative recovery, and, in many cases, avoidance of general anesthesia [3, 4]. Under the guidance of medical imaging, interventionalists deploy guidewires, catheters, balloons, stents, and related devices to reopen stenotic segments and restore perfusion by dilating narrowed arteries and improving regional blood supply.

Despite these advantages, VIS still encounters practical bottlenecks. First, conventional guidewires with pre-shaped passive tips lack intrinsic propulsion and active path-selection capability; therefore, surgeons must continuously perform manual advancement, withdrawal, and axial rotation to navigate the vasculature. Such manual manipulation becomes particularly inefficient in tortuous anatomies and at large-angle bifurcations, where the required operational complexity and time are markedly increased [5]. Second, VIS is commonly conducted under X-ray fluoroscopy, implying that prolonged procedures translate into increased radiation exposure for interventionalists and the clinical team [6, 7]. Third, the overall performance of VIS is highly dependent on operator expertise, while the supply of experienced interventionalists remains insufficient relative to the growing patient demand.

To mitigate these limitations, multiple commercial vascular interventional robotic platforms have been introduced for clinical use in recent years [8, 9]. Representative systems include Siemens Healthcare Corindus' CorPath GRX (USA), which has achieved broad clinical deployment for percutaneous coronary intervention and has obtained both FDA and CE approvals; Robocath's R-One (France), which has received CE approval; and Johnson & Johnson Medical's Magellan system (USA) for peripheral vascular intervention, reported to have been applied in over 1000 clinical cases. Nevertheless, most existing products primarily adopt master–slave teleoperation, where the robot reproduces basic hand motions—advancing, retracting, and rotating the guidewire/catheter—rather than providing fundamentally new navigation modalities. Moreover, these systems typically inherit the intrinsic constraints of pre-shaped passive guidewires; consequently, robust and efficient manipulation at complex bifurcations remains challenging in terms of flexibility, efficiency, and autonomy.

Magnetic fields can safely penetrate biological tissues [10, 11], making them an attractive energy source for wireless actuation of magnetically driven microrobots, with extensive use in biomedical scenarios [12, 13]. By integrating interventional instruments with magnetically actuated microrobotic components, vascular robotic systems can potentially achieve improved maneuverability, controllability, and safety, thereby enabling access to vessels that are otherwise difficult to reach using pure mechanical means [14, 15]. In this context, a variety of guidewire/catheter-based magnetic continuum robots (MCRs) have been proposed to facilitate vascular interventions [16–18], and several research groups have actively explored magnetically manipulated interventional robotic systems.

For instance, Zhao's group at MIT developed a magnetically manipulated neurointervention system that supports both steering and propulsion [9]. The platform comprises a magnetically controllable guidewire, a movable permanent-magnet actuation unit, and a guidewire/catheter advancing module. Choi's group at DGIST (Korea) proposed an electromagnetically controlled microrobotic intervention system that enables teleoperated, real-time manipulation of micro-guidewires; the system integrates magnetic guidewires (MGs), a human-scale electromagnetic actuation setup, and propulsion units, and its navigation capability has been validated through in vitro and in vivo porcine experiments [19]. Nelson's team reported variable-stiffness magnetic catheter robots and their manipulation systems [20, 21], whereas Zhang's group investigated ultrasound-guided

MG robotic systems with moving electromagnetic actuation structures [22, 23]. Xu's group further presented a magnetically controlled guidewire combining a large-workspace magnetic manipulation system with an active propulsion device [24, 25]. Overall, these magnetically manipulated systems offer clear advantages over traditional passive guidewires in flexibility and controllability and can reduce radiation exposure by enabling remote operation. However, the level of autonomy remains limited in many cases, and achieving automated navigation typically requires accurate MG modeling and dedicated trajectory planning and control algorithms.

In this work, we present a MCGRS for vascular intervention, supported by explicit modeling and trajectory planning methods. The main contributions are summarized as follows:

1. The steering kinematics of the MG are formulated based on the constant-curvature assumption. Building on this, a continuum mechanics model is established to predict magnetic-tip deformation by coupling a dipole-field description with a Cosserat-rod formulation.
2. A trajectory planning algorithm is developed to guide the MG along vessel pathways. The proposed models and planning strategy are validated via trajectory-following experiments in three vascular phantoms, demonstrating that the planned commands can drive the magnetic tip to track the desired trajectory with reliable performance.
3. Additional navigation experiments are conducted in a 3D vascular phantom, showing that the MCGRS can be remotely operated to actively steer the MG and reach target regions within complex anatomies.
4. Untethered swimming SMMs are delivered through a catheter in a living rabbit to realize renal artery embolization, thereby supporting the feasibility of the proposed multistage embolization strategy for hard-to-reach vessels.

8.2 Design of the MCGRS

This section first discusses the limitations of conventional passive guidewires and then presents the proposed MCGRS for VIS. Traditional guidewires typically employ a pre-shaped, fixed-curvature tip and thus rely on the interventionalist to manually perform axial rotation and push–pull motions for path selection and forward propulsion. To overcome these constraints, the MCGRS integrates a MG, a magnetic actuation system (MAS), a guidewire/catheter advancer (GCA), a teleoperation console (TC), and an X-ray imaging system. As illustrated in Figure 8.1a, the proposed configuration provides active steering together with controllable propulsion, which facilitates rapid and accurate navigation at complex vascular bifurcations and has the potential to shorten procedure duration.

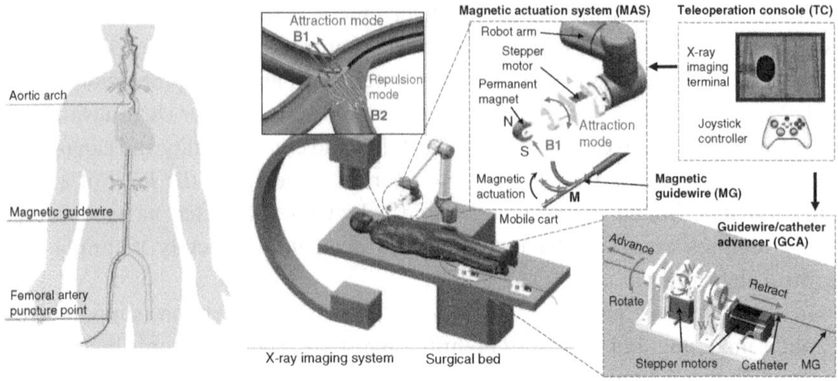

(a) (b)

Figure 8.1 Schematic diagram of interventional surgery and the proposed magnetically controlled guidewire robot system (MCGRS). (a) Schematic diagram of magnetic guidewire (MG) intervention. (b) Conceptual diagram of the future clinical application of the MCGRS.

8.2.1 Design of the MAS

Compared with electromagnets, permanent magnets offer a compact and structurally simple solution because they do not require coil excitation to generate the magnetic field. As a result, issues associated with Joule heating and heat dissipation are avoided, and a higher field strength can be achieved for a comparable device volume. The generated magnetic field depends on the magnet material, geometry, and relative position with respect to the target. Importantly, both the magnitude distribution and the field direction can be regulated by adjusting the permanent magnet pose, thereby enabling controlled deflection of the MG.

A 6-degree-of-freedom (6-DOF) serial robotic arm provides flexible end-effector manipulation and allows arbitrary motions within its reachable workspace. By appropriately selecting the arm size, the operational volume can be scaled to meet the spatial requirements of clinical interventions. In this study, the MAS consists of a mobile cart, a robotic arm, a permanent magnet, and a stepper-motor-based orientation module. An axially magnetized N52-grade cylindrical NdFeB permanent magnet (D50 × L25 mm) was used as the field source. The magnet was mounted on a UR10 robotic arm as the end-effector via 3D-printed connectors, a flexible coupling, and a stepper motor (Figure 8.1b). The robotic arm adjusts the magnet position, whereas the stepper motor controls the magnet orientation, allowing the MG to be deflected to the desired direction and configuration under the external field.

In operation, two working modes are implemented, namely attraction and repulsion, which are switched by rotating the stepper motor to invert the magnet pole orientation (Figure 8.1b, left and middle). In attraction mode, when the

S pole approaches the MG, the resulting magnetic field (denoted as B1) drives the magnetic tip to steer toward the magnet. In repulsion mode, when the N pole approaches the MG, the generated field (denoted as B2) produces an opposite steering tendency, causing the MG to bend away from the magnet. By combining the localized high-gradient field of the permanent magnet with the large reachable workspace of the robotic arm, the MAS enables field generation at arbitrary spatial locations and provides flexible, controllable magnetic steering. This configuration improves system adaptability and is consistent with the practical requirements of clinical deployment.

8.2.2 Design of the GCA

Most existing vascular interventional robotic systems consist of a positioning subsystem and a device pusher. The pusher reproduces the interventionalist's basic actions on guidewires and catheters—advancing, retracting, and rotating—to deliver instruments to the target vessel and support robot-assisted treatment. In this study, a GCA was designed and fabricated based on a continuous propulsion principle. The developed unit provides four degrees of freedom (4-DOFs) and can independently realize advancing, retracting, and rotating operations for both guidewires and catheters (Figure 8.1b).

For linear propulsion, a stepper motor together with a pair of friction wheels was used to clamp and drive the guidewire/catheter. A clamping adjustment mechanism regulates both the wheel gap and the normal force applied to the instrument. This mechanism includes an adjusting spring, base, screw, floating screw, and fixing screw. By rotating the adjusting base and screw, the wheel spacing and clamping force can be finely tuned.

During instrument installation, the adjusting base is manually pressed and rotated clockwise to compress the spring, creating a gap larger than the instrument diameter. The guidewire/catheter is then threaded through the cavities and positioned, after which the adjusting base is released to secure the device. The input/output cavities of the linear propulsion module have a diameter of 5 mm, allowing guidewires/catheters with diameters up to 5 mm. To ensure compatibility with commercial micro-guidewires (typically 0.014 in or 0.36 mm), the minimum supported diameter was set to 0.3 mm. Therefore, the GCA accommodates a broad diameter range from 0.3 to 5 mm, which substantially enlarges the applicable instrument set.

For rotational motion, a stepper motor coupled to a pair of gears was employed to generate rotary propulsion. The driving gear provides input torque, whereas the meshed slave gear transfers the output to the linear propulsion module. With coordinated control of the two stepper motors, one motor actuates the friction wheels to generate linear advancement/retraction, while the other motor drives

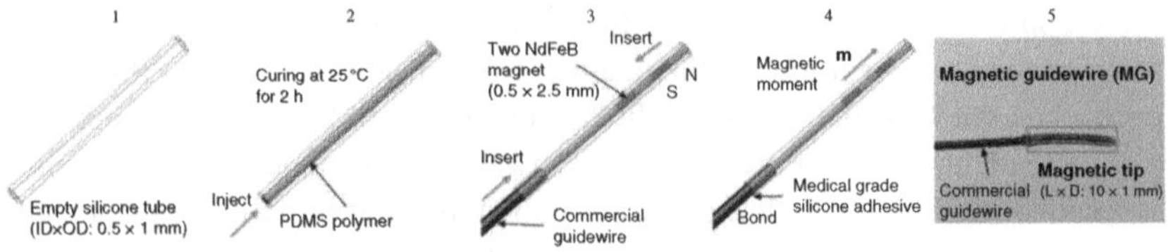

Figure 8.2 Diagram of the fabrication process of the MG.

the meshed cylindrical gears to provide synchronous rotation. The rotational output is transmitted through a rotary connector, thereby imparting axial rotation to the guidewire/catheter. As a result, simultaneous linear and rotational motions can be achieved when required. During intervention, the catheter advancer is placed in the front and the guidewire advancer in the rear to enable concurrent propulsion. The achievable intervention velocity of the guidewire/catheter assembly ranges from 0 to 15 mm/s.

8.2.3 Design of the MG

This chapter proposes an MG composed of a commercial guidewire and a flexible magnetic tip. As shown in Figure 8.2, fabrication begins with a hollow silicone tube (inner/outer diameters: 0.5/1 mm; length $\geq$ 12 mm). A polydimethylsiloxane (PDMS) soft-silicone solution is injected into the tube and cured at 25 °C for 2 h. Subsequently, two axially magnetized N52-grade NdFeB cylindrical magnets (diameter: 0.5 mm; length: 2.5 mm) are sequentially inserted into the distal end of the tube. A 0.035-inch (0.89 mm) commercial guidewire is then inserted from the proximal end with an insertion depth of 2 mm. Finally, the magnetic tip and the guidewire are bonded using medical-grade silicone adhesive. The fabricated MG exhibits a strong magnetic response, a maximum deflection exceeding 90 °, and high flexibility under external magnetic actuation.

8.3 Modeling and Trajectory Planning

This section describes the modeling framework and trajectory planning strategy adopted for the MCGRS. Specifically, a magnetic actuation model is introduced in which the permanent magnet is approximated by an equivalent dipole to compute the magnetic field as well as the resulting force and torque. Based on the constant-curvature assumption, a steering kinematic model is then established to estimate the steering angle of the magnetic tip. Furthermore, a continuum mechanics model is formulated by coupling the dipole-based magnetic loading with a Cosserat-rod description to capture the deformation of the magnetic tip.

Finally, a trajectory planning algorithm is presented to determine the external magnetic motion and the guidewire-advancer velocity that realize a prescribed magnetic-tip trajectory.

8.3.1 Magnetic Actuation Model

For permanent magnets, the dipole approximation is widely adopted because it provides a computationally efficient representation of the magnetic field [11]. In the MAS considered here, the external permanent magnet is modeled as an equivalent point dipole located at its geometric center, yielding a nonuniform magnetic field in space. The magnetic flux density **B** at the target point is given by

$$\mathbf{B} = \frac{\mu_0}{4\pi\|\mathbf{p}\|^3}\left(\frac{3\mathbf{p}\mathbf{p}^T}{\|\mathbf{p}\|^2} - \mathbf{I}\right)\mathbf{M} \tag{8.1}$$

where $\mu_0 = 4\pi \times 10^{-7}$ T m A^{-1} is the vacuum permeability, $\mathbf{p}$ denotes the displacement vector from the external magnet center to the magnetic tip ($\mathbf{p} = \mathbf{p}_t - \mathbf{p}_m$), $\mathbf{I}$ is the identity matrix, and $\mathbf{M}$ is the dipole moment of the external magnet.

As illustrated in Figure 8.3, the MG is actuated by the nonuniform field generated by the external magnet of the MAS. The magnetic tip experiences both a magnetic torque $\mathbf{T_m}$ and a magnetic force $\mathbf{F_m}$. Under the standard dipole-field formulation derived from Maxwell's equations, these quantities can be expressed as

$$\mathbf{T_m} = \mathbf{m} \times \mathbf{B} = \begin{bmatrix} 0 & -m_z & m_y \\ m_z & 0 & -m_x \\ -m_y & m_x & 0 \end{bmatrix} \begin{bmatrix} B_x \\ B_y \\ B_z \end{bmatrix} \tag{8.2}$$

and

$$\mathbf{F_m} = (\mathbf{m} \cdot \nabla)\mathbf{B} = \begin{bmatrix} m_x & m_y & m_z & 0 & 0 \\ 0 & m_y & 0 & m_y & m_z \\ -m_z & 0 & m_x & -m_z & m_y \end{bmatrix} \begin{bmatrix} \frac{\partial B_x}{\partial x} \\ \frac{\partial B_x}{\partial y} \\ \frac{\partial B_x}{\partial z} \\ \frac{\partial B_y}{\partial y} \\ \frac{\partial B_y}{\partial z} \end{bmatrix} \tag{8.3}$$

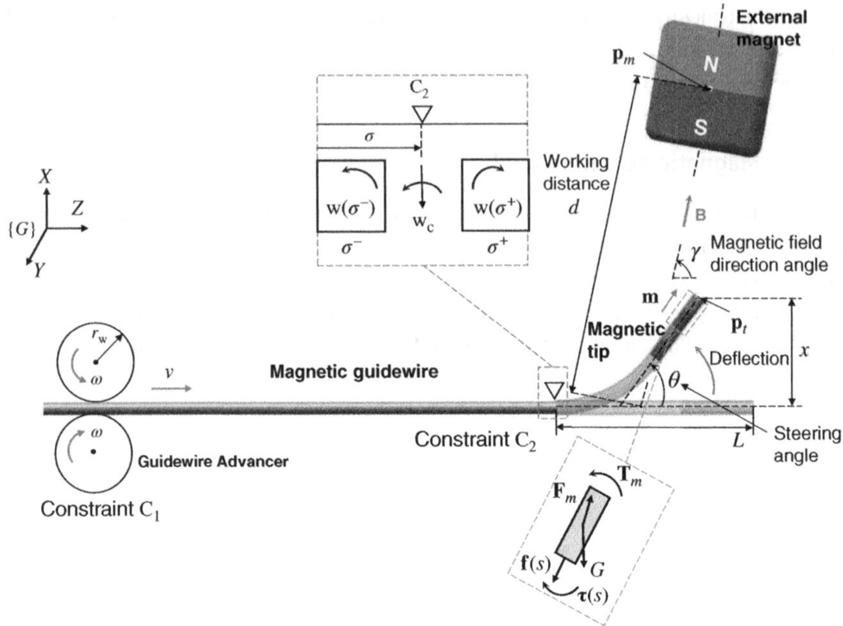

Figure 8.3 Schematic diagram of the overall magnetically actuation kinematic modeling of MG and the analysis schematic of propulsion kinematics of guidewire advancer.

where ∇ denotes the gradient operator and $\mathbf{m}$ is the magnetic moment vector of the tip magnet. The vector $\mathbf{m}$ depends on the pitch angle θ (rotation about the Y-axis) and the yaw angle ϕ (rotation about the X-axis), and can be written as

$$\mathbf{m} = \begin{bmatrix} m_x \\ m_y \\ m_z \end{bmatrix} = \begin{bmatrix} c\theta & 0 & s\theta \\ s\phi s\theta & c\phi & -s\phi c\theta \\ -c\phi s\theta & s\phi & c\phi c\theta \end{bmatrix} \begin{bmatrix} 0 \\ 0 \\ 1 \end{bmatrix} m = \begin{bmatrix} s\theta \\ -c\theta\, s\phi \\ c\theta\, c\phi \end{bmatrix} m \tag{8.4}$$

where m is the magnitude of the magnetic moment, and s and c represent sin(·) and cos(·), respectively.

8.3.2 Steering Kinematic Model

Following prior studies [19, 26], the magnetic tip is modeled as an Euler–Bernoulli beam under the constant-curvature assumption, which enables estimation of the steering angle. The curvature κ is related to the bending moment $\mathbf{T}_b$ and can be written as [22]

$$\frac{d\kappa}{dl} = \frac{1}{EI} \frac{d\|\mathbf{T}_b\|}{dl} \tag{8.5}$$

where EI denotes the effective bending stiffness of the magnetic tip, given by

$$EI = E_t I_t + E_m I_m = \frac{\pi E_t}{64}(d_t^4 - d_m^4) + \frac{\pi E_m}{64} d_m^4 \tag{8.6}$$

in which the subscripts t and m correspond to the silicone tube and the tip magnet, respectively; E_t and E_m are Young's moduli; I_t and I_m are the second moments of area; and d_t and d_m are the outer diameters. Under constant curvature, Eq. (8.5) yields

$$||\mathbf{T}_{\mathrm{b}}|| = \frac{EI\,\theta}{L} \tag{8.7}$$

where θ is the steering angle and L is the total length of the magnetic tip. As shown in the right schematic of Figure 8.2, using the magnitudes of the field $B = ||\mathbf{B}||$ and the moment $m = ||\mathbf{m}||$, the magnetic torque in Eq. (8.2) can be expressed in scalar form as

$$||\mathbf{T}_{\mathrm{m}}|| = mB\sin(\gamma - \theta) \tag{8.8}$$

where γ denotes the direction angle of the external magnetic field. At static equilibrium, the magnetic torque balances the bending moment, i.e.,

$$\mathbf{T}_{\mathrm{b}} = \mathbf{T}_{\mathrm{m}} \tag{8.9}$$

which leads to the implicit relationship for θ:

$$\theta = \frac{mBL\sin(\gamma - \theta)}{EI} \tag{8.10}$$

The steering angle θ in Eq. (8.10) is solved numerically using the Newton–Raphson method [19]. For a given (γ, B), the iteration starts from an initial guess θ_0 and terminates when $|\theta_{n+1} - \theta_n| < 0.001°$, according to

$$\theta_{n+1} = \theta_n - \frac{\theta_n - \frac{mBL\sin(\gamma - \theta_n)}{EI}}{1 + \frac{mBL\cos(\theta_n - \gamma)}{EI}} \tag{8.11}$$

8.3.3 Continuum Mechanics Model

Cosserat elastic rod theory provides an accurate and numerically tractable framework for modeling heterogeneous MCRs undergoing large deflections in spatially varying magnetic fields, while requiring only limited simplifying assumptions [27]. Since the proposed MG exhibits typical continuum-robot characteristics, we integrate a Cosserat-rod formulation with the dipole-field-based magnetic loading model to predict its deformation. As outlined in Figure 8.4a, the resulting model is constructed from three elements: constitutive relations, static equilibrium equations, and boundary conditions.

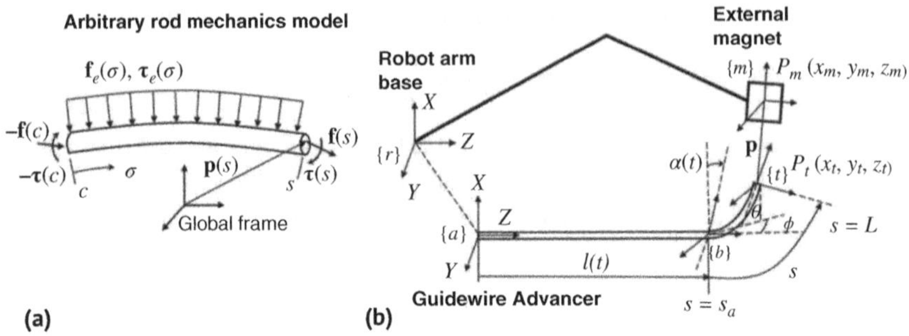

Figure 8.4 (a) Schematic diagram of the Cosserat-rod mechanical model of the magnetic tip. (b) Schematic diagram of the spatial coordinate transformation relationship between MG and other parts of the system.

Constitutive Relations: The configuration of a Cosserat rod along the arc-length parameter s is represented by a homogeneous transformation $\mathbf{T}(s) \in SE(3)$:

$$\mathbf{T}(s) = \begin{bmatrix} \mathbf{R}(s) & \mathbf{p}(s) \\ 0 & 1 \end{bmatrix} \tag{8.12}$$

where $\mathbf{p}(s) \in \mathbb{R}^3$ denotes the centerline position and $\mathbf{R}(s) \in SO(3)$ denotes the material-frame orientation. Their spatial derivatives satisfy

$$\frac{d\mathbf{R}}{ds} = \mathbf{R}(s)\,[\mathbf{u}(s)]_\times \tag{8.13}$$

$$\frac{d\mathbf{p}}{ds} = \mathbf{R}(s)\,\mathbf{v}(s) \tag{8.14}$$

where $\mathbf{v}(s)$ and $\mathbf{u}(s)$ are the strain-rate vectors associated with extension/shear and bending/torsion, respectively, and $[\cdot]_\times$ denotes the skew-symmetric operator:

$$[\mathbf{u}(s)]_\times = \begin{bmatrix} 0 & -u_z & u_y \\ u_z & 0 & -u_x \\ -u_y & u_x & 0 \end{bmatrix} \tag{8.15}$$

The internal force $\mathbf{f}(s)$ and internal moment $\tau(s)$ follow the constitutive laws:

$$\mathbf{f}(s) = \mathbf{R}(s)\,\mathbf{K}_{\text{se}}\left(\mathbf{v}(s) - \mathbf{v}^*(s)\right) \tag{8.16}$$

$$\tau(s) = \mathbf{R}(s)\,\mathbf{K}_{\text{bt}}\left(\mathbf{u}(s) - \mathbf{u}^*(s)\right) \tag{8.17}$$

where $\mathbf{K}_{\text{se}}$ and $\mathbf{K}_{\text{bt}}$ are the shear/extension and bending/torsion stiffness matrices, respectively, and $\mathbf{v}^*(s)$ and $\mathbf{u}^*(s)$ denote the intrinsic strain-rate terms.

Because the commercial guidewire segment is relatively stiff, and the magnetic tip is short, we assume that the magnetic-tip rod undergoes bending and torsion while shear and axial extension are negligible. This reduces the formulation to a Kirchhoff rod model with $\mathbf{v}(s) = \mathbf{v}^*(s) = [0\ 0\ 1]^{\mathrm{T}}$ and $\mathbf{K}_{\mathrm{se}} = 0$. For an axisymmetric elastic rod, the bending/torsion stiffness is written as

$$\mathbf{K}_{\mathrm{bt}} = \mathrm{diag}(EI_{xx},\ EI_{yy},\ GI_{zz}) \tag{8.18}$$

where I_{xx}, I_{yy}, and I_{zz} are the second moments of area about the principal axes, E is Young's modulus, and G is the shear modulus.

Equilibrium Equations: In static equilibrium, the internal force and moment satisfy the nonlinear ordinary differential equations (ODEs) shown in Figure 8.4a:

$$\frac{d\mathbf{f}}{ds} + \mathbf{f}_{\mathrm{e}}(s) = 0 \tag{8.19}$$

$$\frac{d\boldsymbol{\tau}}{ds} + \frac{d\mathbf{p}}{ds} \times \mathbf{f}(s) + \boldsymbol{\tau}_{\mathrm{e}}(s) = 0 \tag{8.20}$$

where $\mathbf{f}_{\mathrm{e}}(s)$ and $\boldsymbol{\tau}_{\mathrm{e}}(s)$ denote the distributed external force and moment per unit length, respectively.

Boundary Conditions: The deformation of the magnetic-tip rod is formulated as a two-point boundary value problem (BVP) over $s \in [s_a,\ L]$ (Figure 8.4b). Because the loads can be discontinuous across a constraint point C_2 at $s = \sigma$, the wrench $\mathbf{w}(s)$ satisfies

$$\mathbf{w}(\sigma^+) = \mathbf{w}(\sigma^-) + \mathbf{w}_{\mathrm{c}}(\sigma) \tag{8.21}$$

where $\mathbf{w}_{\mathrm{c}}$ denotes the constraint wrench.

We assume that a constraint C_2 is located at the base of the magnetic-tip rod, fixing the position and orientation of the centerline at $\mathbf{p}(s_a)$, while allowing rotation about the z-axis. This rotational freedom produces torsional deformation associated with the z-component of the internal moment, $\tau_z(s_a)$. In the base frame $\{b\}$ with $s_a = 0$ (Figure 8.4b), the boundary conditions are:

$$\mathbf{p}(s_a) = 0 \tag{8.22}$$

$$\mathbf{R}(s_a) = \mathbf{R}_z(\alpha) \tag{8.23}$$

$$\mathbf{w}(s_a^+) = \mathbf{w}(s_a^-) + \mathbf{w}_{\mathrm{c}}(s_a) \tag{8.24}$$

$$\mathbf{f}(s_a) = 0 \tag{8.25}$$

$$\tau_z(s_a) = GI_{zz}\big(\mathbf{u}(s_a) - \mathbf{u}^*(s_a)\big)_z \tag{8.26}$$

where $\mathbf{R}_z(\alpha)$ is the rotation matrix about the fixed z-axis, and α is determined by $\tau_z(s_a)$.

At the distal end ($s = L$) in the same frame (Figures 8.3 and 8.4b), the boundary conditions are given by:

$$\mathbf{p}(L) = \mathbf{p}(s_a) + L\,\mathbf{R}(L)\,\mathbf{v}^* \tag{8.27}$$

$$\mathbf{R}(L) = \mathbf{R}_x(\phi)\,\mathbf{R}_y(\theta) \tag{8.28}$$

$$\mathbf{w}(L^+) = \mathbf{w}(L^-) + \mathbf{w}_c(L) \tag{8.29}$$

$$\mathbf{f}(L) = \mathbf{F}_m + \mathbf{G} \tag{8.30}$$

$$\boldsymbol{\tau}(L) = \mathbf{T}_m + \left(0.5\,l_{tm}\,\mathbf{R}(L)\,\mathbf{v}^*\right) \times \mathbf{G} \tag{8.31}$$

where $\mathbf{R}_x(\phi)$ and $\mathbf{R}_y(\theta)$ are rotation matrices about the fixed x- and y-axes, respectively, consistent with the definition of (ϕ, θ) in Eq. (8.4); l_{tm} denotes the length of the tip magnet; $\mathbf{G}$ is the gravitational force; and $\mathbf{F}_m$ and $\mathbf{T}_m$ are the magnetic force and torque computed from the dipole model in Eq. (8.1).

Finally, the deformation of the magnetic tip is obtained by solving the coupled ODEs in Eqs. (8.13), (8.14), (8.19), and (8.20) subject to the boundary conditions in Eqs. (8.22)–(8.31).

8.3.4 Propulsion Kinematic Model

Linear advancement and retraction in the GCA are realized by a stepper motor driving a pair of friction wheels that clamp the guidewire/catheter, enabling accurate regulation of translational motion. Because this linear propulsion directly determines the MG displacement and insertion velocity, the corresponding kinematic relationship is analyzed. As illustrated on the left of Figure 8.2a, the instantaneous propulsion speed can be written as:

$$v(t) = \omega r_w = \frac{\pi r_w}{30\,N_{pr}}\,N_p(t) \tag{8.32}$$

where r_w is the radius of the friction wheel, $N_p(t)$ denotes the actual pulse rate of the motor at time t, and N_{pr} is the number of pulses per motor revolution. For a stepper motor with step angle δ, $N_{pr} = 360/\delta$.

Accordingly, the propulsion displacement over the interval $[0, t]$ is:

$$l(t) = \int_0^t v(\tau)\,dt \tag{8.33}$$

8.3.5 Trajectory Planning

For practical operation of the proposed system, the inverse problem is of primary interest, i.e., computing the external magnet motion and the advancer command from a specified magnetic-tip trajectory. In this subsection, we derive the trajectory planning formulation for the MG.

As shown in Figure 8.4b, the robotic-arm base frame $\{r\}$ is selected as the global frame. The pose of the advancer frame $\{a\}$ with respect to $\{r\}$ is fixed and denoted as $\mathbf{T}_a^r$, where the two frames share parallel orientations and differ only by a constant translation. The transformation from the magnetic-tip distal frame $\{t\}$ to the magnetic-tip base frame $\{b\}$ is $\mathbf{T}_t^b$, and the transformation from $\{b\}$ to $\{a\}$ is $\mathbf{T}_b^a$. Hence, the pose of the magnetic tip in $\{r\}$ is:

$$\mathbf{T}_t^r = \mathbf{T}_a^r\,\mathbf{T}_b^a\,\mathbf{T}_t^b = \begin{bmatrix} \mathbf{R}_t & \mathbf{P}_t \\ 0 & 1 \end{bmatrix} \tag{8.34}$$

with

$$\mathbf{T}_b^a = \begin{bmatrix} \mathbf{R}_z(\alpha) & l(t)\mathbf{v}^* \\ 0 & 1 \end{bmatrix} \tag{8.35}$$

$$\mathbf{T}_t^b = \begin{bmatrix} \mathbf{R}(L) & \mathbf{p}(L) \\ 0 & 1 \end{bmatrix} \tag{8.36}$$

where $\mathbf{P}_t = (x_t, y_t, z_t)$ and $\mathbf{R}_t$ are the position and orientation of the magnetic tip expressed in frame $\{r\}$. The quantities $\mathbf{p}(L)$ and $\mathbf{R}(L)$ are obtained by solving the BVP over $[s_a, L]$ in the base frame $\{b\}$, and $l(t)$ denotes the insertion displacement generated by the advancer.

Let $\mathbf{T}_m^r(\theta)$ denote the pose of the external magnet frame $\{m\}$ with respect to $\{r\}$, parameterized by the robot joint vector θ:

$$\mathbf{T}_m^r(\theta) = \begin{bmatrix} \mathbf{R}_m & \mathbf{P}_m \\ 0 & 1 \end{bmatrix} \tag{8.37}$$

where $\mathbf{P}_m = (x_m, y_m, z_m)$ and $\mathbf{R}_m$ are the magnet center position and orientation matrix in frame $\{r\}$, respectively.

The MAS and the advancer constitute a coupled actuation system. Its input trajectory is the configuration vector $\mathbf{q}(t)$, which collects the insertion displacement and the robot joint coordinates,

$$\mathbf{q}(t) = \begin{bmatrix} l(t) \\ \theta(t) \end{bmatrix} \tag{8.38}$$

while the task-space output is the desired magnetic-tip pose trajectory $\mathbf{r}(t)$ in frame $\{r\}$:

$$\mathbf{r}(t) = \begin{bmatrix} \mathbf{P}_t \\ \mathbf{R}_t \end{bmatrix} \tag{8.39}$$

The dimension of $\mathbf{r}(t)$ (denoted as d) is generally up to six. In this work, since the tip orientation is parameterized by yaw and pitch, d is reduced to five. Let p be the translational DOF of the advancer ($p = 1$) and j be the DOF of the robot arm, then $\mathbf{q}$ has dimension $p + j$.

The overall mapping from configuration space to task space is written as:

$$\mathbf{r} = \mathbf{f}(\mathbf{q}) \tag{8.40}$$

Differentiating Eq. (8.40) yields the velocity relationship through the Jacobian $\mathbf{J}(\mathbf{q}) \in \mathbb{R}^{d\times(p+j)}$:

$$\dot{\mathbf{r}} = \mathbf{J}(\mathbf{q})\,\dot{\mathbf{q}} \tag{8.41}$$

where $\dot{\mathbf{r}}$ is the spatial velocity of the magnetic tip and $\dot{\mathbf{q}}$ is the commanded velocity of the cooperative system. To generate $\mathbf{q}(t)$ for a given $\mathbf{r}(t)$, the incremental control update is computed using the pseudoinverse $\mathbf{J}^{\dagger}(\mathbf{q})$:

$$\delta\mathbf{q} = \mathbf{J}^{\dagger}(\mathbf{q})\,\delta\mathbf{r} = \mathbf{J}^{T}(\mathbf{q})\left(\mathbf{J}(\mathbf{q})\mathbf{J}^{T}(\mathbf{q})\right)^{-1}\delta\mathbf{r} \tag{8.42}$$

After obtaining $\mathbf{q}(t)$ via numerical integration of the propulsion model and the cooperative kinematics (Eqs. (8.32), (8.33), (8.37), and (8.38)), the planned robot motion directly provides the magnet trajectory $[\mathbf{P}_m, \mathbf{R}_m]^T$, and the corresponding advancer velocity is given by $v(t)$.

8.4 Experimental Setup and Validations

This section reports experimental validations of the proposed MCGRS. Trajectory-following studies are conducted in both 2D and 3D vascular phantoms to evaluate navigation accuracy in geometrically complex channels. In addition, in vitro navigation in a 3D phantom is performed to demonstrate feasibility for clinically relevant, hard-to-access vessel paths. Finally, experiments under X-ray fluoroscopy verify that the MCGRS is compatible with standard interventional imaging workflows, thereby underscoring its translational potential. Overall, these results provide experimental support for subsequent automatic control of MGs and contribute to improving the autonomy of magnetically manipulated interventional robotic systems.

8.4.1 Trajectory Following in Three Different Vascular Phantoms with Multiple Angles

Trajectory-following experiments were carried out in three vascular phantoms: 2D Phantom 1 and 3D Phantom 2 with 90° corners, and a tree-like, branched circulatory Phantom 3 with higher geometric complexity (Figure 8.5b). The experiments were implemented on Robot Operating System (ROS Noetic) using four functional node groups: a motion-command generator, a robot-arm controller, a propulsion-motor controller, and a camera-based measurement node.

Two command input modes were supported, including manual teleoperation and automatic execution of preplanned trajectories. In manual mode, the interventionalist inputs commands via a joystick; the robot-arm controller node and the propulsion-motor controller node subscribe to these commands to realize real-time teleoperation. In automatic mode, the motion-command generator node publishes discrete waypoints of the precomputed trajectory, which are subscribed by the two controller nodes to perform closed-loop trajectory following.

The camera measurement node processes images using the OpenCV library and estimates the magnetic-tip pose in real time. The ROS nodes were implemented using a combination of Python and C++, and the full experimental pipeline was executed on Ubuntu 20.04.

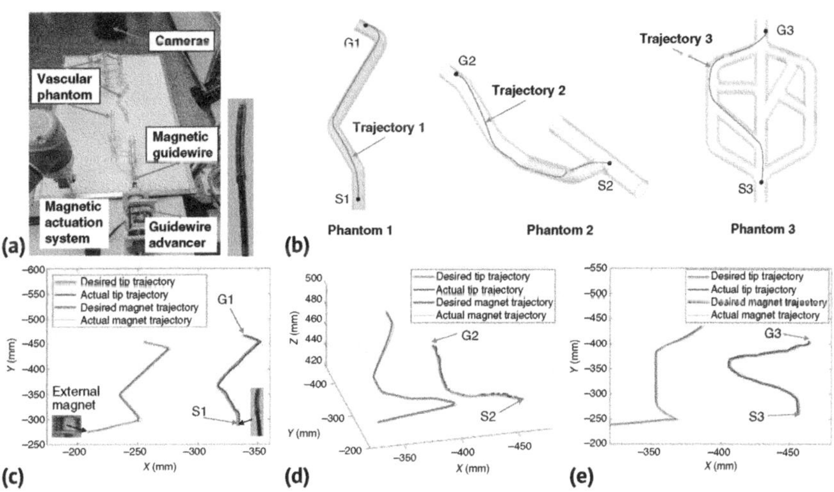

Figure 8.5 Analysis of experimental results of trajectory following. (a) Diagram of the trajectory-following experimental platform. (b) Schematic diagram of three vascular phantoms of different sizes and shapes; the highlighted curves indicate the desired trajectories, with S1–S3 and G1–G3 denoting the starting and target points, respectively. (c–e) Comparison of desired and measured trajectories of the magnetic tip and the external magnet for the three experiments.

The experimental platform is shown in Figure 8.5a. Three continuous, curved vascular phantoms with multiple turns were fabricated via 3D printing using transparent resin. The inner diameters ranged from 10 to 13 mm, and the overall lengths ranged from 160 to 180 mm. The desired trajectories are marked in Figure 8.5b, where S1–S3 denote the start locations and G1–G3 denote the goal locations. These reference paths were discretized to obtain a sequence of desired magnetic-tip positions and orientation angles.

Using the trajectory planning method described in Section 8.3.5, the corresponding external-magnet pose trajectory and the guidewire-advancer propulsion profile were computed in MATLAB from the desired tip trajectory. During experiments, the MAS drives the external magnet to track the planned pose trajectory, while the guidewire advancer provides insertion commands according to the planned propulsion velocity. With coordinated magnetic steering and linear propulsion, the MG tracks the specified straight and turning segments, progressing from S1–S3 to the corresponding targets G1–G3 (Figure 8.5b). As shown in Figure 8.5c–e, the measured trajectories of both the magnetic tip and the external magnet are generally consistent with the planned paths.

Figure 8.6a–c present the desired and measured orientation angles of the magnetic tip and external magnet as functions of time. The reported angles are absolute

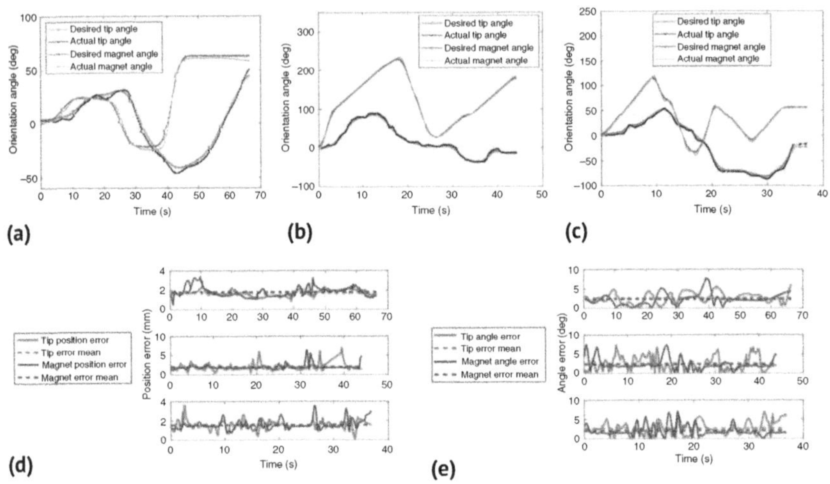

Figure 8.6 (a–c) Desired and measured orientation angles of the magnetic tip and the external magnet over time for the three experiments. (d) Position errors of the magnetic tip and the external magnet. (e) Orientation errors of the magnetic tip and the external magnet.

values referenced to the initial guidewire configuration. In the latter two experiments, the maximum MG deflection approaches 100°, as reflected by the first prominent peak in the tip-angle curve in Figure 8.6b.

Figure 8.6d summarizes the position tracking errors. The mean position errors (mean $\pm$ standard deviation) for the magnetic tip and external magnet are 1.68 ± 0.31 and 1.81 ± 0.52 mm in the first experiment. In the second experiment, they are 1.88 ± 0.96 and 1.99 ± 0.77 mm, respectively. In the third experiment, they decrease to 1.52 ± 0.45 and 1.55 ± 0.46 mm.

Figure 8.6e reports the orientation tracking errors. The mean angle errors for the magnetic tip and the external magnet are $2.36° \pm 1.36°$ and $2.30° \pm 1.47°$ in the first experiment, $2.44° \pm 1.79°$ and $2.29° \pm 1.44°$ in the second experiment, and $2.57° \pm 1.62°$ and $2.07° \pm 1.36°$ in the third experiment, respectively.

Additional analyses of the three trajectory-following trials include time histories of magnet and tip positions, relative separation distances, and the measured tip velocities. The results indicate that higher propulsion speeds are feasible along straight segments, whereas reduced velocities are preferable in turning segments. This speed modulation allows the magnetic tip to respond effectively to the external magnetic field and reach the commanded deflection angles.

The mean propulsion velocities in the three experiments were 2.93, 4.63, and 6.06 mm/s, respectively. Collectively, these experiments demonstrate that when the MAS and advancer execute the planned commands, the MG can be guided to follow the desired trajectories within the fabricated phantoms.

8.4.2 In Vitro Navigation Validation of the MCGRS in a 3D Vascular Phantom

To evaluate the translational feasibility of the proposed MCGRS for interventional applications, in vitro navigation experiments were conducted in a clinically representative 3D vascular phantom, as shown in Figure 8.7a. In addition to the previously described system modules, a customized 3D phantom and a flow pulsation pump (FPP) were procured from a medical-device vendor (TRANDOMED, Ningbo, China). The phantom was designed according to realistic human vessel dimensions and fabricated via 3D printing using transparent silicone (total length: 959 mm). It reproduces an anatomical route extending from the femoral artery through the aortic arch and carotid artery to the intracranial vasculature. For the intracranial segment, four representative pathways, labeled 1–4, were selected for navigation evaluation (Figure 8.7b). The FPP emulates cardiac pumping by providing controllable flow velocity and pulsation frequency, thereby approximating arterial circulation conditions.

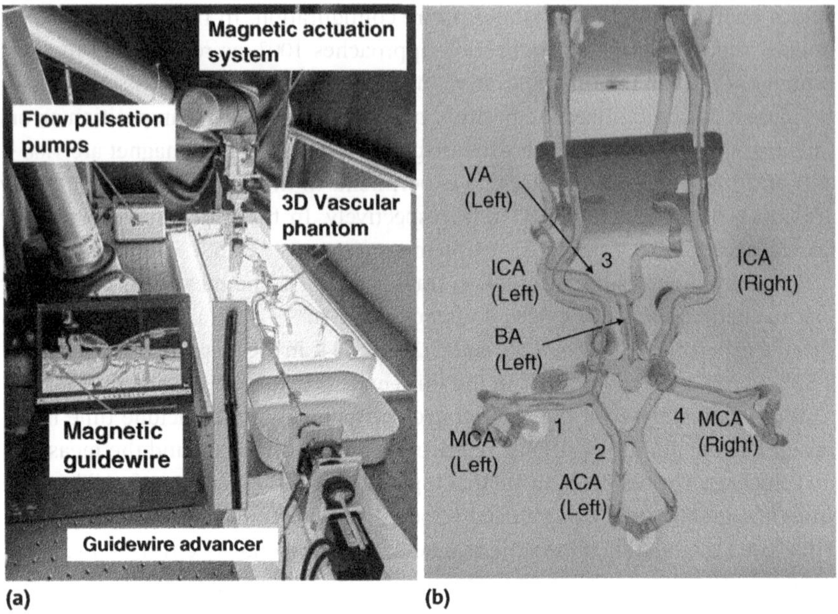

Figure 8.7 In vitro navigation validation of the MCGRS in a 3D vascular phantom. (a) Experimental setup for in vitro evaluation. (b) Four selected vascular pathways (labeled 1–4) in the phantom for assessing navigation capability, with the evaluated routes highlighted.

When the guidewire contacts the compliant vessel wall of the silicone-based phantom, substantial frictional resistance and contact-induced tension can arise, particularly over long insertion distances. This interaction may twist the guidewire shaft and increase the distributed friction along the vessel wall. In practice, the guidewire is therefore advanced while being rotated to mitigate friction and avoid sticking. Nevertheless, excessive propulsive force may lead to abrupt forward motion and uncontrolled "jumping" of the guidewire.

To reduce friction and improve the smoothness of insertion, a blood-mimicking fluid was prepared by mixing a surface lubricant with water at a volumetric ratio of 1 : 100. The mixture was circulated through the phantom using the FPP, which improved sliding conditions, enabled more continuous advancement, and helped extend the service life of the phantom.

For each of the four pathways, the end-magnet pose of the MAS and the propulsion command of the guidewire advancer were teleoperated in real time using an Xbox joystick, thereby coordinating magnetic steering with linear insertion. As summarized in Figure 8.8, the MG was successfully guided through all four pathways and reached the corresponding targets G1–G4.

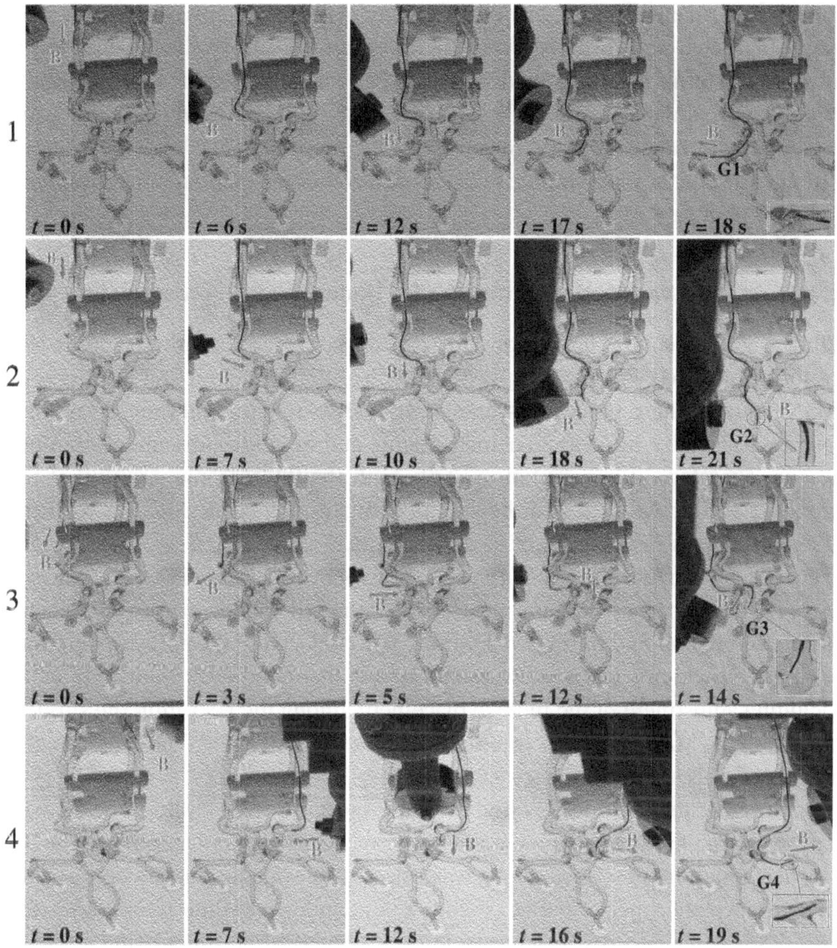

Figure 8.8 Results of in vitro navigation experiments of the MCGRS along the four vascular pathways to targets G1, G2, G3, and G4, respectively. The magnetic-tip states at the four target points are highlighted in the boxed regions.

Navigation in the 3D phantom is nontrivial due to progressively decreasing vessel diameters, frequent turns, short tortuous segments, and the presence of multiple aneurysm-like structures. To ensure that the magnetic tip could respond effectively to changes in the external magnetic field and achieve accurate deflection, relatively low insertion speeds were adopted, with mean propulsion velocities in the range of 1.5–2.5 mm/s. The travel time from the starting position to each target was within 1–2 min (based on the recorded operation time prior to video compression), indicating efficient navigation despite the anatomical complexity.

In clinical practice, the guidewire advancement speed depends on operator skill, vessel diameter, and tortuosity, and therefore no universal velocity standard is available. In terms of procedure duration, experienced interventionalists typically require approximately 1–2 min for a relatively straightforward route and 3–5 min for a highly tortuous pathway. Accordingly, the proposed MCGRS achieves navigation times that are comparable to those of conventional manual intervention for standard pathways.

8.4.3 In Vitro Verification of the MCGRS Under X-ray Fluoroscopy Imaging

Clinical endovascular interventions are typically performed in catheterization laboratories equipped with imaging systems such as digital subtraction angiography (DSA). During these procedures, interventionalists rely on real-time X-ray fluoroscopy to localize intravascular instruments and to guide subsequent manipulations. Therefore, it is essential to verify both the fluoroscopic visibility of the proposed MG under DSA and the overall compatibility of the MCGRS with the DSA workflow, so as to assess clinical feasibility.

For integration testing in the catheterization room of a partner hospital, a mobile cart was developed to support and transport the UR robotic arm together with the host computer and hardware modules. After relocating the entire system into the catheterization laboratory, installation and alignment tests were conducted, including adjustment of the relative placement among the DSA unit, the surgical table, and the MAS, as well as fine-tuning of the DSA tilt angle. The experimental arrangement is shown in Figure 8.9a.

The DSA system used in this study was an Azurion 7M20 (Philips, The Netherlands), a large flat-panel angiography platform. A representative route in the 3D vascular phantom described above was selected to evaluate fluoroscopic visibility and navigation performance (Figure 8.9c). Under real-time fluoroscopy, the MCGRS was teleoperated from outside the catheterization room using an Xbox joystick, and the MG was navigated along the selected path. The corresponding navigation sequence is shown in Figure 8.9b. The magnetic tip remained clearly visible in the fluoroscopic view, as shown in the boxed regions in Figure 8.9, and the overall system layout was compatible with the DSA unit and the surgical table. The mean propulsion velocity of the MG in this experiment was 1.2 mm/s.

During the experiment, the robotic-arm end-effector entered the X-ray beam and partially occluded the phantom in certain views. Nevertheless, the guidewire tip was continuously observable, enabling real-time monitoring of its state. In addition, by adjusting the DSA gantry orientation and coordinating the relative poses of the DSA and MAS, the visibility of the MG tip could be maintained, thereby reducing occlusion-induced interference during fluoroscopic tracking.

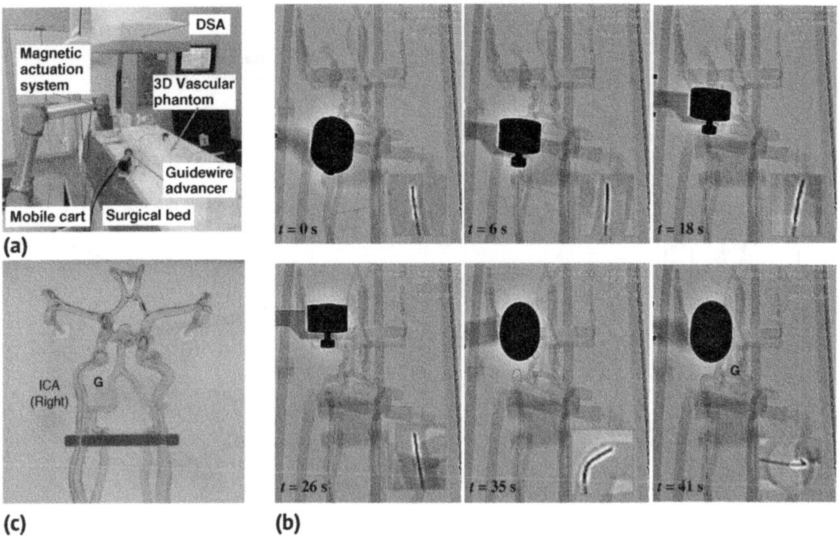

Figure 8.9 In vitro verification of the MCGRS in a 3D vascular phantom under real-time X-ray fluoroscopy imaging. (a) System integration in the interventional catheterization room. (b) Intravascular navigation process under real-time fluoroscopy; the magnetic-tip appearance at key points is highlighted in boxed regions. (c) Selected vascular route for integration testing, where G denotes the target point.

8.5 Strategy Design for Hard-to-reach Vessels

Catheter-based access to the vasculature has fundamentally advanced minimally invasive diagnosis and therapy, offering substantial clinical benefits [28, 29]. Nonetheless, catheter navigation toward distal targets remains challenging because access paths are long, vessels are frequently tortuous, and vascular walls are fragile [30, 31]. As illustrated in Figure 8.10a, catheters are typically introduced via femoral access and advanced from the relatively straight aorta into distal branched vessels with markedly increased geometric complexity. This process requires continuous attention and refined manual skills, and it carries risks of vessel-wall injury caused by contact forces and scraping. Such risks may prolong procedures and increase the amount of contrast agent required, thereby elevating procedural burden and associated complications [32, 33].

A central difficulty is reliable access to distal cortical arteries, which is critical for treating region-specific pathologies, including aneurysms [34, 35], arteriovenous malformations [36], liver tumors [37, 38], vascular tumors [39], and hemorrhage control [40, 41]. Achieving safe catheter advancement through highly

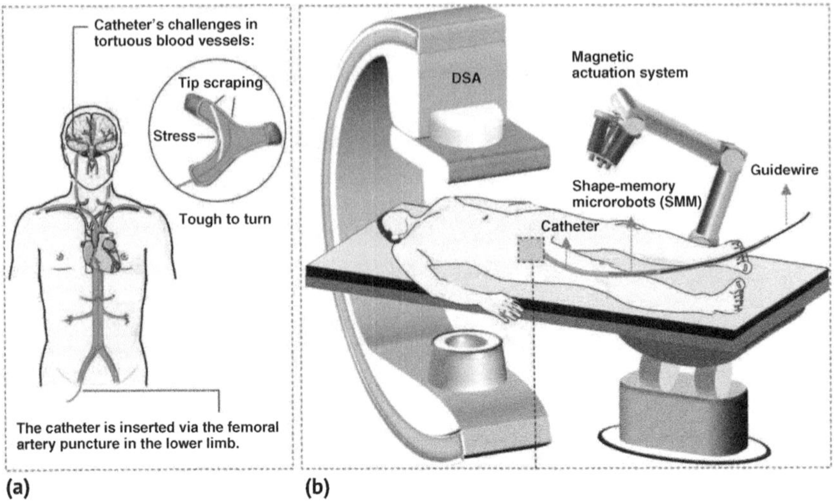

Figure 8.10 Schematic representation of the proposed multistage strategy for targeted embolization. (a) Overview of an interventional route and the associated catheter-navigation challenges in the vascular system. (b) Integrated platform supporting clinical deployment of SMMs, including DSA, a magnetic actuation system, a catheter, and a guidewire.

tortuous anatomies without iatrogenic injury remains difficult for both manual and robotic insertion strategies [42, 43]. In addition, the catheter inner diameter constrains the size and form factor of deliverable therapeutic payloads, which complicates the deployment of coils, stents, microspheres, gelatin sponge particles, and polymeric embolic materials [44]. These considerations motivate the development of more effective tools and strategies for minimally invasive therapies in hard-to-reach regions.

Recent years have seen rapid advances in micro-/nanomachines (MNMs) [45, 46]. Owing to their small size and untethered operation, MNMs can access confined anatomical sites and have shown promise for tasks such as locomotion in restricted spaces, cargo delivery, and targeted interventions [47–49]. MNMs can be actuated remotely using diverse modalities, including magnetic, optical, and chemical approaches [50–52]. Among these, magnetic actuation is particularly attractive because it enables wireless control of locomotion direction and speed while maintaining safe interactions with cells and tissues [50, 53, 54].

Helical micromachines (HMMs), inspired by bacterial flagellar propulsion, represent a typical class of magnetically driven MNMs. Compared with systems that require alternating fields or strong field gradients, HMMs can achieve precise, torque-driven motion under relatively low-strength rotating magnetic fields, which is advantageous for safety and controllability. This capability has stimulated growing interest in biomedical applications. However, existing HMM designs do not yet satisfy all clinical requirements. Further progress is needed in functional integration, real-time localization, the development of magnetic actuation platforms with larger workspaces, and in vivo validation for vascular therapies [55, 56]. Moreover, many demonstrations remain limited to simplified in vitro environments [57], highlighting the need for clinically oriented development that integrates time-efficient delivery, modern imaging guidance, and therapeutic effectiveness.

To address these challenges, we propose an untethered multistage swimming vascular embolization strategy based on SMMs. This strategy aims to achieve time-efficient delivery while enabling access to hard-to-reach vessels for in vivo embolization. The SMMs are fabricated from an organogel matrix containing 20 wt% magnetic particles, providing biocompatibility, radiopacity under X-ray imaging, and rapid thermal and magnetic responses.

An SMM is initially in a linear shape with a diameter of 0.5 mm at 20 °C inside a catheter. After release into the bloodstream, it recovers a preprogrammed helical configuration within 2 s at 38 °C. The helical morphology enables agile swimming in confined and tortuous vessels under rotating magnetic fields. Shape-memory gels have been explored for embolization of vessels and aneurysms [58, 59]; however, shape-memory systems that can rapidly recover at blood temperature while remaining wirelessly actuated are still relatively scarce [60].

Our experiments demonstrate multimodal locomotion of SMMs in carotid-artery models with flow gradients, including upstream swimming against flow and long-distance navigation in human-scale phantoms. The integrated platform (Figure 8.10b) combines DSA, a MAS, a catheter, and a guidewire to support rapid and precise in-body deployment of SMMs. The overall workflow follows the four stages summarized in Figure 8.11: catheter delivery, SMM deformation, magnetic guidance, and embolization.

The procedure starts with catheter insertion through femoral access. The catheter is advanced toward the tortuous region, after which linear SMMs at 20 °C are inserted and pushed by a guidewire. Upon entering the bloodstream, the SMMs transform into helical shapes within 2 s at 38 °C. The helical SMMs are then actuated by a rotating magnetic field to traverse sharp bends or consecutive U-turns and reach the target. Finally, embolization is achieved at the intended vessel or aneurysm site.

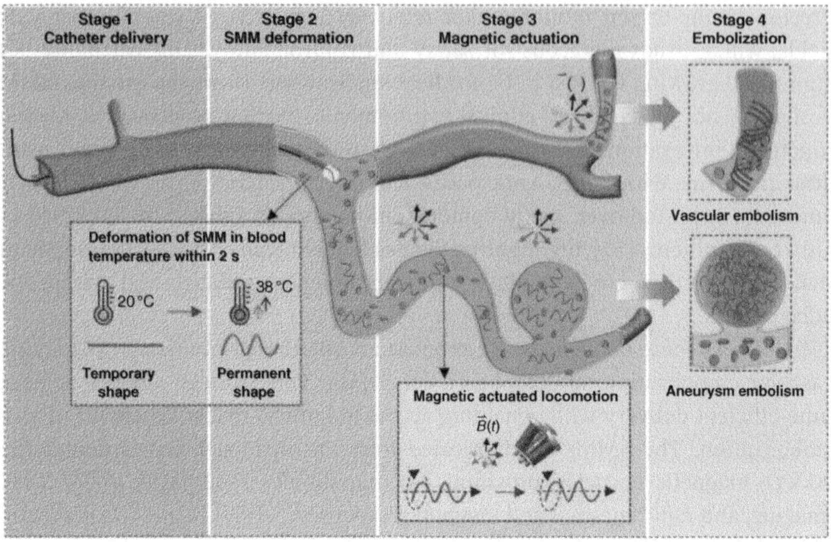

Figure 8.11 Four-step multistage embolization strategy. Step 1—Catheter delivery: the catheter is introduced via femoral access and advanced to the vicinity of tortuous vessels; linear SMMs are inserted and pushed toward the catheter tip by a guidewire. Step 2—SMM deformation: after exiting the catheter, the linear SMMs transform into helical shapes within 2 s at blood temperature. Step 3—Magnetic actuation: helical SMMs swim through sharp turns or multiple U-turns under a rotating magnetic field. Step 4—Embolization: vessel or aneurysm embolization is achieved at the target site.

8.6 In Vivo Validation of the Multistage Embolization Strategy

8.6.1 Experimental Validation of Strategy

Using the renal artery as a representative case, we demonstrate the in vivo performance of the proposed multistage embolization strategy (Figure 8.12). The integrated platform (Figure 8.12a) combines DSA, the MAS, an anesthesia machine, a catheter, and a guidewire.

The renal artery connects the aorta to the kidney (Figure 8.12b). When renal malignancy requires nonsurgical management, angioembolization is a clinically important therapeutic option [61]. As shown in Figure 8.12, the procedure consists of four sequential stages: catheter delivery, microrobot deformation, magnetic actuation, and vascular occlusion. The full process was recorded under real-time DSA fluoroscopy (Figure 8.13), and the duration of each stage is summarized in Table 8.1.

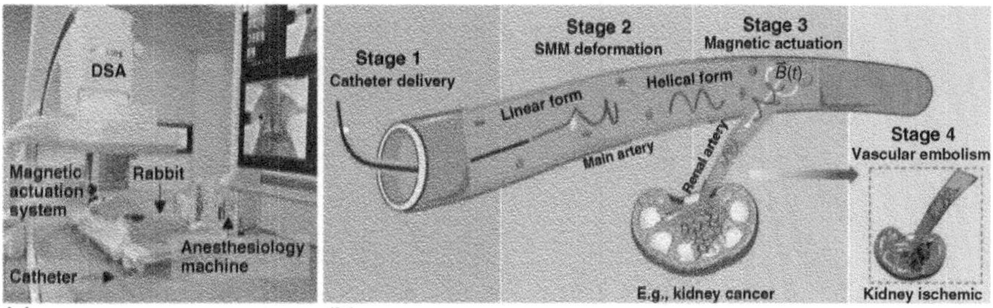

(a) (b)

Figure 8.12 Validation of the multistage strategy for rabbit kidney embolization. (a) Experimental setup for clinical application. (b) Schematic of the multistage kidney embolization strategy in a rabbit.

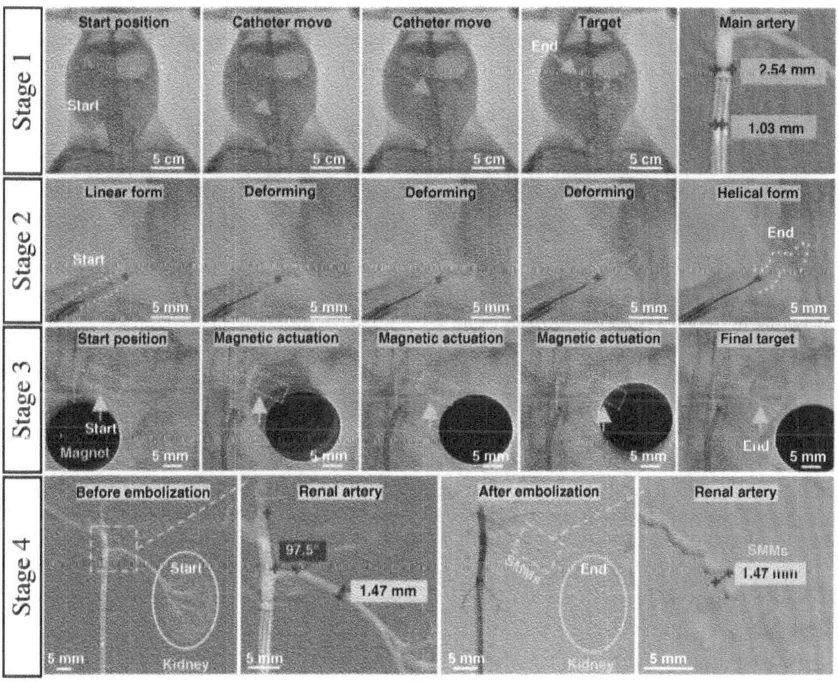

Figure 8.13 DSA chronologically captured the four-stage embolization process in the rabbit. Stage 1: a catheter was inserted via femoral access to the junction between the aorta and renal artery. Stage 2: the linear SMM was pushed out of the catheter and transformed into a helix within 2 s at blood temperature. Stage 3: helical SMMs reached the embolization site under a rotating magnetic field. Stage 4: angiography before and after embolization; disappearance of kidney perfusion indicates complete renal artery embolization.

Table 8.1 Performance in each stage of the multistage embolization procedure.

	Process	Imaging mode (DSA)	Actuation mode	Distance	Time	Speed
Stage 1	The catheter entered through the femoral artery puncture point in the thigh and advanced to the junction of the aorta and renal artery. Linear SMMs were pushed by the guidewire to reach the catheter head	X-ray imaging and angiography	Catheter moving	~17.64 cm	~1 min	2.94 mm/s
Stage 2	SMMs deformed from linear shape to helix at blood temperature	X-ray imaging	Blood temperature	–	~2 s	–
Stage 3	Helical SMMs moving to target position under rotating magnetic field	X-ray imaging	Magnetic field	1.52 cm	~5 s	~3.18 mm/s
Stage 4	Confirmation of vascular embolization	Angiography	–	–	–	–

In Stage 1, a 100-cm catheter was introduced percutaneously through the rabbit's right femoral artery and advanced along the aorta (diameter: 2.46 mm; length: 17.46 mm) to the aortorenal junction in approximately 60 s. A linear SMM (diameter: 0.5 mm) was then advanced through the catheter lumen using a 0.889-mm guidewire, reaching the distal position within 20 s. In Stage 2, the linear SMM exited the catheter and recovered a preprogrammed helical configuration within 2 s at blood temperature. The helical diameter was 1.47 mm. Considering the vessel diameter and locomotion efficiency, the SMM used in this study had a helical diameter of 1.47 mm, a helix angle of 63° (pitch: 2.3 mm), and a total length of 4.6 mm.

In Stage 3, the helical SMM was actuated by a rotating magnetic field, entered the renal artery, and traveled 1.52 cm within 5 s. In Stage 4, angiography clearly visualized renal branching vessels before embolization. After three helical SMMs reached the designated locations in the renal artery, the kidney became invisible in angiography, indicating cessation of renal blood flow and thus complete renal artery embolization.

Embolization efficacy increased as the number of deployed SMMs was raised from 0 to 3, with complete embolization achieved using three SMMs. Deploying more than three SMMs may induce ischemia or necrosis in healthy tissues or organs (e.g., the adrenal gland). In addition, excessive SMM numbers may elevate the mechanical load on vessel walls, disturb hemodynamics, and increase cardiac burden, potentially compromising overall physiological safety.

In summary, the proposed multistage embolization strategy achieved complete kidney embolization within 2 min, covering a delivery distance exceeding 100 cm with an overall turn of 97.5°. Notably, the linear SMM was transported through the catheter by a guidewire at 50 mm/s, which is substantially faster than the maximum in vivo magnetic-actuation speed of SMMs (3.18 mm/s). The catheter effectively serves as a high-throughput delivery channel for the linear SMM, thereby ensuring time efficiency.

8.6.2 Postoperative Observation

Following embolization, the rabbits were monitored for 4 weeks to evaluate both the embolization performance and the biocompatibility of the SMMs. In the experimental group, the left kidney was embolized using SMMs and is marked with cross symbols in Figure 8.14a. Noninvasive ultrasound imaging was performed to assess renal morphology and in vivo perfusion. In B-mode images, the maximal renal cross section is delineated by a white contour (Figure 8.14b), from which the embolized kidney was visibly smaller than the non-embolized kidneys. Color Doppler ultrasound, which visualizes flow direction and magnitude, showed a complete absence of perfusion signals in the embolized kidney (cross-marked region in Figure 8.14b), indicating that the occlusion remained effective after

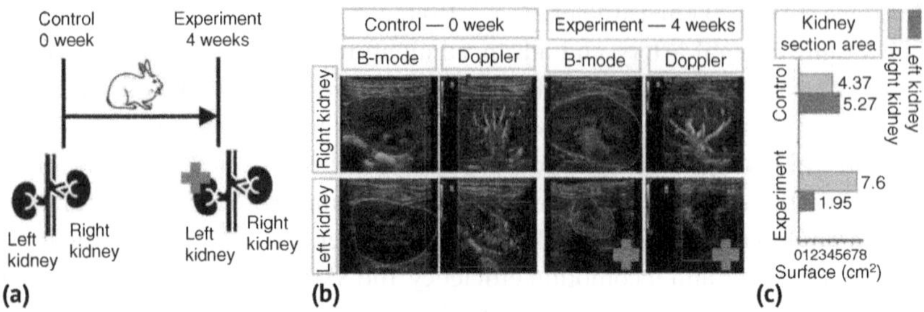

Figure 8.14 Four-week postoperative observation of kidney embolization using SMMs. Cross symbols indicate the embolized kidney. (a) Schematic comparison of a 0-week control rabbit and a 4-week experimental rabbit; the experimental rabbit received multistage embolization in the left kidney for 4 weeks. (b) In vivo assessment using B-mode and Doppler ultrasound. B-mode shows a reduced cross-sectional area of the embolized kidney, while Doppler evaluates blood-flow status and confirms the absence of blood flow in the embolized kidney. (c) The maximal cross-sectional area of the embolized kidney decreased from 5.27 cm^2 to 1.95 cm^2 over 4 weeks.

4 weeks. In contrast, the contralateral and control kidneys exhibited clear hemodynamic signals.

Quantitatively, the maximal cross-sectional area of the embolized kidney decreased from 5.27 to 1.95 cm^2 in B-mode measurements, corresponding to a 62.88% reduction (Figure 8.14c). Meanwhile, the right kidney increased from 4.37 to 7.60 cm^2 (73.74%), consistent with compensatory hypertrophy.

Histological evaluation was performed using hematoxylin and eosin (H&E) staining and Masson's trichrome staining (Figure 8.15a,b). H&E staining indicated pronounced damage to the renal parenchyma in the embolized kidney. The annotated arrows indicate glomerular structures, atrophied capillary collaterals within the glomerulus, and an enlarged glomerular cystic lumen, consistent with capillary atrophy. In comparison, the contralateral unembolized kidney and kidneys in the control group maintained intact tissue architecture.

Masson's staining was used to assess fibrosis, where the positively stained regions reflect fibrotic deposition. The embolized kidney exhibited more extensive positive staining than the other kidneys, indicating a higher degree of fibrosis. Quantitative analysis using ImageJ is summarized in Figure 8.15c. The Masson-positive area in the embolized kidney was significantly larger than that in other kidneys ("***", $P < 0.001$, t-test). No significant difference was observed between the contralateral unembolized kidney and the control group ("ns", $P = 0.05$, t-test).

Figure 8.15d shows histopathological observations at the embolization site. The vascular wall and collagen fibers were closely apposed to the SMMs, suggesting favorable tissue integration and good biocompatibility. As further shown in Figure 8.15e, the embolized kidney weighed 4.8 g, representing a 36% reduction

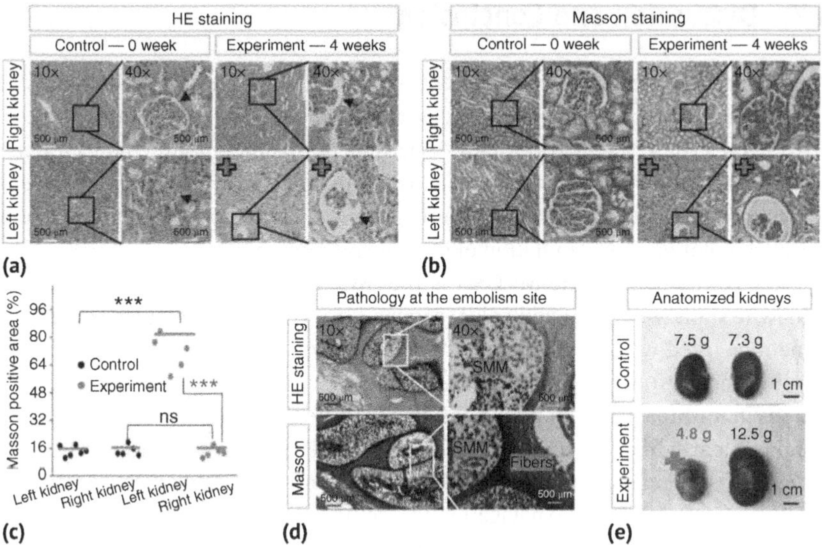

Figure 8.15 (a) H&E staining showing severe damage to the renal parenchyma in the embolized kidney. (b) Masson's staining indicating more pronounced fibrosis in the embolized kidney compared with other kidneys. (c) Quantification of fibrosis from Masson's staining; "***" denotes $P < 0.001$ and "ns" denotes no statistical significance. (d) Histopathological appearance at the renal artery embolization site, showing tight integration between the vascular wall/fibrous tissue and the SMM. (e) Optical images of dissected kidneys.

relative to the 7.5 g left kidney in the control group. In contrast, the right kidney increased from 7.4 to 12.5 g (71.2%), again indicating compensatory growth.

Biocompatibility was further examined by subcutaneously implanting a 20-mm straight SMM in the back of a rabbit for 1 week. Optical inspection and H&E staining showed that the SMM was tightly encapsulated by surrounding tissue, indicating a favorable local tissue response [62, 63].

In summary, 4 weeks after embolization, Doppler ultrasound detected no perfusion signals in the treated kidney, demonstrating that the SMMs could maintain stable occlusion over an extended period. B-mode imaging and gross anatomical inspection confirmed substantial renal atrophy. Histological analyses (H&E and Masson's staining) supported good biocompatibility, with the contralateral kidney remaining healthy and surrounding tissues adhering tightly to the SMMs. These long-term follow-up results validate the non-migration behavior, sustained embolization, and in vivo biocompatibility of the proposed multistage strategy.

8.7 Discussion and Conclusion

This chapter reports two related studies. The first study presents a MCGRS for VIS that provides both active steering and controlled propulsion. The system integrates a MAS and a GCA. The MAS employs a permanent magnet mounted on a robotic arm to generate a spatially controllable magnetic field, while the GCA uses stepper motors and friction-wheel mechanisms to realize linear insertion/retraction as well as rotational manipulation of the guidewire/catheter.

A key component of the MCGRS is the MG, which combines a commercial guidewire shaft with a magnetically responsive tip. The tip is fabricated using a silicone tube filled with a polymer matrix and magnetic particles, enabling flexible bending and a strong magnetic response under external fields. Experimental evaluations in multiple vascular phantoms demonstrate the feasibility of the proposed system. In particular, trajectory-following experiments show that the MCGRS can navigate vessel geometries with multiple turns at an acceptable level of tracking accuracy. Moreover, in vitro navigation in a complex 3D vascular phantom further supports its potential for clinical translation.

The second study introduces an untethered multistage vascular embolization strategy based on SMMs. The SMMs are fabricated from an organogel matrix containing magnetic particles, which provides biocompatibility, radiopacity under X-ray imaging, and rapid thermal and magnetic responses. In vivo validation in rabbits demonstrates that the SMMs can be delivered to the renal artery, achieve embolization, and maintain stable positioning over a 4-week observation period. Postoperative ultrasound and histological analyses confirm sustained occlusion in the embolized kidney, together with favorable tissue compatibility. Overall, these results support the proposed multistage strategy as a promising approach for time-efficient embolization in hard-to-reach vascular regions.

References

1 D. Mozaffarian, et al. Heart disease and stroke statistics—2016 update: a report from the American Heart Association. *Circulation*, 133 (4): e38–e360, 2016.

2 J. Guo, C. Shi, and H. Ren. Ultrasound-assisted guidance with force cues for intravascular interventions. *IEEE Transactions on Automation Science and Engineering*, 16 (1): 253–260, 2018.

3 G. Yang, et al. The grand challenges of science robotics. *Science Robotics*, 3 (14): eaar7650, 2018.

4 M. Cianchetti, C. Laschi, A. Menciassi, and P. Dario. Biomedical applications of soft robotics. *Nature Reviews Materials*, 3 (6): 143–153, 2018.

5 S. A. Menaker, S. S. Shah, B. M. Snelling, S. Sur, R. M. Starke, and E. C. Peterson. Current applications and future perspectives of robotics in cerebrovascular and endovascular neurosurgery. *Journal of NeuroInterventional Surgery*, 10 (1): 78–82, 2018.

6 A. K. Hoshiar, S. Jeon, K. Kim, S. Lee, J. Kim, and H. Choi. Steering algorithm for a flexible microrobot to enhance guidewire control in a coronary angioplasty application. *Micromachines*, 9 (12): 617, 2018.

7 M. Sitti. Miniature soft robots—road to the clinic. *Nature Reviews Materials*, 3 (6): 74–75, 2018.

8 W. Crinnion, et al. Robotics in neurointerventional surgery: a systematic review of the literature. *Journal of NeuroInterventional Surgery*, 14 (6): 539–545, 2022.

9 Y. Kim, et al. Telerobotic neurovascular interventions with magnetic manipulation. *Science Robotics*, 7 (65): eabg9907, 2022.

10 M. Sitti and D. S. Wiersma. Pros and cons: magnetic versus optical microrobots. *Advanced Materials*, 32 (20): 1906766, 2020.

11 Z. Yang and L. Zhang. Magnetic actuation systems for miniature robots: a review. *Advanced Intelligent Systems*, 2 (9): 2000082, 2020.

12 T. Xu, C. Huang, Z. Lai, and X. Wu. Independent control strategy of multiple magnetic flexible millirobots for position control and path following. *IEEE Transactions on Robotics*, 38 (5): 2875–2887, 2022.

13 X. Tang, et al. Magnetic–acoustic sequentially actuated CAR T cell microrobots for precision navigation and in situ antitumor immunoactivation. *Advanced Materials*, 35 (18): 2211509, 2023.

14 J. Choi, J. Hwang, J.-Y. Kim, and H. Choi. Recent progress in magnetically actuated microrobots for targeted delivery of therapeutic agents. *Advanced Healthcare Materials*, 10: 2001596, 2021.

15 F. Soto, J. Wang, R. Ahmed, and U. Demirci. Medical micro/nanorobots in precision medicine. *Advanced Science*, 7 (21): 2002203, 2020.

16 J. Hwang, J. Kim, and H. Choi. A review of magnetic actuation systems and magnetically actuated guidewire-and catheter-based microrobots for vascular interventions. *Intelligent Service Robotics*, 13 (1): 1–14, 2020.

17 Y. Kim, G. A. Parada, S. Liu, and X. Zhao. Ferromagnetic soft continuum robots. *Science Robotics*, 4 (33): eaax7329, 2019.

18 G. Pittiglio, P. Lloyd, T. Veiga, O. Onaizah, C. Pompili, J. H. Chandler, and P. Valdastri. Patient-specific magnetic catheters for atraumatic autonomous endoscopy. *Soft Robotics*, 9 (6): 1120–1133, 2022.

19 J. Hwang, et al. A magnetically controlled soft microrobot for targeted drug delivery. *Advanced Healthcare Materials*, 11: 2102529, 2022.

20 J. Lussi, et al. A submillimeter continuous variable stiffness catheter for compliance control. *Advanced Science*, 8 (18): 2101290, 2021.

21 S. Gervasoni, J. Lussi, S. Viviani, Q. Boehler, N. Ochsenbein, U. Moehrlen, and B. J. Nelson. Magnetically assisted robotic fetal surgery for the treatment of spina bifida. *IEEE Transactions on Medical Robotics and Bionics*, 4 (1): 85–93, 2022.

22 Z. Yang, L. Yang, M. Zhang, Q. Wang, S. Chun Ho Yu, and L. Zhang. Magnetic control of a steerable guidewire under ultrasound guidance using mobile electromagnets. *IEEE Robotics and Automation Letters*, 6 (2): 1280–1287, 2021.

23 Z. Yang, L. Yang, M. Zhang, N. Xia, and L. Zhang. Ultrasound-guided wired magnetic microrobot with active steering and ejectable tip. *IEEE Transactions on Industrial Electronics*, 70 (1): 614–623, 2022.

24 S. Zhang, et al. Design and characteristics of 3D magnetically steerable guidewire system for minimally invasive surgery. *IEEE Robotics and Automation Letters*, 7 (2): 4040–4046, 2022.

25 S. Fu, et al. Path planning and control of a magnetic guidewire in vascular phantoms. In *2022 IEEE International Conference on Real-Time Computing and Robotics (RCAR)*, pages 57–62, Piscataway, NJ, USA, 2022. IEEE.

26 C. Chautems, A. Tonazzini, Q. Boehler, S. H. Jeong, D. Floreano, and B. J. Nelson. Magnetic continuum device with variable stiffness for minimally invasive surgery. *Advanced Intelligent Systems*, 2 (6): 1900086, 2020.

27 D. Lin, J. Wang, N. Jiao, Z. Wang, and L. Liu. A flexible magnetically controlled continuum robot steering in the enlarged effective workspace with constraints for retrograde intrarenal surgery. *Advanced Intelligent Systems*, 3 (10): 2000211, 2021.

28 R. Secoli, et al. Modular robotic platform for precision neurosurgery with a bio-inspired needle: system overview and first *in-vivo* deployment. *PLoS One*, 17 (10): e0275686, 2022.

29 V. M. Pereira, et al. First-in-human, robotic-assisted neuroendovascular intervention. *Journal of NeuroInterventional Surgery*, 12 (4): 338–340, 2020.

30 J. L. Saver, et al. Thrombectomy for distal, medium vessel occlusions: a consensus statement on present knowledge and promising directions. *Stroke*, 51 (9): 2872–2884, 2020.

31 L. Pancaldi, et al. Flow driven robotic navigation of microengineered endovascular probes. *Nature Communications*, 11 (1): 6356, 2020.

32 S. M. Bagheri, M. Ghadamzadeh, and M. Chavoshi. Percutaneous embolization of renal pseudoaneurysms: a retrospective study. *Indian Journal of Urology*, 38 (4): 296–301, 2022.

33 M. S. Alwarraky, M. M. Abdallah, and M. S. Elgharbawy. Clinical outcome and safety of selective renal artery embolization using permanent occlusive agents for acute renal bleeding. *Egyptian Journal of Radiology and Nuclear Medicine*, 51 (1): 205, 2020.

34 K. Rotim, B. Splavski, V. Kalousek, M. Jurilj, and T. Sajko. Endovascular management of intracranial aneurysms on distal arterial branches: illustrative case series and literature retrospection. *Acta Clinica Croatica*, 59 (4): 712–719, 2020.

35 A. Ricci, et al. Cortical aneurysms of the middle cerebral artery: a review of the literature. *Surgical Neurology International*, 8: 117, 2017.

36 M. N. Shah, et al. The relationship of cortical folding and brain arteriovenous malformations. *Neurovascular Imaging*, 2 (1): 13, 2016.

37 S. Young, A. J. Taylor, and T. Sanghvi. Post locoregional therapy treatment imaging in hepatocellular carcinoma patients: a literature-based review. *Journal of Clinical and Translational Hepatology*, 6 (2): 189, 2018.

38 A. Mehta, S. R. Venkat, L. Portelance, and L. G. Feun. Congenital absence of the portal vein complicated by hepatocellular carcinoma in the liver of an adult woman: review of imaging, literature and management. *Hepatoma Research*, 4 (2): 7, 2018.

39 J. Lim, G. Choi, K. Il Joo, H. J. Cha, and J. Kim. Embolization of vascular malformations via *in situ* photocrosslinking of mechanically reinforced alginate microfibers using an optical-fiber-integrated microfluidic device. *Advanced Materials*, 33 (14): 2006759, 2021.

40 Y. Heo, H. L. Kang, and D. H. Kim. Exercise-induced traumatic muscle injuries with active bleeding successfully treated by embolization: three case reports. *Journal of Trauma and Injury*, 35 (3): 219, 2022.

41 L. Masood, Z. A. Khan, M. S. Ahmed, and A. I. Rana. Saving the uterus; our experience in uterine artery embolization for acquired uterine vascular abnormalities. *Journal of Ayub Medical College Abbottabad*, 34 (4 (SUPPL1)): 913–918, 2022.

42 T. Gopesh, et al. Soft robotic steerable microcatheter for the endovascular treatment of cerebral disorders. *Science Robotics*, 6 (57): eabf0601, 2021.

43 A. I. Chen, M. L. Balter, T. J. Maguire, and M. L. Yarmush. Deep learning robotic guidance for autonomous vascular access. *Nature Machine Intelligence*, 2 (2): 104–115, 2020.

44 Y. Yang, et al. Endothelium-mimicking multifunctional coating modified cardiovascular stents via a stepwise metal-catechol-(amine) surface engineering strategy. *Research*, 2020: 9203906, 2020.

45 B. Wang, K. Kostarelos, B. J. Nelson, and L. Zhang. Trends in micro-/nanorobotics: materials development, actuation, localization, and system integration for biomedical applications. *Advanced Materials*, 33 (4): 2002047, 2021.

46 S. Zhao, D. Sun, J. Zhang, H. Lu, Y. Wang, R. Xiong, and K. T. V. Grattan. Actuation and biomedical development of micro-/nanorobots–a review. *Materials Today Nano*, 18: 100223, 2022.

47 T. Xu, J. Yu, C. Vong, B. Wang, X. Wu, and L. Zhang. Dynamic morphology and swimming properties of rotating miniature swimmers with soft tails. *IEEE/ASME Transactions on Mechatronics*, 24 (3): 924–934, 2019.

48 C. Qiu, et al. Advanced strategies for overcoming endosomal/lysosomal barrier in nanodrug delivery. *Research*, 6: 0148, 2023.

49 F. Zhang, et al. Lipid-based intelligent vehicle capabilitized with physical and physiological activation. *Research*, 2022: 9808429, 2022.

50 H. Zhou, C. C. Mayorga-Martinez, S. Pané, L. Zhang, and M. Pumera. Magnetically driven micro and nanorobots. *Chemical Reviews*, 121 (8): 4999–5041, 2021.

51 X. Wang, Z. Ye, S. Lin, L. Wei, and L. Xiao. Nanozyme-triggered cascade reactions from cup-shaped nanomotors promote active cellular targeting. *Research*, 2022: 9831012, 2022.

52 L. Zhang, et al. A magnetic-driven multi-motion robot with position/orientation sensing capability. *Research*, 6: 0177, 2023.

53 L. Wang, et al. Reconfigurable vortex-like paramagnetic nanoparticle swarm with upstream motility and high body-length ratio velocity. *Research*, 6: 0088, 2023.

54 S. Yu, et al. Trimer-like microrobots with multimodal locomotion and reconfigurable capabilities. *Materials Today Advances*, 14: 100231, 2022.

55 Y. Dong, L. Wang, V. Iacovacci, X. Wang, L. Zhang, and B. J. Nelson. Magnetic helical micro-/nanomachines: recent progress and perspective. *Matter*, 5 (1): 77–109, 2022.

56 M. Medina-Sánchez and O. G. Schmidt. Medical microbots need better imaging and control. *Nature*, 545 (7655): 406–408, 2017.

57 T. Li, et al. Bioinspired claw-engaged and biolubricated swimming microrobots creating active retention in blood vessels. *Science Advances*, 9 (18): eadg4501, 2023.

58 Y. Zhang, et al. Radiopaque highly stiff and tough shape memory hydrogel microcoils for permanent embolization of arteries. *Advanced Functional Materials*, 28 (9): 1705962, 2018.

59 B. Liu, et al. Stiffness self-tuned shape memory hydrogels for embolization of aneurysms. *Advanced Functional Materials*, 30 (22): 1910197, 2020.

60 S. Wang, et al. Preshaped 4D photocurable ultratough organogel microcoils for personalized endovascular embolization. *Advanced Materials*, 35 (52): 2308130, 2023.

61 G. Go, et al. Multifunctional microrobot with real-time visualization and magnetic resonance imaging for chemoembolization therapy of liver cancer. *Science Advances*, 8 (46): eabq8545, 2022.

62 L. F. García. Immune response, inflammation, and the clinical spectrum of COVID-19. *Frontiers in Immunology*, 11: 1441, 2020.

63 L. Wang, pH and lipase-responsive nanocarrier-mediated dual drug delivery system to treat periodontitis in diabetic rats. *Bioactive Materials*, 18: 254–266, 2022.

9

Parallel-mobile-coil System for Wired Magnetic Microrobot

This chapter proposes a control framework of a miniature magnetic driller-tipped guidewire (MDG) with an integrated actuation system. Furthermore, a wired magnetic microrobot (WMM) with an ejected helical bullet (HB) mounted on the guidewire is proposed for smoothly steering in lumen environment.

9.1 Introduction

As human lifespan increases and exposure to risk factors grows, the prevalence of Peripheral Vascular Disease (PVD) continues to rise globally. According to statistics, this condition affects approximately 200 million people worldwide [1]. As a minimally invasive treatment, endovascular intervention is broadly applied to manage symptoms of occlusive PVD, such as atherosclerotic obliterans and deep vein thrombosis with severe manifestations. However, current therapeutic approaches face significant limitations. One major issue is the disappointing success rate resulting from the diffuse nature of atherosclerosis and the presence of chronic total occlusions [2]. Additionally, most guidewires/catheters require manual operation at the remote end, heavily relying on the operator's experience and precision [3]. Furthermore, the prolonged duration of these procedures increases radiation exposure for both patients and clinicians, potentially posing health risks [4].

In traditional methods, operators frequently opt for high-stiffness guidewires with tapered tips to advance through clots via pushing. However, this technique entails risks of vascular injury or even perforation. To address these challenges, novel devices aimed at restoring blood vessel patency have emerged. Examples include excimer laser atherectomy, rotational atherectomy, rheolytic thrombectomy, and aspiration thrombectomy. Rotational atherectomy, in particular, has undergone extensive research, utilizing spinning motor-driven tips equipped with blades or burrs to excise or abrade plaques. Several commercial products, such as the Jetstream Atherectomy Device, Phoenix Atherectomy System, and Rotablator System, have received FDA approval [2]. Furthermore, clinical studies have demonstrated enhanced success rates with these devices. Despite these advancements, the aforementioned guidewires and catheters remain passive instruments, offering limited control during their insertion process.

Advancements in interventional device technology have led to the development of various steerable instruments for enhanced maneuverability, encompassing designs based on concentric tubes, tendon actuation, hydraulics, pneumatics, and magnetism [5, 6]. Among these, magnetically driven guidewires/catheters stand out due to their simplicity, compactness, and versatility in operation [7]. These devices incorporate magnetic components at their tips, enabling swift and responsive movements in response to external magnetic fields (EMFs), as demonstrated in studies [8–12]. These continuum manipulators are designed to be compliant, allowing for significant deformation and adept navigation through complex vascular anatomies, as reviewed in [13]. However, their ability to access blocked vessels has received limited attention. Conversely, miniature helical swimmers have proven effective in removing blood clots by converting rotational motion into linear progression and clearing obstructions through mechanical friction [14–17]. Nevertheless, the retrieval issue and multifunctional implementation remain challenging. Combining these two technologies into a unified design could potentially address the aforementioned limitations, reconciling the conflict between manipulation safety and penetration capability observed in traditional guidewires. Preliminary concepts showcasing this integration have been presented in [18–20], where the proposed assemblies of guidewires/catheters exhibit selective capabilities for steering and clot removal. However, further exploration is needed to develop designs that can achieve large steering angles under moderate magnetic field strengths.

This chapter introduces and centers on the development of a robotic approach for treating endovascular occlusive PVD, incorporating magnetic actuation and ultrasound (US) imaging to enhance procedural safety and minimize radiation exposure. Additionally, it advances the implementation of this technology to a human-scale level. Despite their clinical promise, several challenges persist, including: (1) the rapid decrease in magnetic field strength as distance increases [21, 22]; (2) the narrow ultrasonic beam emitted by 2D probes; and (3) the low signal-to-noise ratio (SNR) in US images. To address these, a magnetic driller is affixed to a guidewire via a silicone rubber cannula, enabling simultaneous magnetic guidance and occlusion clearance. An integrated system, adapted from prior research [23], encompasses a motor module, electromagnet module, transducer module, and feeding module. Built upon this system, a user-intuitive control framework facilitates remote operation. A conceptual illustration is depicted in Figure 9.1. A potential intervention procedure is presented to treat occlusive PVD, featured by dual modes, US guidance, and magnetic control.

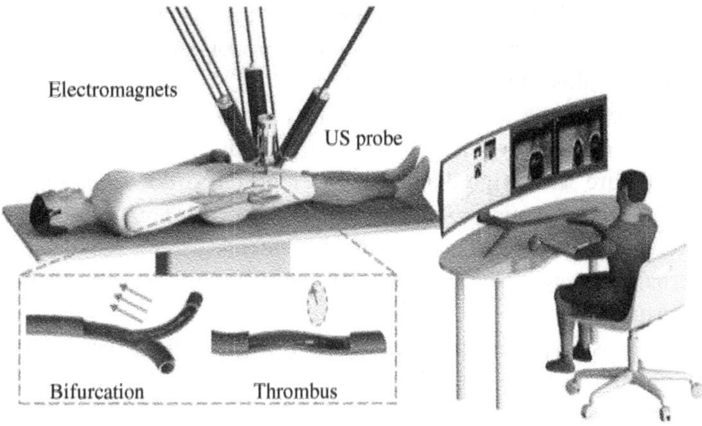

Figure 9.1 Conceptual diagram of the overall system. Three electromagnets and an adjustable US probe are integrated into the parallel mechanism for large-workspace magnetic actuation and US imaging. The MDG has two modes: bending mode under directional magnetic fields for navigation in complex vasculature; drilling mode under rotating magnetic fields for passing through clots.

To this point in time, a diverse range of steerable interventional devices has been conceived to enhance maneuverability, with magnetic actuation being a prominent approach. These designs boast miniaturization, superior flexibility, and economic feasibility [24]. Instances encompass magnetically controllable guidewires [9, 10, 23] and catheters [11, 25–27]. Yet, the efficacy of navigating through tortuous vasculatures, characterized by curves, kinks, loops, and spiral twists, has scarcely been assessed, and it is documented that manipulating such structures can be arduous, occasionally leading to instrument fracture owing to the forces exerted between the wire and the intricate vessel walls [28]. Alternatively, wireless miniature robots have garnered extensive research attention for their potential to access intricate and inaccessible anatomical regions, enabling various medical procedures such as targeted drug administration and clot removal [16, 29, 30]. Nevertheless, controlling these robots in swift blood flows remains a formidable challenge [31]; specifically, there is a risk of the minute robot being displaced during its journey to the lesion site or while returning post-treatment.

Furthermore, there is a need to investigate medical imaging techniques that can reduce X-ray exposure. US imaging stands out as a potential alternative due to its rapid imaging capabilities, absence of ionizing radiation, and widespread availability [32]. However, the reliability of US imaging is heavily reliant on the operator, leading to poor reproducibility. Additionally, its low SNR and constrained field of view pose significant challenges for clear visualization. Prior research has demonstrated the feasibility of robotically controlling diverse microrobots using

US guidance [4, 33–36]. Nevertheless, a solution for concurrently employing multimode magnetic control and US tracking of microrobots in complex tissue-mimicking settings remains an unaddressed area.

In this chapter, we present a novel design for a WMM that combines the advantages of both wired and wireless microrobot technologies. When navigating through vasculature with large diameters and high blood flow, we employ the tethered mode, where the microrobot's forward and backward movements are regulated by a feeding device, while its steering movements are controlled using directional external magnetic fields (D-EMFs). In contrast, for movement within narrowed and tortuous vascular structures, we utilize the untethered mode, in which the ejected HB is propelled wirelessly by rotating external magnetic fields (R-EMFs). Additionally, we have developed a custom system for US-guided multimode control of the WMM across large workspaces. This system involves an initial automatic scan of the vessel, followed by the sequential implementation of the tethered mode, tip ejection, and untethered mode, all with real-time US imaging feedback.

9.2 Robotic Hardware and Control Framework

9.2.1 Robotic Hardware

9.2.1.1 Magnetic Driller-tipped Guidewire

(1) Structure Design: The architecture of the devised MDG is depicted in Figure 9.2a. It comprises a magnetic driller tip, a flexible neck section, and a base shaft. The dimensional specifics of the prototype, as outlined in Table 9.1, are rationalized by considering the typical vessel diameters within the range of 3.0–10.6 mm encountered in most lower limb catheterization procedures, [37]. The components of the driller tip were fabricated using three-dimensional (3D) printing technology (employing RC Series material from EnvisionTEC, Inc.). The helical segment's configuration adheres to the optimal design proposed in [19]. The ball joint-inspired structure allows for agile axial rotation and slight radial flexibility. A cylindrical magnet (CM), radially magnetized, is embedded within the tip's cavity to facilitate simultaneous steering and drilling actions. The tip is mechanically coupled to the cannula through the use of microthreads on the connector and the inherent elasticity of silicone. A spring, made of stainless steel, serves as the linkage between the cannula and the coil, which is further reinforced with elastic adhesive. The neck's stiffness can be varied by adjusting the position of the inner core, resulting in a rotational stiffness that spans from 6.71×10^{-5} to 5.91×10^{-4} Nm/rad. The structural reliability was assessed by applying tensile forces to the tip and repeatedly altering the neck's stiffness.

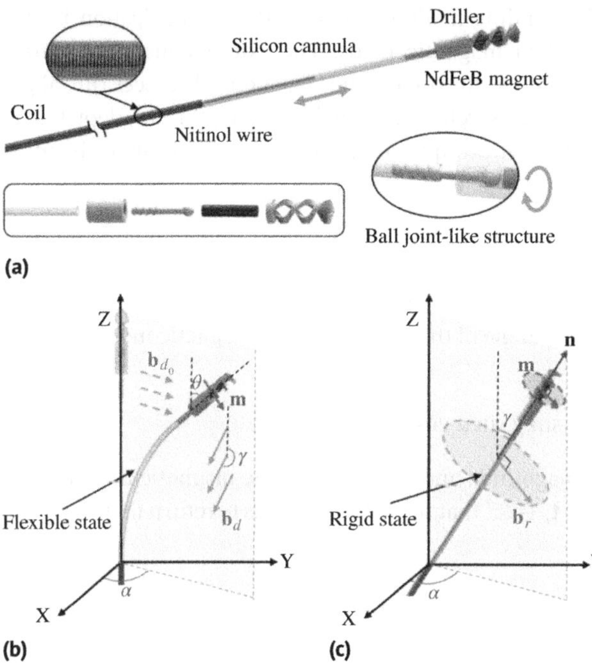

Figure 9.2 Schematic of the MDG. (a) Structural design. (b) Bending mode under directional magnetic fields. $\mathbf{b}_{d_0}$ is the initial directional magnetic field that aligns the magnetization in the bending plane. (c) Drilling mode under rotating magnetic fields. $\mathbf{n} \in \mathbb{R}^3$ is the rotation axis of the field.

Table 9.1 Geometric parameters of the fabricated prototype MDG.

Properties	Dimensions
Outer diameter of driller	1.8 mm
Length of driller	8.5 mm
Outer diameter of NdFeB magnet	1 mm
Length of NdFeB magnet	5 mm
Outer diameter of silicone cannula	900 μm
Length of silicone cannula	30 mm
Outer diameter of spring coil	0.035 in
Length of spring coil	120 mm
Outer diameter of Nitinol wire	260 μm
Length of Nitinol wire	160 mm

(2) Working Mode: The bending mode utilized for flexible navigation is governed by directionally applied magnetic fields. The external magnetic torque exerts an influence that aims to orient the magnetized object in alignment with the external field, as referenced in [36] and [37]. As depicted in Figure 9.2b, the resultant magnetic torque, denoted as $t_m \in \mathbb{R}^3$ within the bending plane, is equivalent to:

$$||\mathbf{t}_m|| = ||\mathbf{m}||||\mathbf{b}_d|| \sin\left(\gamma - \theta - \frac{\pi}{2}\right) \tag{9.1}$$

where γ is the pitch angle of the directional magnetic field $\mathbf{b}_d \in \mathbb{R}^3$, represented by Eq. (9.2); θ is the bending angle of the MDG; m is the magnetic moment of the installed magnet.

$$\mathbf{b}_d = A_m \left[\sin\gamma \cos\alpha \sin\gamma \sin\alpha \cos\gamma\right]^T \tag{9.2}$$

where A_m and α are the magnitude and yaw angle of the magnetic field, respectively. The restoring torque $\mathbf{t}_r \in \mathbb{R}^3$ that the MDG attempts to return to the original state equals:

$$||\mathbf{t}_r|| = EI_A\theta/L_C \tag{9.3}$$

where E and I_A are the elastic modulus and area moment of inertia of the cannula, respectively; L_C is the length from the tip to the bending point. The equilibrium condition reaches when $||\mathbf{t}_m|| = ||\mathbf{t}_r||$, which is used to predict deflection.

The drilling mode is controlled by rotating magnetic fields $\mathbf{b}_r \in \mathbb{R}^3$, which is represented as:

$$\mathbf{b}_r = A_m \begin{bmatrix} \cos\gamma \cos\alpha \cos(2\pi f t) + \sin\alpha \sin(2\pi f t) \\ \cos\gamma \sin\alpha \cos(2\pi f t) - \cos\alpha \sin(2\pi f t) \\ -\sin\gamma \cos(2\pi f t) \end{bmatrix} \tag{9.4}$$

The frequency of magnetic field rotation is represented by f. As shown in Figure 9.2c, the driller rotates along its axis in response to the EMFs, facilitating the removal of blood clots through mechanical abrasion. To ensure stability in the rotational motion and provide support for the drilling action, a rigid configuration can be achieved by extending the inner core to its fullest distal position. It is important to note that repositioning the inner core is not a mandatory step; it is only necessary when the clot is located in a relatively straight section of the vasculature, thereby minimizing any significant impact on the tip's orientation during adjustment.

9.2.1.2 Actuation System

In this chapter, the comprehensive system comprises four distinct modules, as depicted in Figure 9.3. Firstly, the motor module features a bespoke parallel mechanism, with its superior ends linked to three guide rails and inferior ends attached to an endplate. This mechanism is propelled by three stepper motors (specifically, Model: 86BYGH114). Secondly, the electromagnet module encompasses an assembly of three mobile electromagnetic coils, which are actuated by the parallel mechanism and receive power from three programmable power supplies (Model: IT6006C-500-30, manufactured by ITECH Corp.). Thirdly, the transducer module consists of a compact, in-house developed mechanism designed to secure and adjust the orientation of a US probe (Model: 15L4, supplied by Teratech Corp.), allowing for its installation on the endplate. Lastly, the feeding module is a tailored device intended for automated insertion and retrieval of MDGs, capable of accommodating guidewires/catheters with diameters between 0.5 and 5 mm. The advantages of this design include the mobile-coil configuration, which enhances workspace while minimizing energy consumption, and the eye-in-hand setup of the probe, which facilitates efficient US imaging without necessitating additional driven manipulators.

The parallel mechanism's motion and the production of the magnetic field are governed by two distinct microcontrollers, specifically the NI myRIO (National Instruments Inc.). The rotation of the US probe and the feeding of the guidewire

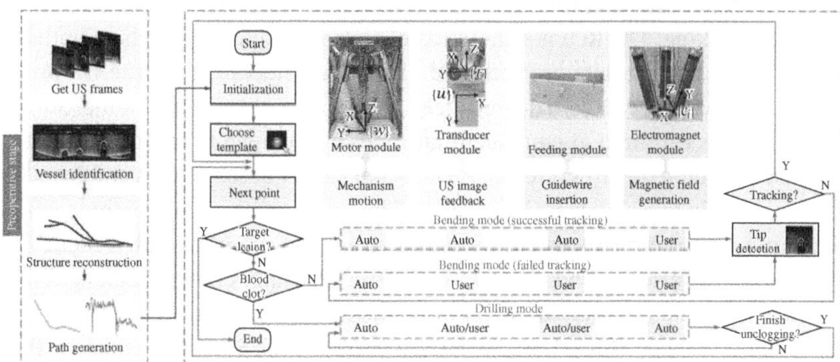

Figure 9.3 Block diagram of the control framework, consisting of the preoperative stage and intraoperative stage. The integrated system has four modules: motor module, transducer module, feeding module, and electromagnet module. Several coordinate systems are defined: the world coordinate system $\mathcal{W}$, the endplate coordinate system $\mathcal{F}$, the US image coordinate system $\mathcal{U}$, and the electromagnet coordinate system $\mathcal{C}_i$. This strategy internally coordinates all modules and provides options of bending mode and drilling mode.

are managed by an Arduino Uno R3 microcontroller. Additionally, US images are captured using a commercial device, the Terason t3200 (Teratech Corp.). A central computer, equipped with an Intel(R) Core(TM) i7-9700KF CPU running at 3.60 GHz and boasting 32 GB of RAM, serves as the hub for coordinating all modules. This coordination is facilitated through a user interface developed using LabVIEW 2017 (National Instruments Inc.) and MATLAB R2020a (MathWorks Inc.), with operational commands input via a joystick. The prototype system features a cylindrical workspace measuring 500 mm in diameter.

9.2.2 Control Framework

A workflow consisting of two phases has been devised, taking into account the characteristics of US imaging and the system's configuration. During the preoperative phase, the region of interest (ROI) containing the lesion is scanned and reconstructed automatically, with pertinent data extracted for the purpose of generating surgical prompts and a reference pathway. Transitioning to the intraoperative phase, the control program orchestrates intricate calculations across the various modules to guarantee that the US probe traverses the vessel and aligns with the MDG in real time, offering operators an intuitive interface for making immediate decisions. The schematic representation of the control framework is illustrated in Figure 9.3.

9.2.2.1 Preoperative Stage

(1) Vessel Tracking: The ROI is scanned as depicted in Figure 9.4a, where both US images and positional data are recorded at regular intervals, with the US probe angle (PA) maintained at zero degrees relative to the xz plane of the frame. Subsequently, frames captured at equivalent latitudes are merged to form comprehensive cross-sectional pictures (OCPs). A Gaussian kernel-based filter is employed to mitigate noise in these images (Figure 9.4). The subsequent portion of this section delves into extracting valuable information from the preprocessed B-mode US data. Discerning arteries and veins within static US images poses challenges due to the low SNR and the presence of additional hypoechoic structures. Various vessel segmentation algorithms have been explored, with the Star–Kalman algorithm demonstrating notable effectiveness for segmenting convex shapes and exhibiting robustness against noise, as reported in [23, 24], and [38].

Building upon this concept, the study incorporates robust ellipse fitting and temporal consistency to enhance vessel tracking and bifurcation identification. The flowchart illustrating the proposed methodology is presented in Figure 9.5. Initially, the operator selects a start seed point S1, located within the central area

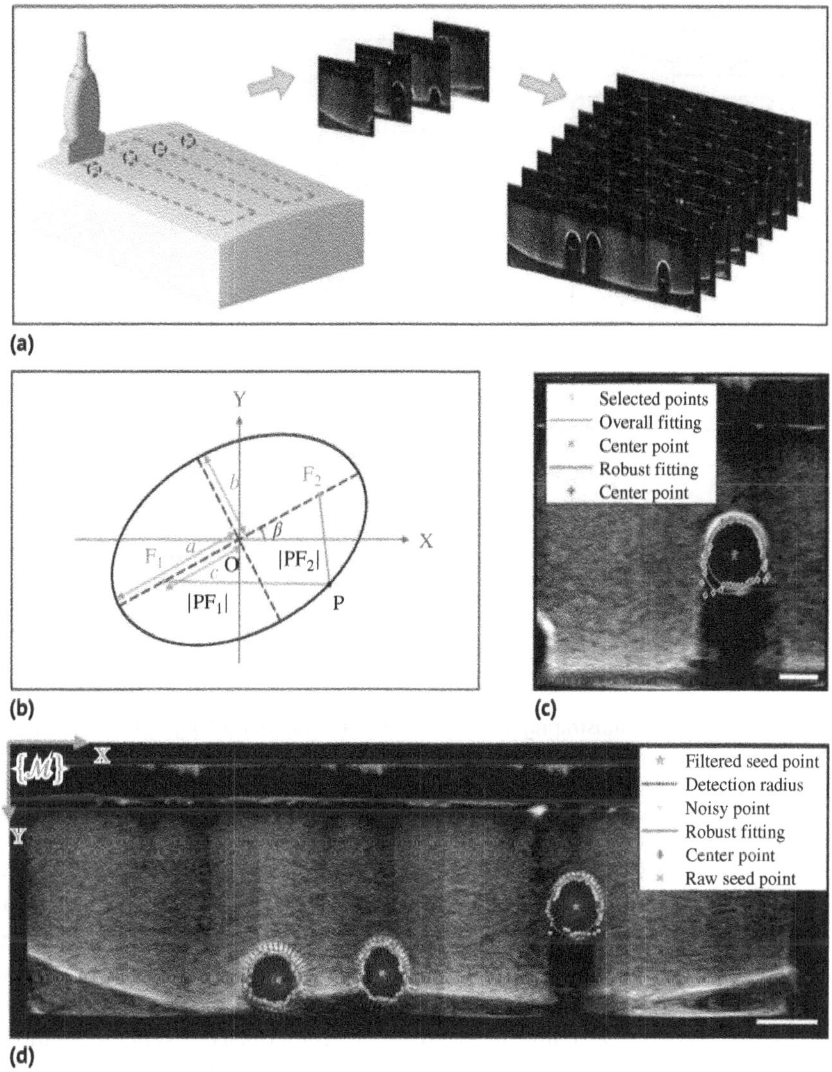

Figure 9.4 Vessel tracking. (a) Schematic of the overall scanning procedure. The dash line indicates the scanning path. It also demonstrates an array of US images recorded at the same latitude and OCPs stitched at intervals. (b) Diagram of a general ellipse model with geometric parameters. (c) Comparison between the overall ellipse fitting and proposed robust ellipse fitting. Scale bar is 5 mm. (d) An example of OCP and vessel tracking result. $\mathcal{M}$ is the map coordinate system. Scale bar is 10 mm.

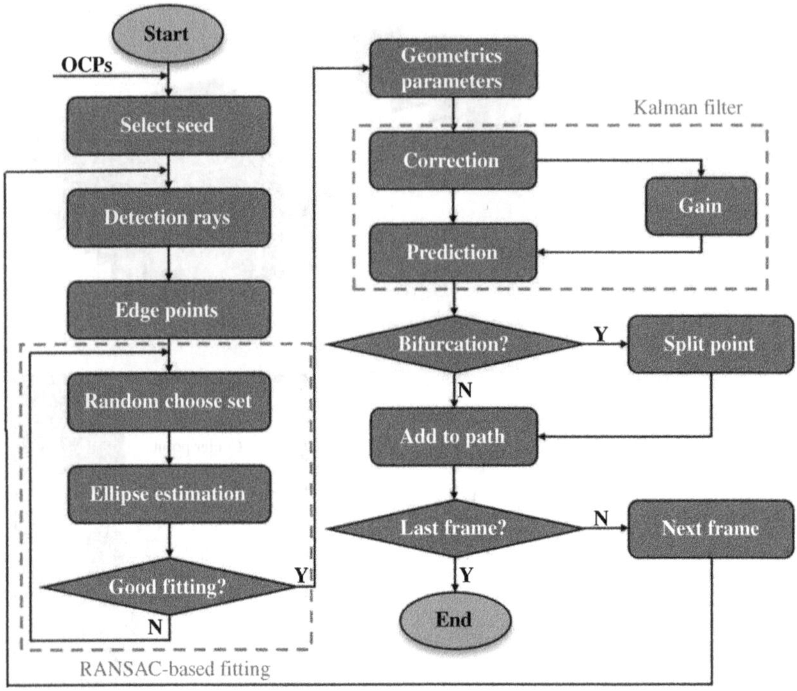

Figure 9.5 Flowchart illustrating the main steps of the vessel tracking method.

of the vessel cavity, in the first overall OCP. Subsequently, one-dimensional (1D) images are sampled along K rays, which are uniformly distributed and centered on the seed point, with appropriate lengths. Potential edge points (E_i, $i = 1$ *to* K) are then detected by computing the normalized cross-correlation (NCC) using a predefined template ($\mathbf{T}_e = [0 \cdots 0\ 1 \cdots 1]$), an array of suitable length consisting of zeros followed by ones, mimicking the vessel boundaries. Ultimately, random sample consensus (RANSAC) is utilized for contour fitting to estimate the center points. In this context, an ellipse profile is adopted, described by a general equation given in Eq. (9.5), which aptly represents the cross-sectional shape of vessels, as exemplified in Figure 9.4b.

$$A_e x^2 + B_e xy + C_e y^2 + D_e x + E_e y + 1 = 0 \tag{9.5}$$

where A_e, B_e, C_e, D_e, E_e are elliptic coefficients. The coordinate of the elliptical center point O equals:

$$\mathbf{p}_O = \left[\frac{B_e E_e - 2C_e D_e}{4A_e C_e - B_e^2} \; \frac{B_e D_e - 2A_e E_e}{4A_e C_e - B_e^2} \right]^T \tag{9.6}$$

Some geometric parameters of the ellipse can be deduced as:

$$\begin{cases} a^2 = \dfrac{2\left(A_e x_O^2 + B_e x_O y_O + C_e y_O^2 - 1\right)}{(A_e + C_e) - \sqrt{(A_e - C_e)^2 + B_e^2}} \\ b^2 = \dfrac{2\left(A_e x_O^2 + B_e x_O y_O + C_e y_O^2 - 1\right)}{(A_e + C_e) + \sqrt{(A_e - C_e)^2 + B_e^2}} \\ c^2 = \dfrac{4\left(A_e x_O^2 + B_e x_O y_O + C_e y_O^2 - 1\right)\sqrt{(A_e - C_e)^2 + B_e^2}}{4A_e C_e - B_e^2} \\ \beta = \operatorname{atan}\left(\dfrac{C_e - A_e - \sqrt{(A_e - C_e)^2 + B_e^2}}{B_e}\right) \end{cases} \tag{9.7}$$

where a denotes the semi-major axis, b represents the semi-minor axis, c signifies the linear eccentricity, and β stands for the inclination angle. The focuses of the ellipse, designated as $F1$ and $F2$, can be derived based on these parameters. For each vascular region, the boundary ellipse is estimated using Algorithm 1 in the Appendix. An illustrative comparison is provided in Figure 9.4c, demonstrating that the proposed robust fitting technique effectively excludes outliers and more accurately captures the true contour of the vessel.

Updating the seed point across frames is essential, as any individual instance of failure or misidentification can compromise the subsequent tracking process. To ensure temporal consistency, a Kalman filter has been implemented, formulated to maintain this consistency over time.

$$\begin{cases} \mathbf{x}(t) = \mathbf{F}\mathbf{x}(t-1) + \mathbf{u}(t) \\ \mathbf{z}(t) = \mathbf{H}\mathbf{x}(t) + \mathbf{v}(t) \end{cases} \tag{9.8}$$

where $\mathbf{x} = \left[\mathbf{p}_O^*; \dot{\mathbf{p}}_O^*\right]$ is the Kalman state vector comprising the position and velocity of the estimated ellipse center; $\mathbf{z} = \mathbf{p}_O$ is the measured center; $\mathbf{F} = \begin{bmatrix} \mathbf{I}_2 & \mathbf{I}_2 \\ \mathbf{0}_2 & \mathbf{I}_2 \end{bmatrix}$ is the state transition matrix; $\mathbf{H} = \begin{bmatrix} \mathbf{I}_2 & \mathbf{0}_2 \end{bmatrix}$ is the observation matrix; and u and v are assumed to be zero-mean white-noise sequences.

In the tracking methodology, incorporate the consideration of bifurcation, which is anticipated based on the morphological attributes of the detected vessel ellipse. Specifically, when the computed eccentricity surpasses a predefined threshold and persists in increasing for three consecutive frames, the current frame is flagged as a potential bifurcation point. Utilizing the elliptical configuration, the seed point is then divided into two separate points, which can subsequently be refined or validated by the operator. The vessel-tracking procedure is detailed in Algorithm 2 in the Appendix. It is worth noting that each point is internally annotated with both nodal and temporal data, facilitating the straightforward tracing of paths from any specific outlet back to the inlet. Additionally, upon the occurrence of bifurcation, the Kalman filter is duplicated to ensure that each branching path maintains its own instance for continued tracking. An exemplary result of vessel tracking is depicted in Figure 9.4d.

(2) Path Generation: The sequence of pertinent central points along a given pathway can be autonomously tracked from the outlet toward the inlet, represented by a novel collection denoted as $\mathcal{R} = \{R_1, R_2, \cdots R_L\}$. Presently, this sequence resides within the $\mathcal{M}$ coordinate system and necessitates transformation into the $\mathcal{W}$ coordinate system:

$$\bar{\mathbf{p}}_{r_i} = {}^{\mathcal{W}}_{\mathcal{M}}\mathbf{T}\,(\mathbf{p}_F, \omega)\,\left[\lambda \mathbf{p}_{r_i}^{\mathcal{M}}; 0; 1\right] \tag{9.9}$$

where $\bar{\mathbf{p}}_{ri} \in \mathbb{R}^4$ is the homogeneous coordinate of reconstructed point R_i; ${}^{\mathcal{W}}_{\mathcal{M}}\mathbf{T} \in \mathbb{R}^{4\times4}$ is the transformation matrix from $\mathcal{M}$ frame to $\mathcal{W}$ frame, related to the endplate position ($\mathbf{p}_F \in \mathbb{R}^3$) and probe angle ($\omega$) when recorded; and λ is the scale factor matching the US image pixel distance and physical distance, proportional to the imaging depth.

Upon subsequent activation, the US probe is engineered to initially sustain a roughly short-axis perspective aligned with the pathway, necessitating an orientation perpendicular to the vessel's centerline. Consequently, the determination of the PA at point R_i is contingent upon this requirement:

$$\mathbf{w}_{u_i} = \begin{cases} \mathbf{w}_z \times \left(\mathbf{p}_{r_{i+1}} - \mathbf{p}_{r_i}\right) & i = 1 \\ \mathbf{w}_z \times \left(\mathbf{p}_{r_{i+1}} - \mathbf{p}_{r_{i-1}}\right) & i = 2 \text{ to } L-1 \\ \mathbf{w}_z \times \left(\mathbf{p}_{r_i} - \mathbf{p}_{r_{i-1}}\right) & i = L \end{cases} \tag{9.10}$$

$$\omega_i' = \text{atan 2}\left(\mathbf{w}_{u_i}(2), \mathbf{w}_{u_i}(1)\right) \tag{9.11}$$

where $\mathbf{w}_z = [0\ 0\ 1]^T$, ω_i' denotes the raw angle measurement obtained from the probe. To enhance the smoothness of the derived sequence, a low-pass filter is incorporated into the process. Subsequently, the reference angle of the probe is denoted by $\Omega = [\omega_1, \omega_2, \cdots \omega_L]$.

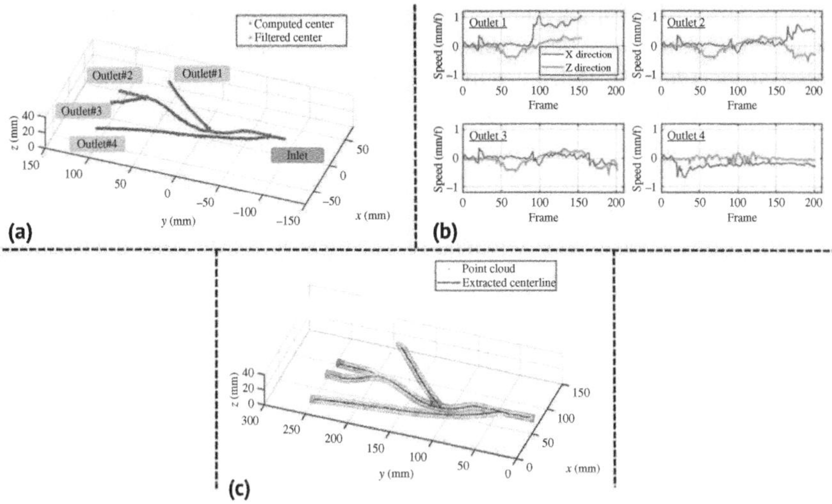

Figure 9.6 Path generation. (a) The reconstructed centerlines form US data. (b) Kalman filter estimated center point velocities of different paths. (c) The extracted point clouds and centerlines using VMTK.

Figure 9.6a displays the reconstructed centerlines derived from ultrasound (US) data, including the raw seed points extracted from each frame and the corresponding seed points updated by Kalman filtering. Additionally, Figure 9.6b illustrates the estimated velocities in both the x-direction and y-direction for the center of each path, aligning with the observed distribution trends. Meanwhile, Figure 9.6c is generated by importing a specifically designed 3D vessel model (created using SolidWorks 2019 by SOLIDWORKS Corp.) into the Vascular Modeling Toolkit, an open-source platform. The similar distribution patterns exhibited in Figure 9.6c underscore the efficacy of the proposed methodology.

9.2.2.2 Intraoperative Stage

(1) Tip Detection: Automated tip detection plays a pivotal role in achieving continuous control. Among the numerous similarity functions designed for image matching, the NCC has gained widespread adoption owing to its resilience to intensity variations and computational efficiency [39]. In our research, we utilize NCC-based template matching. To expedite the search process, the initial recognition phase is confined to a predefined area centered around a reference point. Moreover, we incorporate a pyramidal approach, where both the template and the real-time US image undergo downsampling to a lower resolution through Gaussian pyramids. This allows for a target search strategy that progresses from coarse to fine resolutions.

In scenarios where applications span from preoperative to intraoperative stages, potential vessel displacements may arise. To address this, a hierarchical matching strategy is devised, as detailed in Algorithm 3 Appendix. Initially, the operator selects the vessel region to serve as the vessel template, and subsequently identifies the tip region to function as the tip template. During subsequent detections, the vessel region is first reidentified within a predefined ROI, centered on a reference point and encompassing the vessel with a substantial margin to accommodate potential shifts in location. Following this, the tip is detected within the localized vessel region. The vessel template is updated once per reference point, whereas the tip template remains consistent. Figure 9.7b showcases the results, where the annotated boxes denote the predefined ROI, the re-identified vessel region, and the detected tip, respectively. A more in-depth discussion of these results is provided in Section 9.5.

(2) Operation Strategy: The operational sequence is depicted in Figure 9.3, encompassing both bending and drilling modes. For the purpose of large-workspace operation, concurrent control of all system modules is imperative: the motor module must synchronize the movement of the parallel mechanism with the magnetic device; the transducer module requires precise adjustment of the probe to facilitate real-time imaging; the feeding module must accurately insert the MDG to the designated location; and the electromagnet module must produce the necessary fields for magnetic actuation. Manually inputting all commands can impose a significant burden on the operator. Therefore, a semi-autonomous strategy, which is widely recognized [40], is employed. Prior to the operation, an initialization process is conducted where the reference path, inclusive of the endplate position and US probe orientation, is predetermined as outlined in the preoperative stage.

During the bending mode, the mechanism operates in a point-by-point manner with an adjustable interval, ensuring the US probe maintains a short-axis view above the vessel region. Additionally, the feeding distance of the MDG is automatically calculated as the distance between the current point and the target point. Periodic detection of the MDG within the US image is performed. If the tip is successfully identified, the process proceeds to the subsequent point. Conversely, in the event of tracking failure, the operator can manually rotate the US probe to locate the tip and subsequently adjust the insertion or retraction of the MDG until it becomes visible. It is important to note that the operator retains the discretion to bypass detection and continue with the procedure, as well as to pause at any point for observational or operational purposes (Figure 9.7).

Given the constraints inherent in 2D US imaging, a real-time graphical overlay has been incorporated into the user interface, presenting the reconstructed vessel layout, the current location of the MDG tip, and the intended next target point,

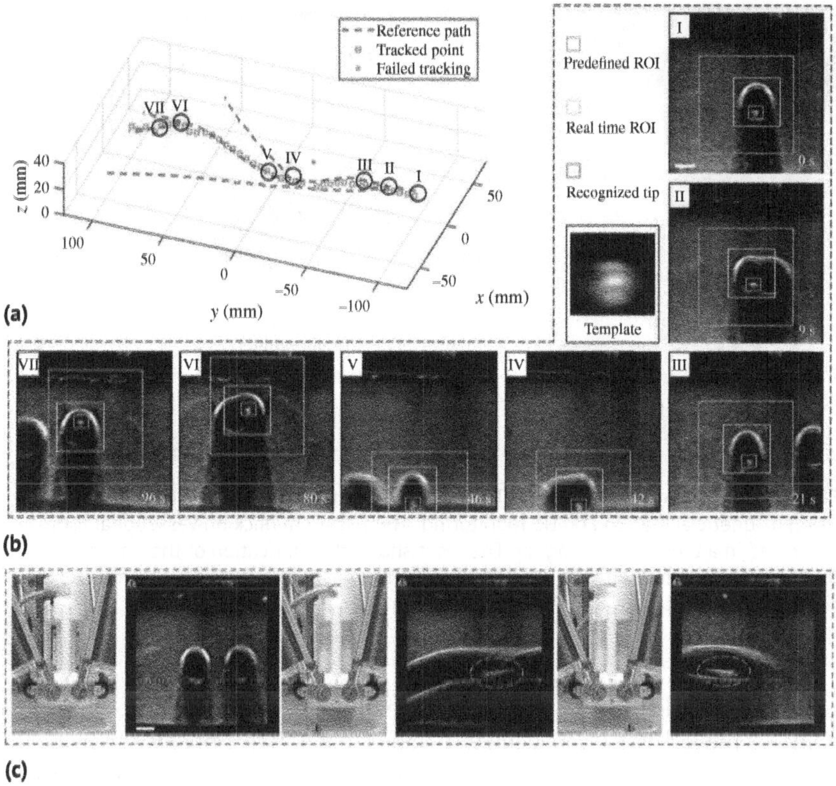

Figure 9.7 Magnetic control of the MDG with US imaging feedback. (a) The reconstructed trajectory for MDG from the Inlet to Outlet 3 in bending mode. (b) US images and tracking results during the intervention, with the relative position indicated in the coordinate plot. Scale bar is 5 mm. (c) Screenshots of the probe configuration and US image when the MDG loses tracking in drilling mode. Arrows indicate the probe orientation, and dashed circles represent the tip position. Scale bar is 5 mm.

as illustrated in Figure 9.8. By leveraging this graphical overlay in conjunction with real-time US surveillance, the operator can exercise precise control over the orientation of the directional magnetic field, thereby guiding the MDG along the predetermined pathway.

In the drilling mode, the mechanism operates with a similar yet finer control interval. A rotating magnetic field, aligned with the vessel axis, is employed. To account for potential disruptions caused by blood clots, the MDG advancement and US probe orientation are initially synchronized with the reference path, allowing for operator-initiated adjustments as required. For instance, MDG retraction

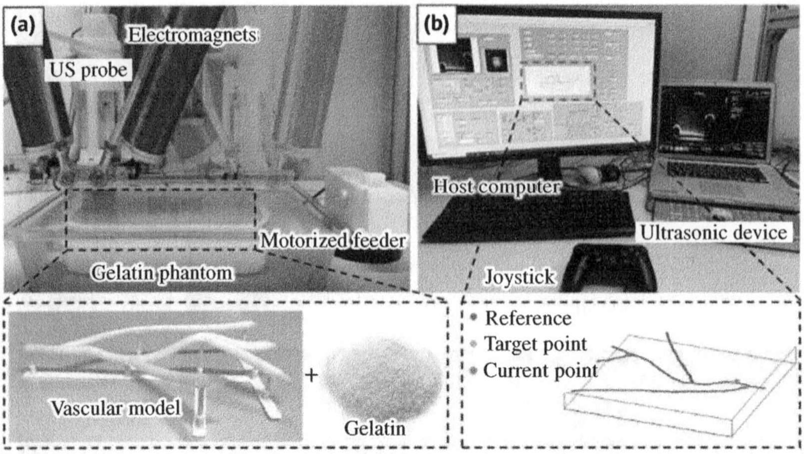

Figure 9.8 Experimental setup. (a) The prototype system for a robotic catheterization using magnetic actuation and US monitoring. The tissue- mimicking vessel phantom is immersed in a tank for US imaging. The inset shows the fabrication of the phantom. (b) The host computer and US imaging equipment. A user interface is programmed for controlling the entire system. The inset is the graphic prompt.

or further insertion may be necessary if the drilling tool encounters obstruction, while US probe rotation can facilitate detailed monitoring. Notably, automatic tip detection is disabled in this mode. Instead, the operator retains the flexibility to maintain the current position, proceed to the next point, or revert to the previous point based on real-time conditions.

Due to the intricate interactions among modules and the nonlinear distribution of the magnetic field, the electric currents supplied to each electromagnet must be meticulously computed to achieve the desired magnetic field configuration. The parallel mechanism design establishes a unique correspondence between the endplate position and the joint space ($J \in \mathbb{R}^3$) [41]. Furthermore, a unit-current grid-field map for each electromagnet ($\Phi_i : \mathbf{p}_D^{C_i} \in \mathbb{R}^3 \rightarrow \hat{\mathbf{b}}_i \in \mathbb{R}^3$) is obtained through a combination of software simulations (employing COMSOL Multiphysics 5.3a, COMSOL Inc.) and sensor calibration (using Model TLE493D W22B6, Infineon Inc.). Based on this information, the real-time currents are determined as follows: Firstly, ascertain the current orientations of all electromagnets. Secondly, utilize the hierarchical tip detection algorithm to identify the location of the MDG ($\mathbf{p}_T^{\mathcal{U}} \in \mathbb{R}^3$) within the US image. Thirdly, employ a look-up-table approach to determine the unit-current magnetic field vector ($\hat{\mathbf{b}}_i \in \mathbb{R}^3$) generated at the tip of each electromagnet, with reference to its respective coordinate frame. Subsequently, compute the powering currents ($\mathbf{i}_p \in \mathbb{R}^3$) using an inverse field algorithm. This process involves multiple coordinate transformation.

$$\hat{\mathbf{b}}_i = \boldsymbol{\Phi}_i \left({}^{\mathcal{C}_i}_{\mathcal{W}}\mathbf{T}(\mathbf{J})^{\mathcal{W}}_{\mathcal{F}}\mathbf{T}(\mathbf{p}_F)^{\mathcal{F}}_{\mathcal{U}}\mathbf{T}(\omega) \left(\lambda \mathbf{p}^{\mathcal{U}}_{T}\right)\right) \tag{9.12}$$

$$\mathbf{i}_p = \left[{}^{\mathcal{W}}_{\mathcal{C}_1}\mathbf{R}(\mathbf{J})\,\hat{\mathbf{b}}_1 {}^{\mathcal{W}}_{\mathcal{C}_2}\mathbf{R}(\mathbf{J})\,\hat{\mathbf{b}}_2 {}^{\mathcal{W}}_{\mathcal{C}_3}\mathbf{R}(\mathbf{J})\,\hat{\mathbf{b}}_3 \right]^{\dagger} \mathbf{b}_{\text{des}} \tag{9.13}$$

where ${}^{\mathcal{C}_i}_{\mathcal{W}}\mathbf{T} \in \mathbb{R}^{4\times4}$ is the transformation matrix from $\mathcal{W}$ to $\mathcal{C}_i$ frame; ${}^{\mathcal{W}}_{\mathcal{C}_i}\mathbf{R} \in \mathbb{R}^{3\times3}$ is the rotation matrix from $\mathcal{C}_i$ to $\mathcal{W}$; † means pseudo-inverse; and $\mathbf{b}_{\text{des}} \in \mathbb{R}^3$ is the desired magnetic field, computed from Eq. (9.2) or (9.4) according to different modes.

9.3 Structure Design and Control Framework of WMM

The MDG presented in Section 9.2 demonstrates robust capabilities in navigating bifurcated vasculatures and penetrating occlusive clots via combined magnetic steering and drilling motion, with US guidance effectively mitigating radiation risks. However, its integrated structure, while reliable for targeted unclogging, faces limitations in accessing extremely tortuous or microscale distal vessels where the tethered design may induce excessive vascular wall contact. To address this accessibility challenge, the WMM in this section advances the magnetic actuation paradigm by introducing a switchable tethered-untethered architecture. Retaining the core advantages of magnetic control and US feedback, this design enables flexible steering in larger vessels via directional fields while allowing controlled ejection of an HB—propelled by rotating fields—to traverse narrow, convoluted segments that are inaccessible to rigid guidewires.

9.3.1 Structure Design

The WMM consists of a commercial guidewire, specifically the HW-25-150 model from Cook Medical, and a modular tip assembly. These two components are interconnected via a silicone cannula segment, as depicted in Figure 9.9a. The modular tip comprises three functional elements: a base frame (BF), a rotary holder (RH), and an HB, all of which are produced through 3D printing techniques. The BF features a lower attachment point for connecting to the guidewire and an upper hollow section designed to house the RH. The RH, in turn, has a lower stem that fits into the BF, two middle openings for the insertion of axially magnetized CMs with alternating polarities, and an upper receptacle meant for the HB.

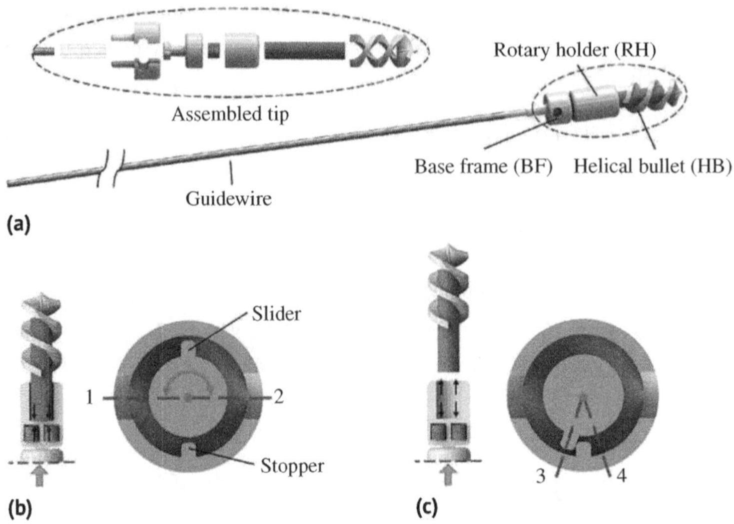

Figure 9.9 Design of the WMM. (a) Detailed structure. Profile charts showing internal states under (b) flexible bending stage and (c) tip ejection stage, where δ equals $0°$ and $180°$, respectively.

The HB is structured with an outer spiral casing and an internally positioned radially magnetized CM, allowing for potential clinical customization. For assembly purposes, the BF and RH are initially manufactured as separate components and then bonded together using medical-grade adhesive.

The BF's cavity incorporates a stopper, while the RH's shaft features a slider. This mating mechanism facilitates limited axial rotation of the RH while concurrently constraining its radial deflection and axial displacement. This assembly is designated as the rotating locker. Additionally, the RH's bucket possesses an inner diameter marginally greater than the outer diameter of the HB, which serves to stabilize its relative positioning while permitting axial rotation and movement. This arrangement is termed the magnetic ejector. The integration of these two structures provides the capability for flexible angular maneuvering in the tethered configuration and facilitates tip ejection for transitioning to the untethered mode.

Figure 9.9b,c present profile charts of the assembled tip module. During flexible angle steering, an EMF is utilized to guarantee the absence of any interference between the slider and the stopper, enabling transitions such as from position 1 to position 2. The HB remains securely within the RH due to the magnetic attraction between the CMs. The combined RH and HB align with the EMF for effective steering. For tip ejection, the EMF is initially applied to cause the slider to engage with the stopper at specific positions, such as position 3 and position 4. Subsequent changes in the EMF direction result in the rotation of the HB alone. Once the intersection angle (δ) reaches a critical threshold, the repulsive force between the CMs ejects the HB from the RH. The ejected HB is then propelled wirelessly, capable of executing forward, steering, and backward movements. Notably, the portion remaining attached to the guidewire exhibits a negligible magnetic moment (approximately 10^{11} times smaller than the assembled module), ensuring it remains unaffected by the EMF during the untethered mode and remains stationary. Upon the return of the HB, internal magnetic forces facilitate automatic reassembly.

9.3.2 Control Framework WMM

1. **System Overview:** Magnetic control at the human scale poses significant challenges owing to the magnetic field's rapid decrease with distance, as highlighted in [22] and [21]. To address this issue, a custom-built system incorporating a parallel manipulator and an electromagnet array has been employed, as described in [42]. Additionally, a 2D US probe (Model: 15L4, Teratech Corp.) is attached to the mobile endplate to provide medical imaging feedback. A motorized feeder has been designed to facilitate the guidewire's forward and backward motion. During operation, the parallel manipulator offers 3-DOFs) for positioning the US probe and electromagnet array close to the device. The US probe itself possesses 1-DOF for adjusting the imaging angle, while the electromagnet array possesses 3-DOFs to generate arbitrary magnetic fields. These modules are synchronously controlled via a user interface (Figure 9.10), with commands input via a joystick, as illustrated in Figure 9.10a.

   Due to the mechanism design, a one-to-one relationship exists between joint space ($\mathbf{Q}$) and the endplate position ($\mathbf{P}_E$) as:

$$\boldsymbol{P}_E = \mathcal{F}(\boldsymbol{Q}) \tag{9.14}$$

   and inverse kinematic can be represented by $\boldsymbol{Q} = \mathcal{F}^{-1}(\boldsymbol{P}_E)$. Moreover, all electromagnets are in the linear region and the generated EMF at the target point ($\mathbf{P}_T$) equals:

$$\begin{aligned}\boldsymbol{B}(\boldsymbol{P}_T, \boldsymbol{P}_E) &= \mathbf{M}_b(\boldsymbol{P}_T, \boldsymbol{P}_E)\boldsymbol{I}_c \\ &= \left[\overline{\boldsymbol{B}}_1(\boldsymbol{P}_T, \boldsymbol{P}_E) \quad \overline{\boldsymbol{B}}_2(\boldsymbol{P}_T, \boldsymbol{P}_E) \quad \overline{\boldsymbol{B}}_3(\boldsymbol{P}_T, \mathbf{P}_E)\right]\boldsymbol{I}_c\end{aligned} \tag{9.15}$$

   where $\overline{\boldsymbol{B}}_n(\boldsymbol{P}_T, \boldsymbol{P}_E)$ is the unit-current magnetic field vector of the n_{th} electromagnet that is calculated according to the real-time $\boldsymbol{P}_T$ and $\boldsymbol{P}_E$; $\boldsymbol{I}_c$ represents the powered electric currents that can be deduced as:

$$\boldsymbol{I}_c = \mathbf{M}_b(\boldsymbol{P}_T, \boldsymbol{P}_E)^{\dagger}\boldsymbol{B}(\boldsymbol{P}_T, \boldsymbol{P}_E) \tag{9.16}$$

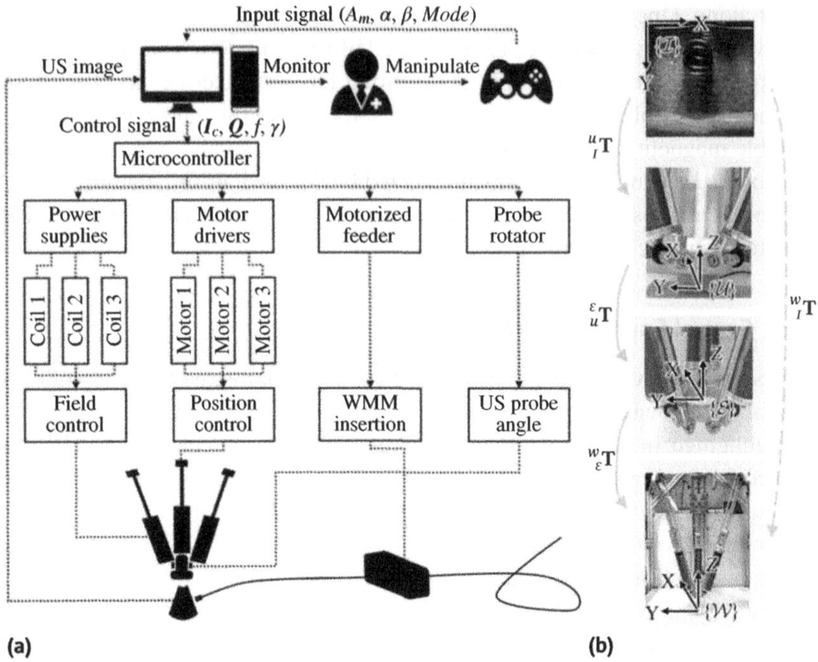

Figure 9.10 Overview of the actuation system. (a) Block diagram showing the system structure and control process. (b) Defined coordinate systems and transformation relationships: US image frame $\mathcal{I}$; US probe frame $\mathcal{U}$; endplate frame $\mathcal{E}$; world frame $\mathcal{W}$.

In addition to motion mechanisms and field production, the application of US data necessitates coordinate transformations. The specified frames are illustrated in Figure 9.13b, and the corresponding conversion relationship has been derived:

$$ {}^{\mathcal{W}}_{\mathcal{I}}\mathbf{T}\,(\boldsymbol{P}_E, \gamma, d) = {}^{\mathcal{W}}_{\mathcal{E}}\mathbf{T}\,(\boldsymbol{P}_E)\,{}^{\mathcal{E}}_{\mathcal{U}}\mathbf{T}(\gamma)\,{}^{\mathcal{U}}_{\mathcal{I}}\mathbf{T}(d) \tag{9.17} $$

where ${}^{\mathcal{W}}_{\mathcal{I}}\mathbf{T}$ is the transformation matrix from $\mathcal{I}$ frame to $\mathcal{W}$ frame, other matrices have a similar definition; γ is the PA; d is the imaging depth that relates to the scale factor converting the pixel length into the physical distance.

2. **Targeted Scanning:** To acquire the centerline, a scanning process is executed along the tubular structure. This procedure solely relies on US image data, encompassing real-time segmentation of the vascular area and estimation of its distribution pattern. Given the low SNR inherent in US imaging, a U-net architecture is employed for robust and real-time identification of the ROI.

This U-net represents a cutting-edge approach in biomedical image segmentation [43]. As depicted in Figure 9.11, the network is structured with an encoder pathway and a decoder pathway, interconnected by skip connections within corresponding layers. A gelatin phantom, featuring tubular internal structures, is utilized for network training. The input dataset comprises 300 US images (256 × 256 pixels), primarily captured in a transverse view, while the output dataset consists of corresponding segmentation masks, with vessel regions manually delineated. Subsequent to segmentation, additional image processing steps are implemented to eliminate outliers and extract descriptive features ($\mathcal{D}_u$) from the ROI.

$$\mathcal{D}_u = (\mathbf{h}_u, \boldsymbol{a}_u, \boldsymbol{e}_u, n_u) \tag{9.18}$$

The variable $\mathbf{h}_u$ represents the central points, which are calculated through the identification of centroids and the application of coordinate transformations as described in Eq. (9.17). $\boldsymbol{a}_u$ signifies the cumulative areas, while $\boldsymbol{e}_u$ denotes the eccentricities of ellipses that share the same second-moments as the segmented regions. The symbol nu stands for the number of regions. In instances where nu exceeds one or both $\boldsymbol{a}_u$ and $\boldsymbol{e}_u$ exhibit a consistent increase over three consecutive frames, the operator must intervene by selecting a centroid via an image click, thereby determining the target branch for scanning at bifurcations. Consequently, the final descriptive feature set, denoted as $\mathcal{D}_f$, comprises a single ROI and is recorded as $\mathcal{D}_f = (\mathbf{h}_u, \boldsymbol{a}_u, \boldsymbol{e}_u)$.

A Kalman filter is designed to estimate the distribution trend that can be expressed as:

$$\begin{cases} \boldsymbol{x}^n = \mathbf{A}\boldsymbol{x}^{n-1} + \boldsymbol{\varsigma}^n \\ \boldsymbol{y}^n = \mathbf{C}\boldsymbol{x}^n + \boldsymbol{\eta}^n \end{cases} \tag{9.19}$$

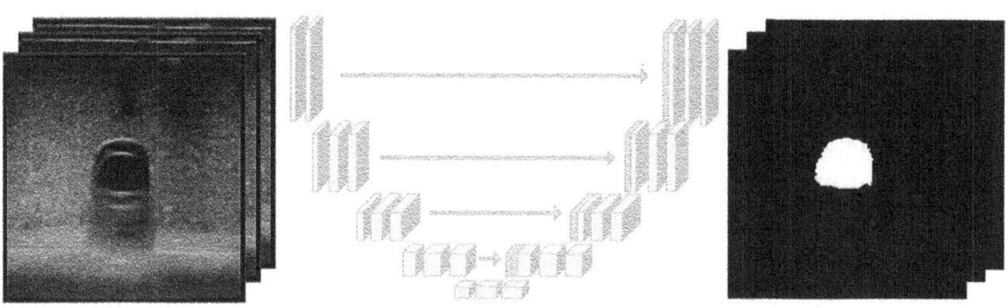

Figure 9.11 U-net architecture with input US images ($\mathbf{U}_i$) and output segmentation masks ($\mathbf{U}_s$).

the state vector $\boldsymbol{x}^n = \left[\boldsymbol{h}_f^*; \dot{\boldsymbol{h}}_f^*\right]$ comprises the estimated position and velocity of the center point in three dimensions, denoted as $\left[\boldsymbol{h}_f^*; \dot{\boldsymbol{h}}_f^*\right]$. The observed center point position is represented by $\boldsymbol{y}^n = \boldsymbol{h}_f$. The system dynamics are governed by a constant state transition matrix $\mathbf{A} = \begin{bmatrix} \mathbf{I}_3 & \mathbf{I}_3 \\ \mathbf{0}_3 & \mathbf{I}_3 \end{bmatrix}$, and an observation matrix $\mathbf{C} = \begin{bmatrix} \mathbf{I}_3 & \mathbf{0}_3 \end{bmatrix}$. Both ζ^n and η^n are sequences of zero-mean white noise, accounting for uncertainties. Subsequently, this updated state information is utilized to determine the motion of the parallel manipulator.

$$\boldsymbol{P}_E^n = \boldsymbol{P}_E^{n-1} + g\dot{\boldsymbol{h}}_f^* \tag{9.20}$$

where g is the defined scanning interval.

During targeted scanning procedures, the US probe is engineered to sustain an approximately transverse orientation, meaning that the image plane aligns perpendicularly with the tube's centerline. Consequently, the initial PA can be determined based on this alignment.

$$\gamma^* = \text{atan 2}\left(\dot{\boldsymbol{h}}_f^*(2), \dot{\boldsymbol{h}}_f^*(1)\right) - \pi/2 \tag{9.21}$$

Since derivation is susceptible to noise, a low-pass filter is implemented to mitigate angle jumps and thereby smooth the γ^* values, enabling the determination of γ. This point-by-point scanning process continues iteratively until the target location is reached, as illustrated in Figure 9.12a. The sequence of center points and PAs acquired during the scanning process is documented and referred to as the reference path R.

3. **US-guided Control:** To facilitate magnetic control guided by US in a large workspace, it is necessary to coordinate multiple simultaneous motions: the endplate of the parallel manipulator should glide along the tubular framework, the US probe should scan in synchronization with the WMM, and the electromagnet array should produce the necessary electromagnetic fields (EMFs) for effective control. In this conceptual framework, the mechanical movements are automated, allowing the operator to solely focus on determining the direction of real-time EMFs and issuing commands as required.

   (1) Imaging Mode: Both short-axis (transverse) and long-axis (longitudinal) US image views are incorporated to monitor the status of the WMM. Figure 9.13 depicts the illustration of US imaging. In the tethered configuration, the transverse view is employed, where the image plane of the US probe intersects the centerline of the tubular structure at a right angle. The assembled tip of the WMM appears as a highly echogenic white dot in the US image (more prominent than the shaft guidewire due to the integrated contrast markers, or CMs), facilitating the visualization of the WMM's progression. Furthermore, the guidewire's constraint facilitates straightforward

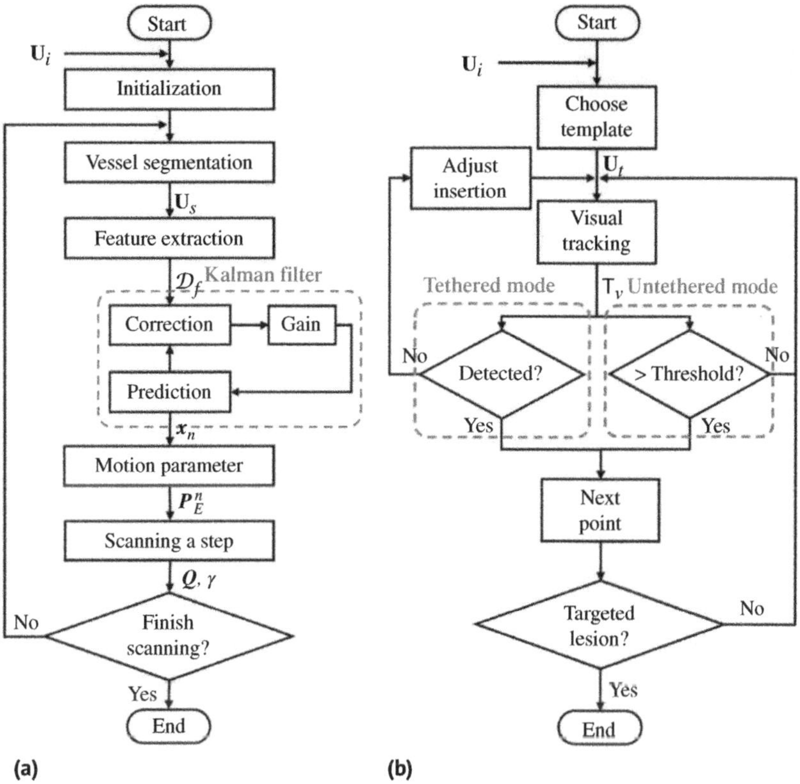

Figure 9.12 Flowcharts of the control framework. (a) Targeted scanning procedure. (b) US-guided control procedure, including two modes.

control over the WMM's axial motion. Hence, the transverse view is a suitable option, given its robustness to relative movements between the probe and the structure under examination.

During the tip ejection phase and in the untethered mode, the longitudinal view is utilized, where the image plane of the US probe is aligned with the centerline of the tubular structure. In this view, the assembled tip appears as a bright, hyperechoic block with indistinct edges within the US image. A drawback of this perspective is its susceptibility to losing the target with minimal radial movements, owing to the narrow ultrasonic beam. Nevertheless, given that the progression of the ejected HB may be influenced by environmental factors (such as fluid flow rate and vessel inclination), this view is advantageous for observing the relative positioning between the US probe and the HB, thereby enabling appropriate motion compensation to ensure stable tracking.

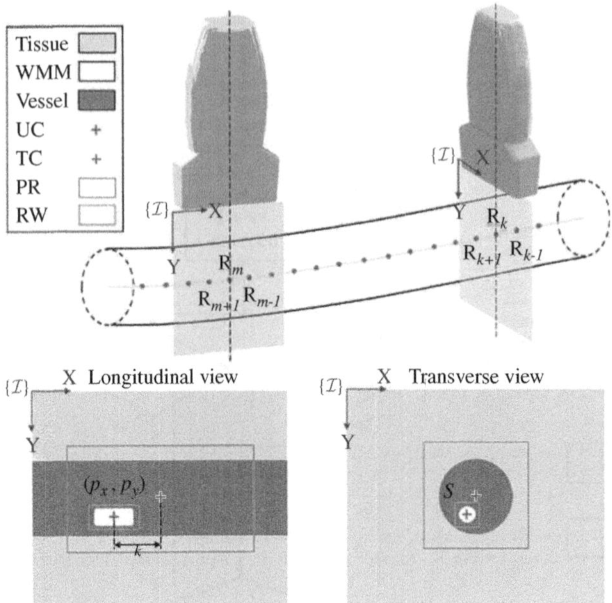

Figure 9.13 Illustrations of US imaging during control. The abstract diagrams represent longitudinal and transverse views, including the tissue area, WMM zone, vessel region, US image center (UC), tracked image center (TC), predefined ROI (PR), and recognized WMM (RW).

(2) Visual Tracking: To track the WMM in environments with noise, the template matching algorithm is adopted. The NCC criterion is utilized to assess the similarity between images, noted for its robustness against intensity variations and its computational simplicity [39]. Optimal matching is achieved when the NCC value (denoted as V) attains its maximum.

$$V = \frac{\sum_{i=1}^{I}\sum_{j=1}^{J}\left(U_t(i,j)-\mu_t\right)\left(U_m(i,j)-\mu_m\right)}{\sqrt{\sum_{i=1}^{I}\sum_{j=1}^{J}\left(U_t(i,j)-\mu_t\right)^2}\sqrt{\sum_{i=1}^{I}\sum_{j=1}^{J}\left(U_m(i,j)-\mu_m\right)^2}} \tag{9.22}$$

where $U_t(i, j)$ and $U_m(i, j)$ are intensities of (i, j) pixels of the selected template and cropped matching images, respectively; $\mu_t = \frac{1}{IJ}\sum_{i=1}^{I}\sum_{j=1}^{J}U_t(i,j)$ and $\mu_m = \frac{1}{IJ}\sum_{i=1}^{I}\sum_{j=1}^{J}U_m(i,j)$ are average intensities of U_t and U_m, separately. To enhance both the speed and precision of tracking, the initial search is confined to a rectangular zone encompassing the vessel region with an added safety margin. A coarse-to-fine searching strategy is employed, which involves down sampling both images to lower resolutions using Gaussian pyramids. The tracking outcome is represented as $\mathcal{T}_v = \{S, p_x, p_y\}$, where S signifies the tracking score, and p_x and p_y denote the coordinates of the tracking center in pixel space.

(3) Control Strategy: Considering the distinct motion and imaging characteristics of the WMM across various stages, various strategies are devised with the aid of R. The control flowchart for these strategies is depicted in Figure 9.12b. In the tethered operational mode, the process proceeds on a point-by-point basis. The predefined path serves as a guide for manipulating the electromagnet array, adjusting the angle of the US probe, and inserting the guidewire. The operator selects the appropriate D-EMF based on real-time US imagery to steer the WMM toward the intended branch. Prior to initiating this mode, a template image is selected by marking an area on the screen. Visual tracking is executed at each designated reference point. Successful detection of the WMM triggers the system to automatically proceed to the next point, executing synchronized probe movement and guidewire insertion. In case of WMM loss, the operator remotely adjusts the feeder motor to control the WMM's forward and backward motion until it reappears in the image. This sequence is repeated iteratively until the desired position for tip ejection is reached.

During the tip ejection phase, the operator oversees the process. Initially, the US probe is oriented to a longitudinal view for enhanced visibility (Figure 9.13). Subsequently, the position of the assembled tip is adjusted toward one edge of the US image to ensure adequate space for the ejection, accomplished either by modifying the probe's position or retracting the guidewire. Following this, an R-EMF is applied to initiate the tip ejection and is subsequently deactivated once the ejection is complete. As a final step, the probe's position is fine-tuned to position the ejected HB centrally within the US image.

The untethered operational mode employs in plane visual servoing techniques. The predefined path serves as guidance for manipulating the electromagnet array, adjusting the angle of the US probe, and orienting the direction of the R-EMF. Prior to initiating this mode, the operator selects a template image, similar to previous steps. Continuous visual tracking is performed throughout. Upon detecting the HB and recognizing that the x-direction deviation ($\varkappa$) surpasses a predetermined threshold, the US probe advances to the subsequent reference points until the target is centered within the image. This allows the US probe to dynamically align with the HB's movement, regardless of any variations in speed due to external factors. Manual adjustments are made to the US probe's out-of-plane motion to ensure optimal imaging. This sequence is iteratively repeated until the intended lesion is reached.

9.4 Experimental Results and Discussion

In this section, the MDG is first demonstrated to validate the proposed control strategy; moreover, the CCD imaging method is utilized to validate the function and the performance of the proposed design, showing an intuitional state of the WMM during the whole procedure. The US imaging method is then employed to verify the proposed control framework, making a step toward clinical applications.

9.4.1 Demonstration of the MDG

(1) Characterization: To validate the kinematic model, directional magnetic fields are utilized, with a Helmholtz coil serving as the characterization tool. During experimentation, the MDG is positioned at varying lengths, taking into account the practical distance between the tip and bending point when navigated through confined spaces. The intensity of the magnetic field is maintained at 10 mT, aligning with subsequent operational requirements. The yaw angle of the magnetic field is varied in increments of 10°, ranging from 90° to 300°. Furthermore, the cannula length is adjusted to 10 mm, 20 mm, and 30 mm for testing. A comparison of the theoretically calculated bending angle with the experimentally measured bending angle is presented in Figure 9.14. The results indicate a maximum discrepancy of 8.4° and an average error of 1.24°.

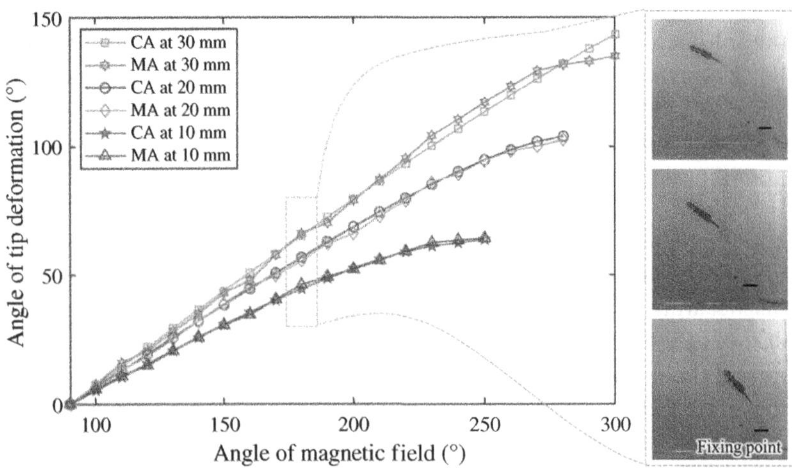

Figure 9.14 Comparison between the calculated angle (CA) and the measured angle (MA) with different cannula lengths under 10 mT magnetic field. The insets are captured when the angle of the magnetic field equals 180°.

To assess the drilling mode, rotating magnetic fields are employed, with frequencies spanning from 1 to 30 Hz. The findings reveal that when subjected to a magnetic field of 0.5 mT in magnitude, the integrated magnetic driller tip maintains continuous rotation in sync with the external fields, without experiencing any deviation. Consequently, the friction arising from the ball joint-like configuration can be successfully mitigated, enabling the advanced MDG to execute drilling actions within blood clots.

(2) Bending Mode for Magnetic Navigation: Optical imaging techniques are utilized to provide a clear visualization of the MDG's behavior under the influence of EMFs. The prototype of the MDG is displayed in Figure 9.15a. A laser-cut triple-layer Polymethyl Methacrylate (PMMA) channel model, featuring branched and tortuous pathways, has been designed. The dimensions of this model are 115 mm by 230 mm by 5 mm, with channel widths varying between 5 and 9 mm. Various pathways are identified and labeled in Figure 9.15b. The experimental arrangement is depicted in Figure 9.15c, where a Charge-coupled Device (CCD) camera, tasked with real-time monitoring and mechanism tracking, is mounted onto the endplate. Throughout the experiment, the operator controls the insertion of the MDG and the orientation of the magnetic field using a joystick. Simultaneously, a parallel mechanism automatically adjusts its movement in tandem with the MDG, based on visual feedback. Representative screenshots captured by the in situ camera at key positions are presented in Figure 9.15d. The results demonstrate that the proposed MDG can be agilely manipulated using magnetic navigation in confined and tortuous environments.

(3) Drilling Mode for Unclogging Motion: Clogged channel environments are constructed to test the drilling mode. The stiffness of agar with various concentrations (0.2, 0.3, 0.4, 0.5, 0.6, 0.7, 0.8, 0.9, 1.1 and 1.4 wt%) are measured by the mechanical tester (Model: MACH-1 V500 C, Biomomentum Inc.). Results are shown in Figure 9.15e, where dash lines are reference stiffness range of a real blood clot given in [44]. In this study, the artificial blood clot is prepared with 0.8 wt% agar as [17], with a similar stiffness of a 14-days thrombus. The red dye colored agar liquid is first injected into the silicone tube, and then cooled to 4°C and kept for 1 h in the refrigerator. The phosphate-buffered saline (PBS) solution is then filled into the rest of the tube. During the drilling motion, the MDG is driven by the rotating magnetic field with a magnitude of 10 mT and rotation frequency of 8 Hz. Figure 9.15f shows the penetration of the MDG through an artificial clot with a length of 30 mm. Results show that the rotation of the driller tip enables the guidewire to go through the clogged channel, without which it keeps stuck at the clot.

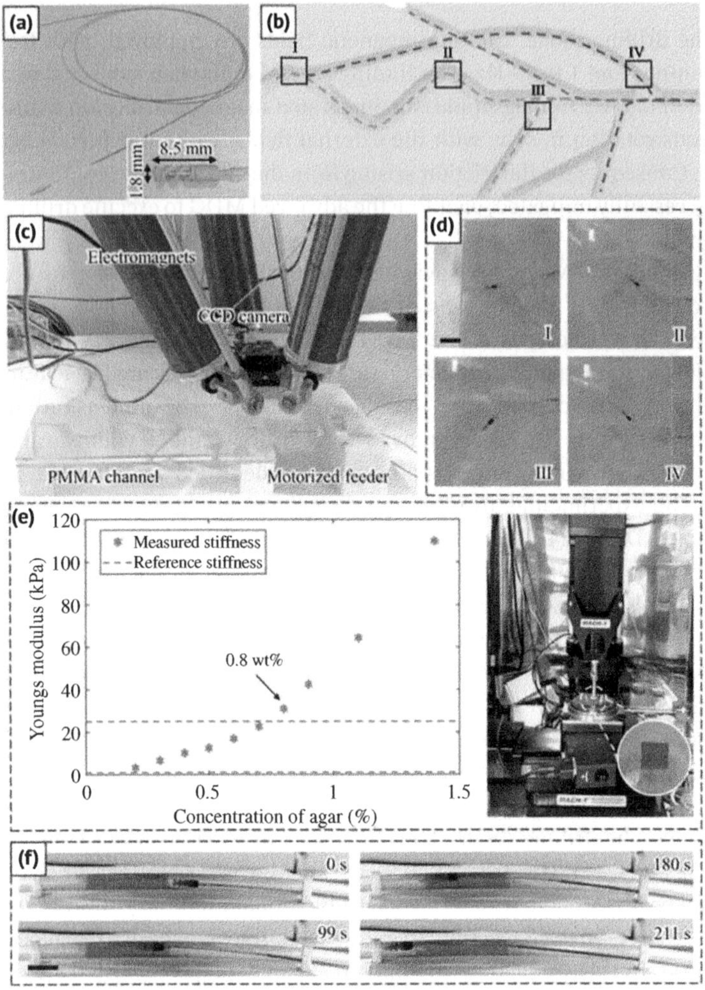

Figure 9.15 Function validation. (a) The fabricated MDG prototype. (b) Schematic diagram of the PMMA channel model. Dash lines demonstrate the desired paths. (c) Experimental setup with optical camera. (d) Images captured by the eye-in-hand camera at bifurcations in the bending mode for magnetic navigation. Scale bar is 10 mm. (e) The measured Young's modulus of artificial blood clots with different concentrations of agar. (f) Images captured by a third camera in the drilling mode for unclogging motion. Scale bar is 10 mm.

9.4.2 Validation of Control Strategy

(1) Experimental Setup: Figure 9.11 illustrates the experimental configuration designed to verify the comprehensive workflow. At the endplate, a rotatable US probe is installed for acquiring US images, while a motor-driven feeder, anchored on the table, facilitates the insertion of MDGs. A phantom resembling tissue, composed of a 3D-printed vascular model embedded in a mixture of 10w% gelatin and 1.5w% cellulose, is submerged in a container filled with PBS solution. The operational height of the entire system is finely tuned to ensure optimal imaging. All components interact with the main computer via a preprogrammed user interface, which also features a real-time graphical guide to aid in navigation. The operator issues commands through a joystick controller.

(2) Intervention on a Vessel Phantom: The scanning process within the ROI is executed with a recording interval of 1 mm and a row spacing of 30 mm, struck to maintain a harmony between computational precision and efficiency. The operator determines the starting point, scanning distance, and number of rows based on practical considerations. Subsequently, vessel tracking and environmental reconstruction procedures are carried out. Figures 9.4 and 9.5 present the computed outcomes for the fabricated phantom, derived from five separate trials with varying initial seed points. The standard path is established as the average position across all these trials. The overall average positional error for the path is 0.28 mm, with the highest positional error observed at a single point being 1.65 mm, occurring at vessel bifurcations.

The flexible magnetic navigation system, described in Section 9.2, is utilized for navigation. To simulate real-world conditions, a location disturbance is introduced by randomly displacing the phantom by several millimeters. Prior to initiating the procedure, the inlet and outlet points of the path are identified to facilitate automatic generation of reference points. Subsequently, the MDG is inserted at an appropriate distance to ensure clear visibility in the US image. The operator selects the vessel region and the tip ROI. Once initialized, the process proceeds step-by-step. Four distinct paths are designed, with each path being repeated three times. Across all trials, the MDG successfully reaches the intended target region. The average time taken to reach different outlets is 103.8, 118.5, 121.6, and 94.7 s, respectively. Figure 9.7a displays the recovered tip position during one trial from Inlet to Outlet #3. Representative real-time US images and detection results at key positions are presented in Figure 9.7b. Additionally, a magnetic field-free catheterization experiment is conducted, where only the insertion motion of the MDG is controlled. This experiment is performed five times, with the MDG reaching Outlet 2 but getting stuck at bifurcations two and three times, respectively. These results underscore the enhanced controllability offered by magnetic navigation.

To validate the ability to cross occlusions under US monitoring, artificial blood clots are prepared within a vascular model. Currently, in the experimental setting, the presence of a clogged region is assessed by observing the state of the MDG, as the clot appears nearly transparent in US images, unlike in clinical applications where the occlusive region can be predetermined through medical examination. During the experiment, if the MDG tip loses tracking and further insertion results in stacking, the procedure is reversed to the previous point with a slight interval until the tip reappears in the US image. At this point, the magnetic field is transitioned from a directional field to a rotating field, and the MDG is manipulated using the framework described in Section 9.2. For a more detailed observation, the angle of the US probe can be adjusted. Figure 9.7c illustrates an instance where the MDG tip falls behind the imaging plane, showing the transverse view (excluding the target), longitudinal view A (with a PA of approximately +90°), and longitudinal view B (with a PA of approximately −90°). The MDG is able to successfully traverse the clogged region.

9.4.3 Validation of the WMM

(1) Model Calibration: The prototype of the WMM is depicted in Figure 9.16a. The assembled tip measures 2.8 mm in outer diameter and 12.5 mm in length, while the attached guidewire has dimensions of 0.025 inches in outer diameter and 1500 mm in length. Figure 9.16b presents a comparison of the computed and experimentally measured bending angles of the WMM under various D-EMFs with a magnitude of 9 mT. This comparison includes various lengths of deformation segments. The maximum bending error observed is 4.5°, with an average error of 1.9°, which is deemed acceptable for applications utilizing a semi-autonomous strategy. Figure 9.16c showcases the tip ejection performance of the WMM. For ease of operation, the tip ejection is initiated by a R-EMF with its axis of rotation aligned with the centerline, followed by a command to deactivate the field. Each set of parameters is tested six times, and the results are categorized as failed ejection, low success rate, medium success rate, and high success (HS) rate, based on the number of successful ejections (0, 1–2, 3–4, and 5–6 times, respectively). The results indicate that a low-frequency, medium-strength R-EMF is optimal for tip ejection. When the EMF magnitude is low (e.g., 1–5 mT), the external magnetic torque is insufficient to alter the existing internal attraction state, resulting in failed tip ejection. Conversely, at high R-EMF frequencies (e.g., above 6 Hz), the HB may be captured by the release hub (RH) during ejection due to rapidly alternating intersection angles, leading to a decreased success rate. Additionally, the environment (air or liquid) does not affect the releasability of the tip but does influence the ejected distance. Figure 9.16d illustrates the motion speed

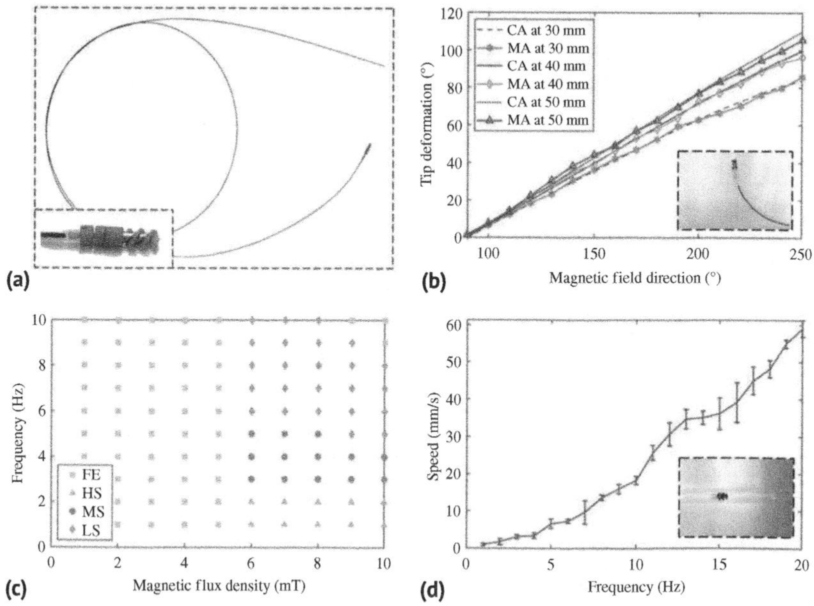

Figure 9.16 Calibration of the WMM. (a) WMM prototype. (b) Comparison between the calculated angle (CA) and the measured angle (MA) with different deformation segments in the tethered mode under D-EMFs. (c) Phase diagram of the tip ejection stage. (d) Speed of the ejected HB under R-EMFs.

of the ejected HB under various R-EMFs. The frequency is varied from 1 to 20 Hz in increments of 1 Hz, and each test is repeated three times to ensure accuracy.

(2) Function Test: Three distinct tubular models, crafted using 3D printing technology, serve as platforms to verify various operational modalities. The initial model, depicted in Figure 9.17a, features five branches that are uniformly dispersed within a 3D space. This configuration allows for seamless insertion of the WMM into any channel, facilitated by D-EMFs. The outcomes demonstrate impressive maneuverability when operated in the tethered configuration. The second model, presented in Figure 9.17b, adopts a looping structure. Upon reaching the intended position, tip ejection of the WMM is activated. Subsequently, R-EMFs are employed to propel the HB wirelessly through the intricate, tortuous pathways. If this entire sequence were attempted in the tethered mode, involving continuous insertion of the WMM while using D-EMFs to guide its orientation, the risk of the WMM becoming lodged within the circular section would increase significantly. The results underscore the superior flexibility achieved in the untethered mode of operation.

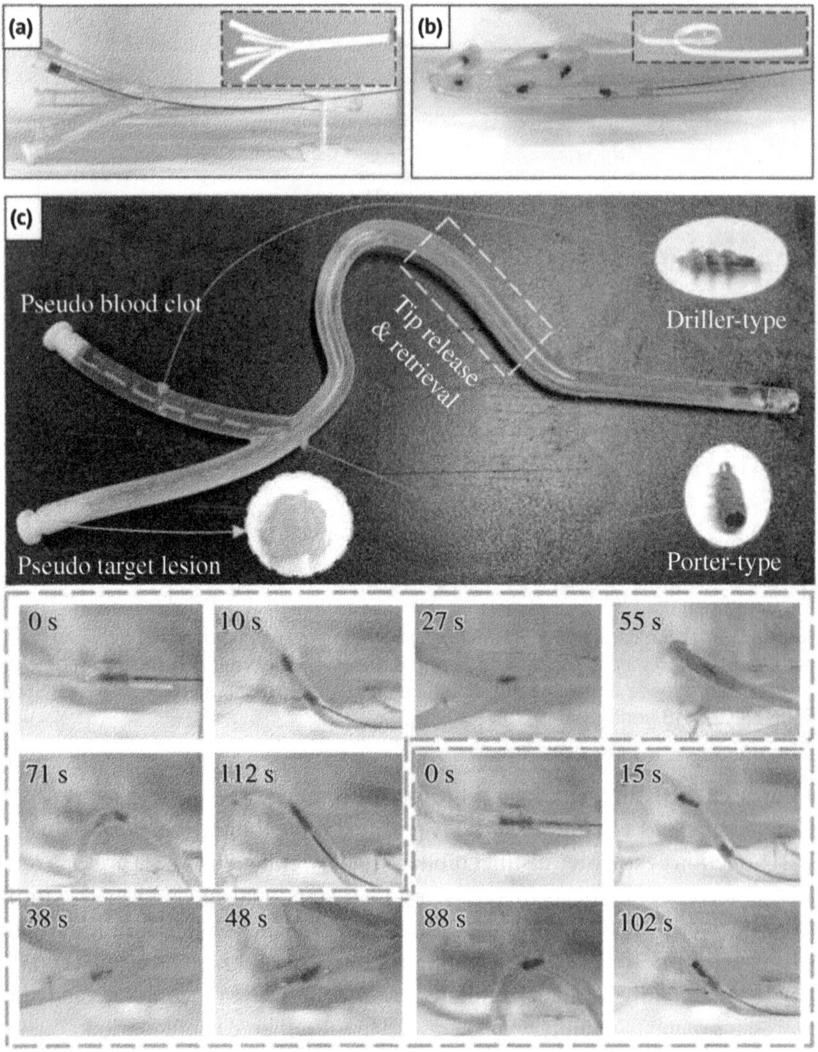

Figure 9.17 Validation of the WMM. (a) Test of the tethered mode in a model with multiple 3D branches. (b) Test of the untethered mode in a model with looping structures. (c) Demonstration of the driller-type design for blood clot removal and the porter-type design for targeted drug delivery.

The third model, illustrated in Figure 9.17c, incorporates a kinking segment and two branches. One of these branches is filled with a 0.2 w% solution of agar, mimicking artificial blood clots, while the other branch contains cotton soaked in an acidic solution, designed to emulate the acidic environment of a target lesion or tumor. To address these specific conditions, two distinct types of functionalized HBs are devised: a driller-type HB, equipped with a tapered tip for effective unclogging, and a porter-type HB, featuring a cavity filled with a cross-linked, pH-responsive hydrogel containing methylthionine chloride, which serves as a surrogate for a loaded drug. During the operational sequence, the driller-type HB is initially attached and propelled into the upper branch. Once the blood clots are removed, it reverses its direction and is retrieved by the robotic hand (RH). Following this, the WMM is withdrawn, and the porter-type HB is installed (Figure 9.17). This second HB is then ejected and activated within the lower branch for drug delivery. Upon completion of its task, it returns to its starting point.

(3) Vascular Model Demonstration: A full-scale model of a leg artery, crafted with physiological accuracy based on authentic computed tomography (CT) data, is utilized for experimental purposes. This model measures 715 mm in length, 239 mm in width, and 100 mm in height, with internal diameters varying between 3 and 20 mm. The experimental arrangement, depicted in Figure 9.18, includes a CCD camera installed on the endplate to provide optical feedback of the procedure. Various routes (indicated by dashed lines) are preplanned, including routes through the renal artery, iliac artery, and femoral artery. The WMM is introduced through a designated insertion point. Additionally, representative screenshots captured at key positions throughout the experiment are presented.

Initially, the tethered mode of operation is employed. The WMM navigates sequentially through point I, proceeds to the abdominal aorta at point II, and then shifts to the left renal artery at point III. It subsequently accesses the posterior and anterior segmental arteries at points IV and V, respectively. A comparable procedure is followed for intervention in the right renal artery. Subsequently, the focus shifts to the lower limb. The WMM repositions itself back to the bifurcation area

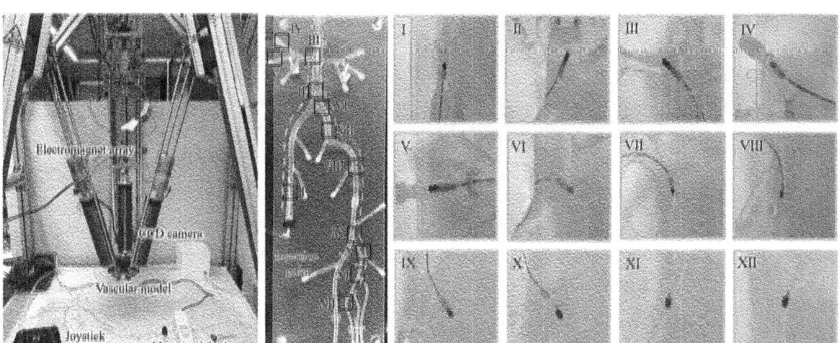

Figure 9.18 Experimental setup with respect to a life-size leg artery model and screenshots of the WMM at representative positions during the intervention.

and aligns toward the iliac artery at point VI, continuing through points VII and VIII. Preparation for tip ejection is initiated as it approaches the relatively constricted vessel at point IX. At point X, the R-EMF is activated to eject the HB. Ultimately, the untethered mode is engaged to reach deeply situated lesions, as demonstrated at points XI and XII.

9.4.4 Intervention Studies on a Phantom

(1) Scanning Along the Vessel: For in vitro experimentation, a tissue-mimicking phantom is constructed. Within this phantom, a bifurcating silicone tubular structure is anchored to emulate blood vessels, and a mixture comprising 10 w% gelatin and 1 w% cellulose is infused to mimic background tissue noise. The experimental configuration, illustrated in Figure 9.19, features a rotatable US probe affixed to the endplate for monitoring purposes. Initially, the operator determines the position and orientation of the US probe to ensure that the vessel is captured in a transverse US image. Subsequently, automated target scanning commences. At each scanning location, the vascular region is delineated by a pretrained U-net model, allowing for real-time adjustments to scanning parameters. A scanning interval of 2 mm is adopted. The trained model efficiently segments the vascular area within 100 ms. B-mode US images are acquired at a rate of 30 fps, while control commands and actuation system status updates are exchanged at a frequency of 50 Hz. The average scanning velocity achieved in these experiments is 1.24 mm/s.

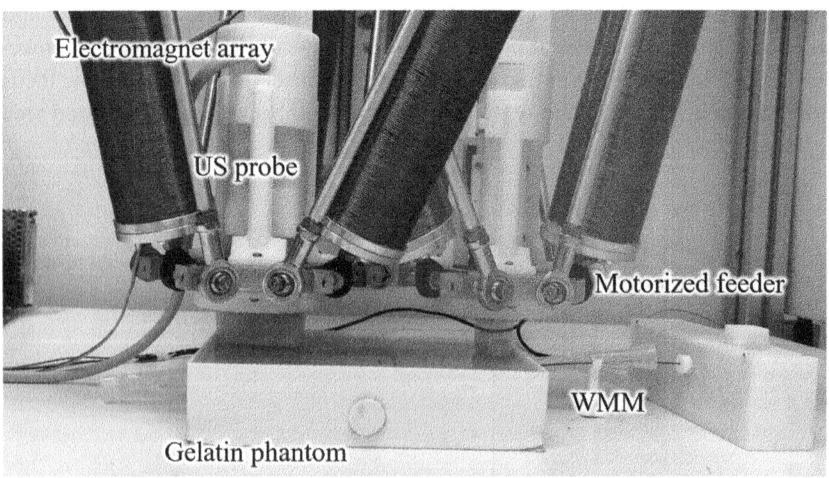

Figure 9.19 Experimental setup with respect to the gelatin phantom, merged US probe configurations for both transverse and longitudinal views.

Figure 9.20a illustrates the fluctuations in the computed PAs and the cumulative vessel areas during a scanning session along both the left and right branches of a trial. Figure 9.20b presents the reconstructed centerlines derived from the US data. Meanwhile, Figure 9.20c showcases the centerlines computed by the Vascular Modeling Toolkit, an open-source tool, utilizing a bespoke 3D silicone tubular model designed in SolidWorks 2019 (SOLIDWORKS Corp.). The maximum discrepancy and average deviation between the VMTK-computed and US-derived centerlines are 5.1 and 1.8 mm, respectively, with partial contribution from the post-curing deformation of the soft phantom. This underscores the efficacy of the proposed scanning technique. Examples of vessel segmentation during scanning, accompanied by annotated descriptive features, are exhibited in Figure 9.20d.

(2) Control Under US Guidance: Magnetic manipulation of the WMM is conducted under US imaging, with no apparent adverse interactions noted between magnetic field manipulation and US imaging. Figure 9.20e exhibits the chosen template image alongside three illustrative screenshots captured in the tethered operation mode, highlighting the outcomes of visual tracking. Our proposed

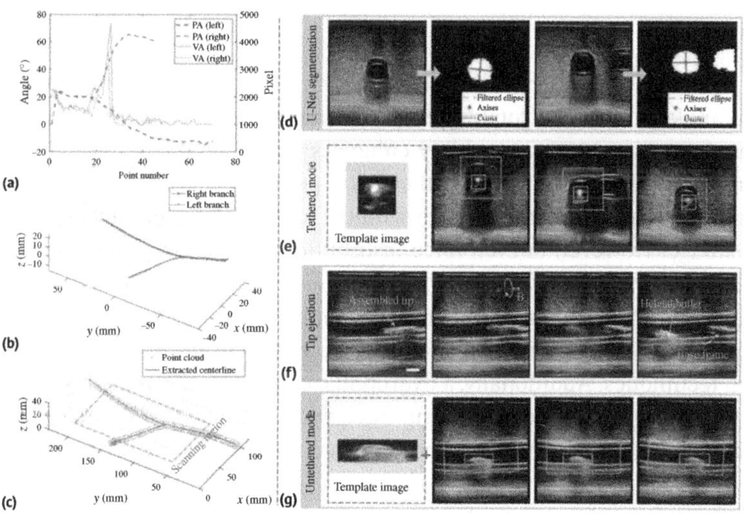

Figure 9.20 Demonstration of the US-guided control. (a) Plot of the estimated probe angle (PA) and detected vessel area (VA). (b) Plot of recovered centerlines. (c) Extracted centerlines and point clouds using VMTK. (d) Examples of original US images and segmented vessel regions. (e) Template and representative US images in the tethered mode using transverse view. The annotated boxes indicate the predefined ROI and the recognized WMM, respectively. (f) Representative US images during the tip ejection stage. (g) Template and representative US images in the untethered mode using longitudinal view.

approach synchronizes the US probe's movement and guidewire insertion to facilitate sequential tracking, enabling the use of D-EMFs to effectively navigate the WMM toward a designated channel. Subsequently, tip ejection occurs in the latter segment of the right branch, as demonstrated in Figure 9.20f, where the US image transitions to a longitudinal view for intricate examination. The four frames depicted illustrate the stages of pre-ejection waiting, intersection angle adjustment, HB ejection, and post-ejection landing, respectively, all clearly discernible under US imaging. Transitioning to the untethered mode, Figure 9.20g displays the selected template image and three representative screenshots. In this mode, our strategy allows for wireless propulsion of the HB via R-EMFs, with the US probe's motion dynamically adjusted based on the HB's real-time position to ensure continuous tracking and actuation. The template matching algorithm achieves WMM detection within 50 ms. The experiments yield average moving speeds of 1.82 and 1.37 mm/s for the tethered and untethered modes, respectively.

9.5 Discussion and Conclusion

9.5.1 Discussion

The MDG in question features a diameter of 1.8 mm, capable of bending up to a maximum angle of 135° and rotating at speeds as high as 30 Hz during drilling operations. Its active steering capability enhances maneuverability through the vascular network, thereby minimizing the risk of perforation [45]. In the drilling mode, the tip's orientation tends to align with the axis of EMFs, minimizing vessel wall damage. However, uneven force distribution may cause lateral contact, posing a risk of scratching, which can be addressed by incorporating a catheter with a protective tip or refining the control mechanism. To cater to smaller vascular applications, the overall size can be further decreased using advanced microfabrication techniques. Additionally, a hydrophilic coating on the exterior surface can reduce insertion friction, enhancing navigational efficiency. Furthermore, coating the drilling tip with diamond powder can augment its cutting force and effectiveness.

In this research, artificial thrombus is employed to test the drilling functionality, recognizing that the physical properties and composition of natural blood clots vary based on their age and other conditions. To bolster the MDG's penetration capacity, particularly when dealing with older clots, it can be paired with catheters designed to administer thrombolytic agents like tissue plasminogen activator (tPA). Enhanced convective flow and mechanical friction can expedite clot dissolution in this process. Furthermore, catheters with specialized functionalities, including balloon occlusion, debris aspiration, and clot retrieval capabilities, can be integrated to serve diverse purposes.

Across all trials, the proposed strategy achieved a perfect success rate of 100%, with an average operational speed of 2.14 mm/s, which aligns competitively with existing methodologies [23]. The procedure's duration can be optimized by tweaking system parameters and enhancing operator training. Currently, the operation maintains a fixed working height and incorporates a thin liquid layer to facilitate US imaging. To accommodate imaging on irregular human body surfaces, force sensors and hybrid control approaches can be seamlessly integrated into the system. The US-guided process is tailored to minimize radiation exposure, while X-ray fluoroscopy remains an option for further interventions, such as angioplasty and stent placement, when deemed necessary.

The prototype of the WMM has undergone validation in a phantom mimicking the leg artery. By applying the proposed principle, the structural parameters can be reconfigured to suit various operational contexts. Moreover, leveraging advanced microfabrication techniques, the overall size of the device can be reduced, enabling its application in smaller vasculatures, such as those in the cerebral region.

The WMM under consideration features a design that incorporates a releasable and retrievable HB, merging the advantages of both tethered and untethered microrobots. However, this design necessitates a compromise in terms of miniaturization of the assembled tip, potentially leading to reduced steerability when compared to simpler, either solely wired or solely wireless, configurations. As the insertion length augments, a slight twisting of the guidewire may occur during the tip ejection phase. This twist is tolerable and results in an extended tip release duration but does not impede the ultimate ejection outcome.

Despite the tip release exhibiting a HS rate, the retrieval efficiency requires enhancement. Specifically, in scenarios where the diameter of the tip release area significantly exceeds that of the WMM, there is a risk of the HB failing to align with the retrieval bucket during retraction and instead being captured by the side of the retrieval head (RH). This issue could potentially be mitigated by incorporating a funnel-shaped, transformable shell onto the BF. Additionally, if debris removed during the process migrates into and obstructs the bucket of the RH, it may adversely impact the retrieval process.

The tethered configuration exhibits satisfactory performance in conditions of high blood flow. Nevertheless, such flow dynamics can potentially affect the untethered configuration. In the majority of vascular areas characterized by narrow and tortuous anatomies, the blood flow is manageable and can be successfully navigated. For regions that necessitate wireless access despite experiencing high blood flow rates, a balloon catheter can be advanced over the guidewire and inflated to temporarily obstruct blood flow.

9.5.2 Conclusion

This section introduces a potential therapeutic approach for treating occlusive PVD, utilizing EMFs for control and radiation-free US imaging for monitoring. The MDG integrates a magnetic driller at its distal end to facilitate both magnetic navigation and occlusion clearance, with mode switching achieved through the application of distinct magnetic fields. The system incorporates a mobile electromagnet array and an eye-in-hand US probe, enabling flexible magnetic field generation and US imaging within a human-scale workspace. The control strategy is designed in two stages to alleviate operator workload: the preoperative stage reconstructs the environment to address the challenges posed by noisy US images, while the intraoperative stage coordinates all system modules internally, allowing users to primarily input magnetic field directions. Experiments with an optical camera demonstrate the MDG's flexible navigation under directional magnetic fields and effective occlusion clearance under rotating magnetic fields. Additionally, US imaging tests on a tissue-mimicking phantom validate the feasibility of the control strategy. To date, we have achieved preliminary verification of US-guided catheterization with combined magnetic navigation and occlusion clearance. Future research will concentrate on in vivo validation.

Besides, this section introduces an innovative intravascular magnetic microrobot featuring multiple actuation modes and explores its control mechanism guided by medical imaging feedback. The introduced WMM offers flexible maneuverability in tethered mode, facilitates tip ejection transition through magnetic triggering, and can be propelled wirelessly in untethered mode. A custom-built electromagnetic system is employed to facilitate large-workspace, multimodal US imaging and magnetic actuation. The proposed control strategy encompasses targeted scanning prior to WMM intervention, addressing noisy imaging challenges, considering motion characteristics of different modes, ensuring system module cooperation, and providing an intuitive user interface. A comprehensive series of experiments, including model calibration, functional testing, and demonstration in a vascular model, have been conducted to validate the structural effectiveness and enhanced flexibility of the design. Additionally, the US-guided control framework has been tested on a tissue-mimicking phantom, demonstrating its potential for preliminary clinical applications. The future endeavors will focus on reducing the overall size of the WMM through the utilization of advanced micromachining technologies. Furthermore, it is aimed to carry out in vivo validations to assess its performance.

Appendix Additional Algorithm

Algorithm 1: RANSAC-based Ellipse Fitting

Require: recognized edge point set $\mathcal{E} = \{E_1, E_2, \dots, E_K\}$
Ensure: ellipse coefficient matrix $\mathbf{P}^\star = [A_e^\star \;\; B_e^\star \;\; C_e^\star \;\; D_e^\star \;\; E_e^\star]$

max#Inlier $\leftarrow 0$
while $\#Iteration \leq$ max#Iteration **do**
    $\mathcal{E}' \leftarrow$ Random_Choose_Point($\mathcal{E}, N$) ▷ Subset with N edge points $\mathcal{E}'$
    $\mathbf{P} \leftarrow$ Least_Square_Ellipse_Estimation($\mathcal{E}'$)
    $a, c, F_1, F_2 \leftarrow$ Compute_Geometry_Parameter($\mathbf{P}$)
    if $\frac{a}{c} < \delta_e$ & $B_e^2 - 4A_eC_e < 0$ **then** ▷ Eccentricity threshold: δ_e
        $\#Inlier \leftarrow 0$
        for $i = 1$ to N **do**
            $d_1 \leftarrow$ Euclidean_Distance(E_i, F_1)
            $d_2 \leftarrow$ Euclidean_Distance(E_i, F_2)
            if $|d_1 + d_2 - 2a| < \delta_d$ **then** ▷ Distance threshold: δ_d
                $\#Inlier \leftarrow \#Inlier + 1$
            end if
        end for
        if $\#Inlier >$ max#Inlier **then**
            max#Inlier $\leftarrow \#Inlier$
            $\mathbf{P}^\star \leftarrow \mathbf{P}$
        end if
    end if
    $\#Iteration \leftarrow \#Iteration + 1$
end while
if max#Inlier $< \delta_i$ **then** ▷ Inlier threshold: δ_i
    $\mathbf{P}^\star \leftarrow [\,]$
end if

Algorithm 2: Modified Star-Kalman Recognition

Require: processed OCP sequence $I_{m1}, I_{m2}, \dots, I_{mM}$
Ensure: center point set $\mathcal{G} = \{G_1, G_2, \dots, G_L\}$

$S_{p1} \leftarrow$ ClickSeedPoint (I_{m1}) ▷ The ith seed list: S_{pi}
while $\#OCP \le$ max#OCP **do**
    $i \leftarrow \#OCP$
    $\#Seed \leftarrow 1$
    while $\#Seed \le$ max#Seed(S_{pi}) **do**
        $S_c \leftarrow$ ExtractPoint ($\#Seed, S_{pi}$) ▷ Current seed: S_c
        for $j \leftarrow 1$ **to** K **do**
            $r_j \leftarrow$ GenerateDetectionRay($S_c, 2\pi j/K$)
            $D_j \leftarrow$ Sample1DImage(I_{m_i}, r_j)
            $E_j \leftarrow$ MaximumNCCValue(D_j, T_e)
            if NCCValue(E_j) $< \delta_n$ **then** ▷ NCC threshold: δ_n
                $E_j \leftarrow [\,]$
            end if
        end for
        $\mathbf{P}^\star \leftarrow$ RansacEllipseFitting($\mathcal{E}$)
        $a, c, x_O, y_O \leftarrow$ ComputeGeometryParameter($\mathbf{P}^\star$)
        $x(i) \leftarrow$ UpdateKalmanFilter($x(i-1), z(i)$)
        if MeetSplitCondition(c/a) **then**
            $O_1, O_2 \leftarrow$ SplitPoint(O, F_1, F_2)
            $S_{p(i+1)} \leftarrow S_{p(i+1)} \cup \{O_1, O_2\}$
        else
            $S_{p(i+1)} \leftarrow S_{p(i+1)} \cup \{O\}$
        end if
        $\#Seed \leftarrow \#Seed + 1$
    end while
    $\mathcal{G} \leftarrow \mathcal{G} \cup S_{pi}$ ▷ Element in seed list renamed as G_n
    $\#OCP \leftarrow \#OCP + 1$
end while

Algorithm 3: Hierarchical Tip Detection

Require: real-time US image I_r; tracking interval t_r; reference path set $\mathcal{R} = \{R_1, R_2, \dots, R_L\}$
Ensure: tracked tip position P_t^U
$T_v \leftarrow$ Select_Vessel_Region(I_r) ▷ Vessel template: T_v
$T_t \leftarrow$ Select_Tip_Region(I_r) ▷ Tip template: T_t
while #RefPoint $\leq$ max#RefPoint **do** ▷ Point number: L
    $i \leftarrow$ #RefPoint
    $[I_o, C_o] \leftarrow$ Initial_Region_Of_Interest(I_r, R_i) ▷ Defined ROI: I_o, ROI corner point: C_o
    $[I_v, C_v, g_v] \leftarrow$ NCC_Template_Matching(I_o, T_v) ▷ Matched vessel image: I_v, vessel corner point: C_v, vessel matching score: g_v
    $[I_t, C_t, g_t] \leftarrow$ NCC_Template_Matching(I_v, T_t) ▷ Matched tip image: I_t, tip corner point: C_t, tip matching score: g_t
    if $g_v > \delta_s$ **and** $g_t > \delta_s$ **then** ▷ Matching score threshold: δ_s
        $P_t^U \leftarrow$ Recover_Tip_Position(C_o, C_v, C_t)
        $T_v \leftarrow I_v$
        #RefPoint $\leftarrow$ #RefPoint + t_r
    end if
end while

References

1 D. G. Kokkinidis and E. J. Armstrong. Current developments in endovascular therapy of peripheral vascular disease. *Journal of Thoracic Disease*, 12 (4): 1681, 2020.

2 N. I. Akkus and A. Abdulbaki. Rotational atherectomy in peripheral arterial interventions. In *Textbook of Catheter-Based Cardiovascular Interventions: A Knowledge-Based Approach*, pages 1377–1388, 2017. Springer.

3 J. Sheng, X. Wang, T.-M. L. Dickfeld, and J. P. Desai. Towards the development of a steerable and MRI-compatible cardiac catheter for atrial fibrillation treatment. *IEEE Robotics and Automation Letters*, 3 (4): 4038–4045, 2018.

4 F. Langsch, S. Virga, J. Esteban, R. Göbl, and N. Navab. Robotic ultrasound for catheter navigation in endovascular procedures. In *2019 IEEE/RSJ International Conference on Intelligent Robots and Systems (IROS)*, pages 5404–5410, 2019. IEEE.

5 H. Rafii-Tari, C. J. Payne, and G.-Z. Yang. Current and emerging robot-assisted endovascular catheterization technologies: a review. *Annals of Biomedical Engineering*, 42: 697–715, 2014.

6 X. Hu, A. Chen, Y. Luo, C. Zhang, and E. Zhang. Steerable catheters for minimally invasive surgery: a review and future directions. *Computer Assisted Surgery*, 23 (1): 21–41, 2018.

7 J. Sikorski, A. Denasi, G. Bucchi, S. Scheggi, and S. Misra. Vision-based 3-D control of magnetically actuated catheter using bigmag—an array of mobile electromagnetic coils. *IEEE/ASME Transactions on Mechatronics*, 24 (2): 505–516, 2019.

8 J. Edelmann, A. J. Petruska, and B. J. Nelson. Magnetic control of continuum devices. *The International Journal of Robotics Research*, 36 (1): 68–85, 2017.

9 S. Jeong, H. Choi, G. Go, C. Lee, S. Y. Ko, J.-O. Park, and S. Park. Feasibility study on magnetically steerable guidewire device for percutaneous coronary intervention. *International Journal of Control, Automation and Systems*, 15 (1): 473–479, 2017.

10 S. Jeon, et al. A magnetically controlled soft microrobot steering a guidewire in a three-dimensional phantom vascular network. *Soft Robotics*, 6 (1): 54–68, 2019.

11 P. Lloyd, A. K. Hoshiar, T. da Veiga, A. Attanasio, N. Marahrens, J. H. Chandler, and P. Valdastri. A learnt approach for the design of magnetically actuated shape forming soft tentacle robots. *IEEE Robotics and Automation Letters*, 5 (3): 3937–3944, 2020.

12 D. Lin, J. Wang, N. Jiao, Z. Wang, and L. Liu. A flexible magnetically controlled continuum robot steering in the enlarged effective workspace with constraints for retrograde intrarenal surgery. *Advanced Intelligent Systems*, 3 (10): 2000211, 2021.

13 J. Hwang, J.-y. Kim, and H. Choi. A review of magnetic actuation systems and magnetically actuated guidewire-and catheter-based microrobots for vascular interventions. *Intelligent Service Robotics*, 13 (1): 1–14, 2020.

14 J. Nam, W. Lee, J. Kim, and G. Jang. Magnetic helical robot for targeted drug-delivery in tubular environments. *IEEE/ASME Transactions on Mechatronics*, 22 (6): 2461–2468, 2017.

15 I. S. Khalil, et al. Mechanical rubbing of blood clots using helical robots under ultrasound guidance. *IEEE Robotics and Automation Letters*, 3 (2): 1112–1119, 2018.

16 J. Leclerc, H. Zhao, D. Bao, and A. T. Becker. *In vitro* design investigation of a rotating helical magnetic swimmer for combined 3-d navigation and blood clot removal. *IEEE Transactions on Robotics*, 36 (3): 975–982, 2020.

17 K. T. Nguyen, B. Kang, E. Choi, J.-O. Park, and C.-S. Kim. High-frequency and high-powered electromagnetic actuation system utilizing two-stage resonant effects. *IEEE/ASME Transactions on Mechatronics*, 25 (5): 2398–2408, 2020.

18 C.-H. Yu and S. H. Kim. Multifunctional robotic guidewire system using spiral-type magnetic microrobot with magnetic manipulation. *Journal of Magnetics*, 21 (4): 616–621, 2016.

19 N. Kim, S. Lee, W. Lee, and G. Jang. Development of a magnetic catheter with rotating multi-magnets to achieve unclogging motions with enhanced steering capability. *AIP Advances*, 8 (5): 056708, 2018.

20 J. Park, W. Lee, J. Kim, E. Jung, N. Kim, and G. Jang, Selective separating and assembling motion control for delivery and retrieval of an untethered magnetic robot in human blood vessels. *AIP Advances*, 9 (12): 125136, 2019.

21 Z. Yang and L. Zhang. Magnetic actuation systems for miniature robots: a review. *Advanced Intelligent Systems*, 2 (9): 2000082, 2020.

22 J. J. Abbott, E. Diller, and A. J. Petruska. Magnetic methods in robotics. *Annual Review of Control, Robotics, and Autonomous Systems*, 3 (1): 57–90, 2020.

23 Z. Yang, L. Yang, M. Zhang, Q. Wang, S. C. H. Yu, and L. Zhang. Magnetic control of a steerable guidewire under ultrasound guidance using mobile electromagnets. *IEEE Robotics and Automation Letters*, 6 (2): 1280–1287, 2021.

24 C. Heunis, J. Sikorski, and S. Misra. Flexible instruments for endovascular interventions: improved magnetic steering, actuation, and image-guided surgical instruments. *IEEE Robotics Automation Magazine*, 25 (3): 71–82, 2018.

25 C. Chautems, A. Tonazzini, Q. Boehler, S. H. Jeong, D. Floreano, and B. J. Nelson. Magnetic continuum device with variable stiffness for minimally invasive surgery. *Advanced Intelligent Systems*, 2 (6): 1900086, 2020.

26 W. Lee, J. Nam, J. Kim, E. Jung, N. Kim, and G. Jang. Steering, tunneling, and stent delivery of a multifunctional magnetic catheter robot to treat occlusive vascular disease. *IEEE Transactions on Industrial Electronics*, 68 (1): 391–400, 2020.

27 D. Lin, N. Jiao, Z. Wang, and L. Liu. A magnetic continuum robot with multi-mode control using opposite-magnetized magnets. *IEEE Robotics and Automation Letters*, 6 (2): 2485–2492, 2021.

28 G. Datta. Broken guidewire–a tale of three cases. *Indian Heart Journal*, 67: S49–S52, 2015.

29 W. Lee, J. Nam, J. Kim, E. Jung, and G. Jang. Effective locomotion and precise unclogging motion of an untethered flexible-legged magnetic robot for vascular diseases. *IEEE Transactions on Industrial Electronics*, 65 (2): 1388–1397, 2017.

30 T. Xu, Y. Guan, J. Liu, and X. Wu. Image-based visual servoing of helical microswimmers for planar path following. *IEEE Transactions on Automation Science and Engineering*, 17 (1): 325–333, 2019.

31 Y. Alapan, U. Bozuyuk, P. Erkoc, A. C. Karacakol, and M. Sitti. Multifunctional surface microrollers for targeted cargo delivery in physiological blood flow. *Science Robotics*, 5 (42): eaba5726, 2020.

32 Q. Wang, L. Yang, J. Yu, P. W. Y. Chiu, Y.-P. Zheng, and L. Zhang. Real-time magnetic navigation of a rotating colloidal microswarm under ultrasound guidance. *IEEE Transactions on Biomedical Engineering*, 67 (12): 3403–3412, 2020.

33 K. J. Boskma, S. Scheggi, and S. Misra. Closed-loop control of a magnetically-actuated catheter using two-dimensional ultrasound images. In *2016 6th IEEE International Conference on Biomedical Robotics and Biomechatronics (BioRob)*, pages 61–66, 2016. IEEE.

34 T. Wang, W. Hu, Z. Ren, and M. Sitti. Ultrasound-guided wireless tubular robotic anchoring system. *IEEE Robotics and Automation Letters*, 5 (3): 4859–4866, 2020.

35 C. M. Heunis, Y. P. Wotte, J. Sikorski, G. P. Furtado, and S. Misra. The ARMM system-autonomous steering of magnetically-actuated catheters: Towards endovascular applications. *IEEE Robotics and Automation Letters*, 5 (2): 705–712, 2020.

36 Z. Yang, L. Yang, M. Zhang, C. Zhang, S. C. H. Yu, and L. Zhang. Ultrasound-guided catheterization using a driller-tipped guidewire with combined magnetic navigation and drilling motion. *IEEE/ASME Transactions on Mechatronics*, 27 (5): 2829–2840, 2021.

37 R. Lorbeer, A. Grotz, M. Dörr, H. Völzke, W. Lieb, J.-P. Kühn, and B. Mensel. Reference values of vessel diameters, stenosis prevalence, and arterial variations of the lower limb arteries in a male population sample using contrast-enhanced MR angiography. *PloS One*, 13 (6): e0197559, 2018.

38 A. M. Zakrzewski and B. W. Anthony. Noninvasive blood pressure estimation using ultrasound and simple finite element models. *IEEE Transactions on Biomedical Engineering*, 65 (9): 2011–2022, 2017.

39 M. Kaya, E. Senel, A. Ahmad, and O. Bebek. Visual tracking of biopsy needles in 2D ultrasound images. In *2016 IEEE International Conference on Robotics and Automation (ICRA)*, pages 4386–4391, 2016. IEEE.

40 E. Fosch-Villaronga, P. Khanna, H. Drukarch, and B. H. Custers. A human in the loop in surgery automation. *Nature Machine Intelligence*, 3 (5): 368–369, 2021.

41 Z. Yang, L. Yang, and L. Zhang. Autonomous navigation of magnetic microrobots in a large workspace using mobile-coil system. *IEEE/ASME Transactions on Mechatronics*, 26 (6): 3163–3174, 2021.

42 L. Yang, X. Du, E. Yu, D. Jin, and L. Zhang. Deltamag: an electromagnetic manipulation system with parallel mobile coils. In *2019 International Conference on Robotics and Automation (ICRA)*, pages 9814–9820, 2019. IEEE.

43 O. Ronneberger, P. Fischer, and T. Brox. U-net: convolutional networks for biomedical image segmentation. In *Medical Image Computing and Computer-assisted Intervention–MICCAI 2015: 18th International Conference, Munich, Germany, October 5-9, 2015, Proceedings, Part III 18*, pages 234–241, 2015. Springer.

44 E. Mfoumou, J. Tripette, M. Blostein, and G. Cloutier. Time-dependent hardening of blood clots quantitatively measured *in vivo* with shear-wave ultrasound imaging in a rabbit model of venous thrombosis. *Thrombosis Research*, 133 (2): 265–271, 2014.

45 X. Hu, L. Cao, Y. Luo, A. Chen, E. Zhang, and W. Zhang. A novel methodology for comprehensive modeling of the kinetic behavior of steerable catheters. *IEEE/ASME Transactions on Mechatronics*, 24 (4): 1785–1797, 2019.

10

Conclusion and Outlook

10.1 Design of Magnetic Actuation System

When employing mobile actuation systems to drive miniature robots, a mechanical platform is required to translate and rotate the coils or permanent magnets. Both permanent magnets and electromagnetic coils can be implemented to generate magnetic fields.

1. Systems with a Single Magnetic Field Magnet or Electromagnetic Coil: Single permanent magnets or electromagnetic coils are commonly employed as end-effectors to generate magnetic fields in mobile systems. A typical configuration involves attaching a permanent magnet to a robotic arm, as illustrated in Figure 10.1g. Through precise modeling and closed-loop control strategies, this system has achieved five-degree-of-freedom (5-DOF) actuation of an untethered miniature robot [1]. To enable time-varying magnetic field generation, the spherical-actuator-magnet manipulator (SAMM) was developed. This system controls the orientation of a spherical permanent magnet via three omniwheels in a closed-loop configuration, allowing continuous rotation around any axis, as shown in Figure 10.1h [2]. The SAMM platform has successfully demonstrated simultaneous actuation and tracking of a helical capsule endoscope using a rotating magnetic field [3]. The advanced robotics for magnetic manipulation (ARMM) system, depicted in Figure 10.1i, utilizes a single electromagnetic coil for navigating magnetic catheters [4, 5]. Furthermore, Ankon Technology Company has commercialized a system featuring a single permanent magnet for active control of capsule endoscopes during gastrointestinal tract examinations.

   The architectures of mobile systems equipped with a single permanent magnet or electromagnetic coil are typically characterized by structural simplicity, which facilitates the derivation of solutions for coil motion dynamics. Meanwhile, when high-degree-of-freedom (high-DOF) mechanical platforms

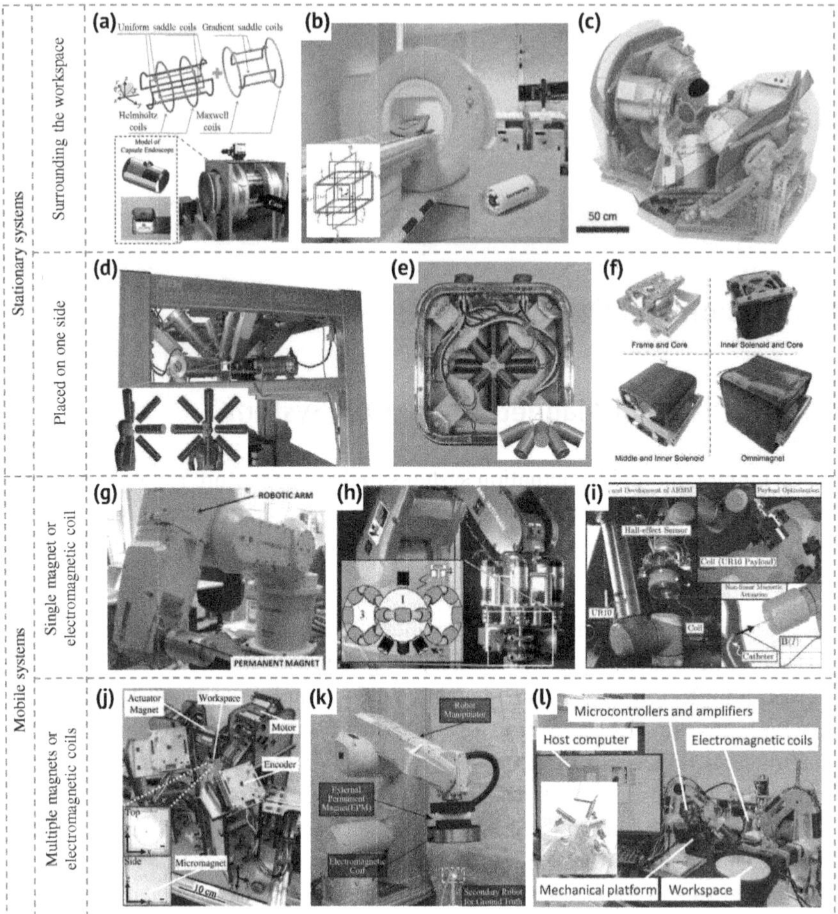

Figure 10.1 Designs of magnetic actuation systems.

are employed to adjust magnet positions, these systems exhibit significant potential in addressing complex clinical scenarios, such as obstacle avoidance (e.g., avoiding patients and surgeons) and seamless integration with imaging devices. Compared to stationary designs, scaling these systems for clinical deployment may be more straightforward by expanding the operational range of the mechanical platform. However, inherent limitations persist. The inertial properties of the magnet or coil assembly impose constraints on the frequency range and variability of the generated time-varying magnetic fields.

2. Systems with Multiple Magnets or Electromagnetic Coils: Mobile systems can also utilize multiple magnets or electromagnetic coils to generate required magnetic fields. For instance, by employing eight rotating permanent magnets, the required magnetic flux density and gradient have been achieved to

enable 5-DOF control over the position and orientation of a microrobot, as demonstrated in Figure 10.1j [6]. Combining an electromagnetic coil with a permanent magnet can eliminate singularity regions during the pose estimation of magnetic robots, thereby enhancing closed-loop control performance, as shown in Figure 10.1k [7–9]. The DeltaMag system employs three mobile coils actuated by a parallel mechanism to generate time-varying magnetic fields [10]. Furthermore, the RoboMag system in Figure 10.1l features three mobile electromagnetic coils with adjustable positions, enabling long-distance actuation of robots while avoiding workspace obstructions and improving energy efficiency [11, 12]. The Niobe system and the latest Genesis RMN system from Stereotaxis Inc. incorporate two permanent magnets that rotate about their center of mass, driven by robotic arms. These systems can steer catheters for minimally invasive diagnostic and ablation procedures.

Mobile systems equipped with multiple magnets or electromagnetic coils offer flexibility as a key advantage. By leveraging mechanical structures, the coils or magnets can be positioned in close proximity to the patient while tracking robots within an expanded operational space. Scaling these designs for clinical applications is feasible by enhancing the motion range of the mechanical platform. The use of multiple mobile magnets or coils enables the system to generate programmed magnetic fields with wide bandwidth [11, 13]. However, the control strategies for such systems are inherently complex. Real-time computation of magnetic fields requires consideration of both the positions of magnets/coils and the dynamic movements of the robots. Well-designed actuation strategies are essential to fully exploit the system's redundancies.

3. Calculation of Demanded Parameters for Magnetic Actuation: To generate the required magnetic torque and force using mobile systems, the parameters requiring control encompass the relative positions $\mathbf{p}_i$ and orientations $\hat{\mathbf{m}}_{s,i}$ of the mobile magnets or coils, as well as the electrical currents I_i when mobile coils are employed. A viable approach for calculating these parameters is through local linearization of the system, which is mathematically expressed as

$$\begin{bmatrix} \delta\tau \\ \delta\mathbf{F} \end{bmatrix} = \sum_{i=1}^{n} \mathbb{J}_i \begin{bmatrix} \delta\mathbf{p}_i \\ \delta\hat{\mathbf{m}}_{s,i} k_i I_i \end{bmatrix} = [\mathbb{J}_1 \dots \mathbb{J}_n] \begin{bmatrix} \delta\mathbf{p}_1 \\ \delta\hat{\mathbf{m}}_{s,1} k_1 I_1 \\ \dots \\ \delta\mathbf{p}_n \\ \delta\hat{\mathbf{m}}_{s,n} k_n I_n \end{bmatrix}$$
$$= \mathbb{J}\delta\mathbf{u} \tag{10.1}$$

where u is the vector of input parameters of the system, $\mathbb{J}_i$ is the Jacobian matrix, k_i is a constant for compensation so that $\hat{\mathbf{m}}_{s,i}k_i$ equals the magnetic moment of the magnet or the moment of the coil when $I_i = 1\,\text{A}$. Besides linearization, the calculation process can also be conducted regarding specific designs and objectives, such as the equation-solving method for the DeltaMag system [10] and the particle swarm optimization (PSO)-based method for the RoboMag system [11].

10.2 Integration Strategy of Imaging and Actuation System

10.2.1 System Framework

A typical framework for an integrated system designed for in vivo medical applications is depicted in Figure 10.2. The core modules of this system encompass the magnetic actuation system, miniature robots, imaging devices, and a host computer. The magnetic actuation system is responsible for generating the required magnetic fields to enable miniature robots to perform designated tasks. These magnetic fields include rotating magnetic fields [14], oscillating magnetic fields, and magnetic field gradients [15]. Magnetic forces are typically employed for specific operational tasks, such as drug release and biopsy procedures [16]. The miniature robot functions as an end-effector to execute required operations within the human body. Imaging devices continuously monitor the status of the miniature robots and provide real-time feedback to the host computer, which serves as the central coordinator for the entire operational workflow. The host computer processes data acquired from imaging devices to calculate the position of the miniature robot and environmental information. A user interface is integrated to facilitate convenient operation for surgeons, allowing real-time execution of commands. Based on the robot's status and model, environmental information, and

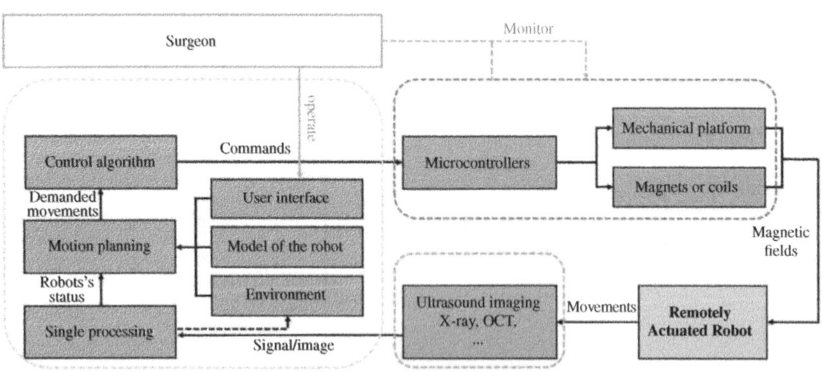

Figure 10.2 General framework of image-integrated magnetic actuation systems.

instructions from the user interface, the host computer plans the robot's motion trajectory. Subsequently, the control algorithm computes the necessary control parameters to achieve the planned motion and transmits commands to the magnetic actuation system. Throughout the procedure, the surgeon retains the highest operational priority, with the authority to monitor the system in real time and override its operations if deemed necessary.

Integration between magnetic actuation systems and localization devices constitutes a critical step in system design. From a hardware perspective, this integration imposes stringent requirements on system configurations. The hardware-level integration necessitates precise alignment of technical parameters, spatial coordination, and signal synchronization between the actuation components and localization sensors to ensure accurate control and real-time feedback during medical interventions. This interoperability is essential for achieving closed-loop control, where magnetic field generation is dynamically adjusted based on the robot's position and environmental feedback obtained through localization systems.

Four categories of magnetic actuation systems can be identified:

- stationary systems surrounding the workspace;
- stationary systems placed on one side;
- mobile systems with a single magnet or coil;
- mobile systems with multiple magnets or coils.

Within the category of mobile systems with multiple magnets or coils, mechanical dexterity varies with the mechanical design. Herein, a criterion is proposed to evaluate system dexterity and estimate compatibility with different imaging devices. The average mechanical degrees of freedom (DOFs) shared by each magnet or coil, $\bar{N}_{\text{dof}}$, is defined as

$$\bar{N}_{\text{dof}} = \frac{N_{\text{mech}}}{n_c}, \tag{10.2}$$

where N_{mech} denotes the total number of mechanical DOFs of the system, representing the cumulative spatial movement capabilities of the magnets or coils, and n_c denotes the number of magnets or coils. For instance, the DeltaMag system [10] comprises three coils actuated by a 3-DOF parallel mechanism, yielding $\bar{N}_{\text{dof}} = 1$. The BigMag system [17] integrates six movable coils with two rotational DOFs, resulting in $\bar{N}_{\text{dof}} = 1/3$. The RoboMag system [11] employs three movable coils, each driven by a 4-axis robotic arm, leading to $\bar{N}_{\text{dof}} = 4$. When the relative positioning of magnets or coils remains constant, they function equivalently to a single magnetic source. If a 6-axis robotic arm transports a single magnet or coil [7] with fixed relative positions, then $n_c = 1$ and $\bar{N}_{\text{dof}} = 6$. This metric facilitates systematic discussion of integration strategies for distinct design paradigms (Table 10.1).

Table 10.1 Comparison of system configurations for imaging and tracking devices.

	Surrounding	One side	Mobile (1 src.)	Mobile (multi)
Encircled imaging	Size↓, quality↓	Size↓, quality↓	Workspace↓	$\bar{N}_{dof} \leq 4$: workspace↓; $\bar{N}_{dof} > 4$: workspace↓
C-shaped imaging	View↓	View↓	✓*	$\bar{N}_{dof} \leq 3$: view↓; $\bar{N}_{dof} > 3$: ✓
Probe imaging	✓	✓	✓	✓
Sensor-array tracking	Array size↓	✓†	✓†	✓†

$\bar{N}_{dof}$: average mechanical DOF per magnet/coil. * Except large mechanical structures.
† Consider interference from actuation fields.

10.3 Application of Image-integrated Mobile Magnetic Actuation System

Image-integrated magnetic actuation systems are extensively employed in the automatic control of small-scale robots. Control methodologies utilizing magnetic fields for medical miniature robots have been the subject of extensive research. At the millimeter scale, closed-loop control with imaging feedback has been implemented on various robots, encompassing magnetically steered catheters [13, 18], tethered capsule colonoscopes [7], magnetic capsule robots [1], and untethered helical robots [19–22], as well as independent control of multiple miniature robots [23, 24]. At the micrometer scale, visually servoing a magnetic microrobot in a low Reynolds number environment has been carried out within the OctoMag system, with the objective of performing intraocular surgeries, as illustrated in Figure 10.3a. As depicted in Figure 10.3b, motion control of a miniature gripper has been achieved under ultrasound imaging feedback for the navigation of a magnetic gradient-controlled microrobot within blood flow [25]. A field-of-view (FoV) tracking scheme has been proposed for 3D position control of a microrobot using magnetic force [26]. An adaptive sliding-mode controller integrated with a nonlinear disturbance observer has been proposed to compensate for the hysteresis nonlinearity of the magnetic actuation system, thereby generating a more precise magnetic field for accurate microrobotic manipulation [27]. Fully automatic formation control, motion control, and pattern control of microrobotic swarms can be executed using a statistics-based algorithm [28], a fuzzy logic controller [29], or a genetic algorithm [30].

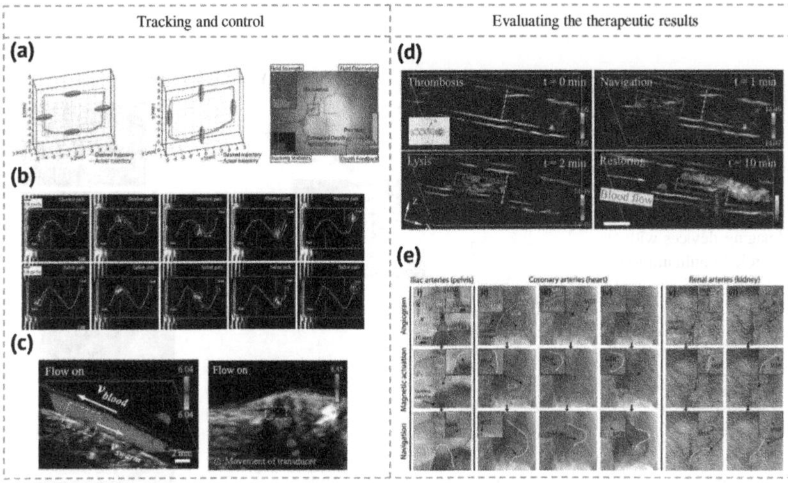

Figure 10.3 Applications of miniature robots with image-integrated magnetic actuation systems: (a) visual servoing of a microrobot, (b) closed-loop control of a miniature gripper, (c) tracking microrobotic swarm in blood flow, (d) thrombolysis, and (e) cardiovascular intervention.

The integrated system also enables in vivo biomedical therapies and monitoring. Magnetic systems integrated with Doppler ultrasound imaging can drive the motion of helical miniature robots, while simultaneously monitoring their locomotion and position. The thrombolysis outcomes can also be evaluated, as demonstrated in Figure 10.3d [31]. Ultrasound imaging combined with a magnetic actuation system has facilitated real-time tracking and navigation of microrobotic swarms using B-mode images [32] and color Doppler images [33], as illustrated in Figures 10.3c and 10.4c1, respectively. Fluorescence image-guided magnetic actuation enables the in vivo navigation of magnetic biohybrid microrobots for targeted therapy [14]. Fluoroscopy has been employed to determine the position and orientation of magnetically steered guidewires, providing visual feedback during cardiovascular interventions, as shown in Figure 10.3e [34]. Additionally, fluoroscopy can assess the efficacy of magnetically actuated mechanical thrombolysis by monitoring the distribution of contrast agents across the occluded site, as depicted in Figure 10.4b1 [35]. Digital subtraction angiography using fluoroscopy, in conjunction with a stationary magnetic actuation system, has been implemented to accumulate microrobotic swarms in porcine kidneys for in vivo embolization (Figure 10.4) [36].

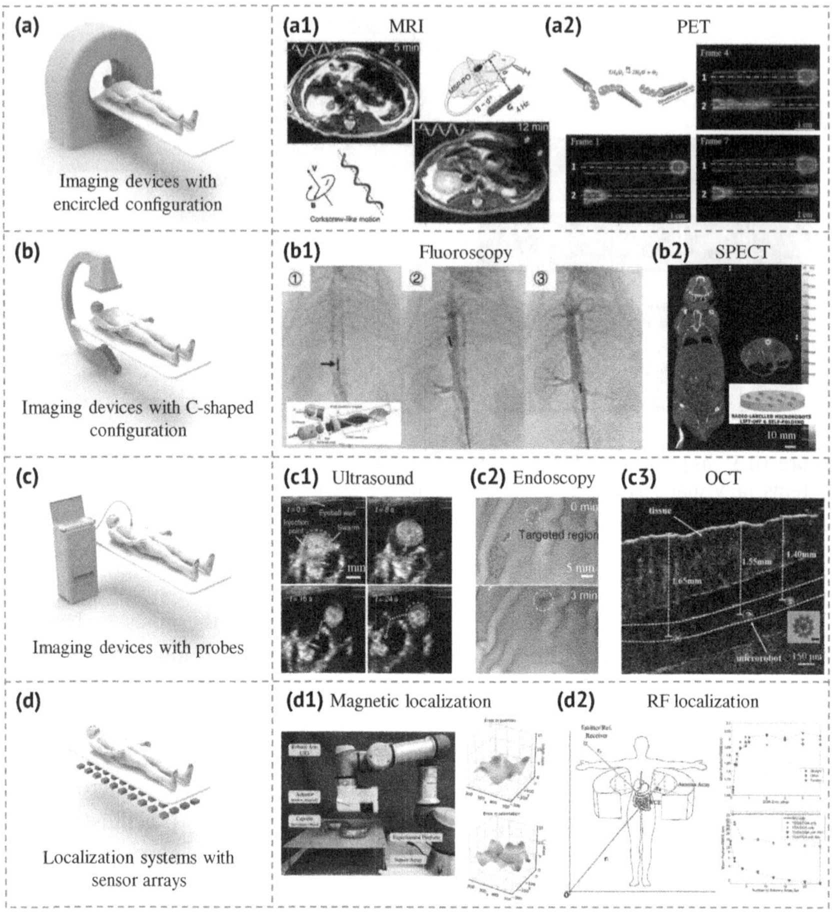

Figure 10.4 Imaging and localization devices for miniature robots. (a) Medical imaging devices with encircled configuration. (a1) Tracking of biohybrid helical microrobots using MRI. (a2) Imaging of micromotors under PET. (b) Imaging devices with C-shaped configuration. (b1) Thrombolysis using a tethered helical microrobot guided by X-ray fluoroscopy to reach the thrombus region, drill through the thrombus, and be retracted from the body. (b2) Imaging of hydrogel-based magnetic microrobots with drug release function under SPECT. (c) Imaging devices with probes. (c1) Tracking of a microrobotic swarm using ultrasound imaging inside a bovine eyeball. (c2) Delivery and observation of magnetic stem cell spheroid microrobots using endoscopy. (c3) Tracking of a magnetic microrobot using OCT inside the portal vein of mice. (d) Localization using a sensor array. (d1) Simultaneous localization and actuation of a capsule robot with a magnetic sensor array. (d2) Localization of a capsule robot using a circular RF sensor array and an IMU.

References

1 A. W. Mahoney and J. J. Abbott. Five-degree-of-freedom manipulation of an untethered magnetic device in fluid using a single permanent magnet with application in stomach capsule endoscopy. *The International Journal of Robotics Research*, 35 (1–3): 129–147, 2016.

2 S. E. Wright, A. W. Mahoney, K. M. Popek, and J. J. Abbott. The spherical-actuator-magnet manipulator: a permanent-magnet robotic end-effector. *IEEE Transactions on Robotics*, 33 (5): 1013–1024, 2017.

3 K. M. Popek, T. Hermans, and J. J. Abbott. First demonstration of simultaneous localization and propulsion of a magnetic capsule in a lumen using a single rotating magnet. In *2017 IEEE International Conference on Robotics and Automation (ICRA)*, pages 1154–1160, 2017. IEEE.

4 C. M. Heunis, Y. P. Wotte, J. Sikorski, G. P. Furtado, and S. Misra. The ARMM system-autonomous steering of magnetically-actuated catheters: towards endovascular applications. *IEEE Robotics and Automation Letters*, 5 (2): 705–712, 2020.

5 J. Sikorski, C. M. Heunis, F. Franco, and S. Misra. The armm system: an optimized mobile electromagnetic coil for non-linear actuation of flexible surgical instruments. *IEEE Transactions on Magnetics*, 55 (9): 1–9, 2019.

6 P. Ryan and E. Diller. Magnetic actuation for full dexterity microrobotic control using rotating permanent magnets. *IEEE Transactions on Robotics*, 33 (6): 1398–1409, 2017.

7 A. Z. Taddese, P. R. Slawinski, M. Pirotta, E. De Momi, K. L. Obstein, and P. Valdastri. Enhanced real-time pose estimation for closed-loop robotic manipulation of magnetically actuated capsule endoscopes. *The International Journal of Robotics Research*, 37 (8): 890–911, 2018.

8 J. C. Norton, et al. Intelligent magnetic manipulation for gastrointestinal ultrasound. *Science Robotics*, 4 (31): eaav7725, 2019.

9 P. R. Slawinski, N. Simaan, A. Z. Taddese, K. L. Obstein, and P. Valdastri. Sensitivity ellipsoids for force control of magnetic robots with localization uncertainty. *IEEE Transactions on Robotics*, 35 (5): 1123–1135, 2019.

10 L. Yang, X. Du, E. Yu, D. Jin, and L. Zhang. Deltamag: an electromagnetic manipulation system with parallel mobile coils. In *2019 International Conference on Robotics and Automation (ICRA)*, pages 9814–9820, 2019. IEEE.

11 X. Du, M. Zhang, J. Yu, L. Yang, P. W. Y. Chiu, and L. Zhang. Design and real-time optimization for a magnetic actuation system with enhanced flexibility. *IEEE/ASME Transactions on Mechatronics*, 26 (3): 1524–1535, 2020.

12 X. Du, L. Yang, J. Yu, K. F. Chan, P. W. Y. Chiu, and L. Zhang. Robomag: a magnetic actuation system based on mobile electromagnetic coils with tunable working space. In *2020 5th International Conference on Advanced Robotics and Mechatronics (ICARM)*, pages 125–131, 2020. IEEE.

13 J. Sikorski, A. Denasi, G. Bucchi, S. Scheggi, and S. Misra. Vision-based 3-D control of magnetically actuated catheter using bigmag—an array of mobile electromagnetic coils. *IEEE/ASME Transactions on Mechatronics*, 24 (2): 505–516, 2019.

14 X. Yan, et al. Multifunctional biohybrid magnetite microrobots for imaging-guided therapy. *Science Robotics*, 2 (12): eaaq1155, 2017.

15 J. Li, et al. Development of a magnetic microrobot for carrying and delivering targeted cells. *Science Robotics*, 3 (19): eaat8829, 2018.

16 D. Son, H. Gilbert, and M. Sitti. Magnetically actuated soft capsule endoscope for fine-needle biopsy. *Soft Robotics*, 7 (1): 10–21, 2020.

17 J. Sikorski, I. Dawson, A. Denasi, E. E. Hekman, and S. Misra. Introducing bigmag—a novel system for 3D magnetic actuation of flexible surgical manipulators. In *2017 IEEE International Conference on Robotics and Automation (ICRA)*, pages 3594–3599, 2017. IEEE.

18 J. Edelmann, A. J. Petruska, and B. J. Nelson. Magnetic control of continuum devices. *The International Journal of Robotics Research*, 36 (1): 68–85, 2017.

19 J. Leclerc, H. Zhao, D. Bao, and A. T. Becker. *In vitro* design investigation of a rotating helical magnetic swimmer for combined 3-D navigation and blood clot removal. *IEEE Transactions on Robotics*, 36 (3): 975–982, 2020.

20 A. Oulmas, N. Andreff, and S. Régnier. 3D closed-loop swimming at low Reynolds numbers. *The International Journal of Robotics Research*, 37 (11): 1359–1375, 2018.

21 S. Xu, J. Liu, C. Yang, X. Wu, and T. Xu. A learning-based stable servo control strategy using broad learning system applied for microrobotic control. *IEEE Transactions on Cybernetics*, 52 (12): 13727–13737, 2021.

22 T. Xu, C. Huang, Z. Lai, and X. Wu. Independent control strategy of multiple magnetic flexible millirobots for position control and path following. *IEEE Transactions on Robotics*, 38 (5): 2875–2887, 2022.

23 S. Chowdhury, W. Jing, and D. J. Cappelleri. Towards independent control of multiple magnetic mobile microrobots. *Micromachines*, 7 (1): 3, 2015.

24 M. Salehizadeh and E. Diller. Three-dimensional independent control of multiple magnetic microrobots via inter-agent forces. *The International Journal of Robotics Research*, 39 (12): 1377–1396, 2020.

25 S. Scheggi, K. K. T. Chandrasekar, C. Yoon, B. Sawaryn, G. Van De Steeg, D. H. Gracias, and S. Misra. Magnetic motion control and planning of untethered soft grippers using ultrasound image feedback. In *2017 IEEE International Conference on Robotics and Automation (ICRA)*, pages 6156–6161, 2017. IEEE.

26 L. Zheng, Y. Jia, D. Dong, W. Lam, D. Li, H. Ji, and D. Sun. 3D navigation control of untethered magnetic microrobot in centimeter-scale workspace based on field-of-view tracking scheme. *IEEE Transactions on Robotics*, 38 (3): 1583–1598, 2021.

27 Y. Jia, L. Zheng, D. Dong, Y. Wang, and D. Sun. Robust navigation control of a microrobot with hysteresis compensation. *IEEE Transactions on Automation Science and Engineering*, 19 (4): 3083–3092, 2021.

28 L. Yang, J. Yu, and L. Zhang. Statistics-based automated control for a swarm of paramagnetic nanoparticles in 2-d space. *IEEE Transactions on Robotics*, 36 (1): 254–270, 2019.

29 J. Yu, L. Yang, X. Du, H. Chen, T. Xu, and L. Zhang. Adaptive pattern and motion control of magnetic microrobotic swarms. *IEEE Transactions on Robotics*, 38 (3): 1552–1570, 2021.

30 H. Xie, X. Fan, M. Sun, Z. Lin, Q. He, and L. Sun. Programmable generation and motion control of a snakelike magnetic microrobot swarm. *IEEE/ASME Transactions on Mechatronics*, 24 (3): 902–912, 2019.

31 Q. Wang, X. Du, D. Jin, and L. Zhang. Real-time ultrasound doppler tracking and autonomous navigation of a miniature helical robot for accelerating thrombolysis in dynamic blood flow. *ACS Nano*, 16 (1): 604–616, 2022.

32 J. Yu, D. Jin, K.-F. Chan, Q. Wang, K. Yuan, and L. Zhang. Active generation and magnetic actuation of microrobotic swarms in bio-fluids. *Nature Communications*, 10 (1): 5631, 2019.

33 Q. Wang, et al. Ultrasound doppler-guided real-time navigation of a magnetic microswarm for active endovascular delivery. *Science Advances*, 7 (9): eabe5914, 2021.

34 J. Hwang, et al. An electromagnetically controllable microrobotic interventional system for targeted, real-time cardiovascular intervention. *Advanced Healthcare Materials*, 11 (11): 2102529, 2022.

35 K. T. Nguyen, et al. Guide-wired helical microrobot for percutaneous revascularization in chronic total occlusion *in-vivo* validation. *IEEE Transactions on Biomedical Engineering*, 68 (8): 2490–2498, 2020.

36 J. Law, et al. Microrobotic swarms for selective embolization. *Science Advances*, 8 (29): eabm5752, 2022.

Index

f

Printed and bound by CPI Group (UK) Ltd, Croydon, CR0 4YY

17/09/2026

14964234-0006